Advances in MICROBIAL ECOLOGY
Volume 15

ADVANCES IN MICROBIAL ECOLOGY

Sponsored by the International Committee on Microbial Ecology
(ICOME), a committee of the International Union of
Microbiological Societies (IUMS) and the International Union of
Biological Sciences (IUBS)

EDITORIAL BOARD

Bernhard Schink
Universität Konstanz
Konstanz, Germany

Warwick F. Vincent
Université Laval
Québec, Canada

David Ward
Montana State University
Bozeman, Montana

A Continuation Order Plan is available for this series. A continuation order will bring delivery of each new volume immediately upon publication. Volumes are billed only upon actual shipment. For further information please contact the publisher.

Advances in
MICROBIAL ECOLOGY

Volume 15

Edited by
J. Gwynfryn Jones
Freshwater Biological Association
Ambleside, Cumbria, England

PLENUM PRESS • NEW YORK AND LONDON

The Library of Congress cataloged the first volume of this title as follows:

Advances in microbial ecology. v. 1–
New York, Plenum Press ©1977–
v. ill. 24 cm.
Key title: Advances in microbial ecology, ISSN 0147-4863
1. Microbial ecology—Collected works.

QR100.A36	576'.15	77-649698

ISBN 0-306-45559-5

© 1997 Plenum Press, New York
A Division of Plenum Publishing Corporation
233 Spring Street, New York, N. Y. 10013

http://www.plenum.com

All rights reserved

10 9 8 7 6 5 4 3 2 1

No part of this book may be reproduced, stored in a retrieval system, or transmitted in any form or by any means, electronic, mechanical, photocopying, microfilming, recording, or otherwise, without written permission from the Publisher

Printed in the United States of America

Contributors

Robert T. Anderson, Department of Civil and Environmental Engineering, University of Massachusetts, Amherst, Massachusetts 01003

Douglas E. Caldwell, Department of Applied Microbiology and Food Science, University of Saskatchewan, Saskatoon, Saskatchewan, S7N 5A8 Canada

Donat-P. Häder, Institut für Botanik und Pharmazeutische Biologie, Friedrich-Alexander-Universität, D-91058 Erlangen, Germany

Ann E. Hajek, Department of Entomology, Cornell University, Ithaca, New York 14853

David L. Kirchman, College of Marine Studies, University of Delaware, Lewes, Delaware 19958

Darren R. Korber, Department of Applied Microbiology and Food Science, University of Saskatchewan, Saskatoon, Saskatchewan, S7N 5A8 Canada

John R. Lawrence, National Hydrology Research Institute, Saskatoon, Saskatchewan, S7N 3H5 Canada

Derek R. Lovley, Department of Microbiology, University of Massachusetts, Amherst, Massachusetts 01003

Jan Molin, Institute of Organization and Industrial Sociology, Copenhagen Business School, DK-2200N Blaagaardsgade 23B, Denmark

Søren Molin, Department of Microbiology, Technical University of Denmark, DK-2800 Lyngby, Denmark

Toshi Nagata, Ocean Research Institute, University of Tokyo, Nakano, Tokyo 164, Japan

Jörg Overmann, Institut für Chemie und Biologie des Meeres, Universität Oldenburg, D-26111 Oldenburg, Germany

Gideon M. Wolfaardt, Department of Applied Microbiology and Food Science, University of Saskatchewan, Saskatoon, Saskatchewan, S7N 5A8 Canada

Preface

This is the third volume of *Advances in Microbial Ecology* to be produced by the current editorial board. I would, therefore, like to take this opportunity to thank my co-editors for all their efforts, particularly in maintaining a balance of subject matter and geographical distribution of the contributions.

Volume 15 is no exception in that we have a balance between the prokaryotic and eukaryotic organisms and a range of subject matter from applied ecology through process ecology to ecological theory. The response from our readers has been encouraging in the sense that the breadth of coverage is much appreciated, particularly by teachers and postgraduate/postdoctoral researchers. However, we still strive to improve our coverage and particularly to move wider than the North America/Europe axis for contributions. Similarly, we would like to see coverage of the more unusual microbes, perhaps a chapter devoted to the ecology of a particular species or genus. There must exist many ecological notes on "rarer" organisms that have not found their way into the standard textbooks or taxonomic volumes; properly compiled these could provide valuable information for the field ecologist.

Ecological theory has, until recently, been the domain of the "macroecologist." Recent advances in molecular techniques will ensure that the microbial ecologist will play a more significant role in the development of the subject. We shall not, therefore, change our policy of encouraging our contributors to speculate, permitting them sufficient space to develop their ideas.

Our current timetable includes the production of Volume 16 before the Eighth International Symposium on Microbial Ecology in Halifax, Nova Scotia. Doubtless the Editorial Board will be scouring that meeting for contributors. However, in the meantime, no one should hesitate to discuss any proposed chapter with us.

J. Gwynfryn Jones, Editor
Bernhard Schink
Warwick F. Vincent
David Ward

Contents

Chapter 1

Effects of UV Radiation on Phytoplankton

Donat-P. Häder

1. Stratospheric Ozone and Solar UV Radiation	1
1.1. The Ozone Layer	1
1.2. Solar UV Radiation	2
1.3. Penetration of Solar UV-B Radiation into the Water Column	3
2. Phytoplankton	3
2.1. Phytoplankton and Global Carbon Cycles	4
2.2. Global and Vertical Distribution of Phytoplankton	4
2.3. The Biological Food Web	6
3. UV-B Effects on Phytoplankton	6
3.1. UV-B Effects on Motility, Orientation, and Vertical Distribution	7
3.2. Cyanobacteria	10
3.3. Metabolism and Development	14
3.4. Targets of UV-B Radiation	16
3.5. Protective Strategies	17
4. Consequences of UV-B Damage in Aquatic Ecosystems	18
5. Conclusions	18
References	19

Chapter 2

CASE: Complex Adaptive Systems Ecology

Jan Molin and Søren Molin

1. Microbial Ecology	27
1.1. *In Situ* Studies	27

2. CASE 29
   2.1. A Paradigmatic Platform 29
3. SCIO—Parameters for Description 38
   3.1. Development of an Analytical Model 38
   3.2. *In Situ* Parameter Analysis 42
   3.3. SCIO in General 49
4. Behavior and Organization 50
   4.1. Levels of Analysis 50
   4.2. Interference as an Experimental Approach 54
5. Identity 62
   5.1. Contextual Understanding 62
   5.2. Coupling and Regulation 64
   5.3. The Four Analytical Windows for Identification 66
   5.4. The Windows and Laboratory Biofilms 69
6. CASE Study—Research Designs 69
   6.1. Naturalistic Inquiry Paradigm 70
   References 73

Chapter 3

Roles of Submicron Particles and Colloids in Microbial Food Webs and Biogeochemical Cycles within Marine Environments

Toshi Nagata and David L. Kirchman

1. Introduction 81
2. Definition and Determination of Colloids and
   Submicron Particles 83
3. Occurrences of Submicron Particles and Colloidal Organic Matter
   in Marine Environments 84
4. Turnover Rates of DOC and Colloids 86
5. Sources of Submicron and Colloidal Particles 88
6. Sinks of Submicron and Colloidal Particles 90
   6.1. Bacterial Degradation and Its Constraints 90
   6.2. Consumption of Colloidal Particles by Protozoan and
        Metazoan Grazers 94
   6.3. Aggregation 97
7. Summary and Future Challenges 97
   References 98

Chapter 4

Do Bacterial Communities Transcend Darwinism?

Douglas E. Caldwell, Gideon M. Wolfaardt, Darren R. Korber, and
John R. Lawrence

1. Introduction 105
2. Conceptual Barriers to Community-Level Thought and
   Experimentation 109
   2.1. Organismic Selection Theory 110
   2.2. Germ Theory 115
   2.3. Selective Enrichment Theory 118
   2.4. Reductionism 120
3. Community Theory 122
   3.1. Are There Community-Level Reproductive Strategies? ... 122
   3.2. A Mini Gaia Hypothesis 124
   3.3. Individual Traits vs. Community Traits 124
   3.4. Testing the Mini Gaia and Community-Level Strategy
        Hypothesis 125
4. Methods of Community-Level Cultivation 126
   4.1. The Chemostat, Nutristat, and Gradostat 128
   4.2. The Dual-Dilution Continuous Culture 133
   4.3. Rototorque Annular Bioreactor 133
   4.4. Continuous-Flow Slide Cultures 134
   4.5. The Microstat 135
5. Methods of Community-Level Analysis 139
   5.1. Scanning Confocal Laser Microscopy (SCLM) and
        Fluorescent Molecular Probes 139
   5.2. Genetic Analyses 153
   5.3. Fatty Acid Methyl Ester Profiles 155
   5.4. Substrate Utilization Profiles 156
   5.5. Other Technologies 157
6. Selected Examples of Community-Level Bacterial
   Associations 157
   6.1. Degradative Microbial Consortia 157
   6.2. Anaerobic Digestor Granules 162
7. Criteria Used in the Isolation and Study of Communities . 165
8. Conclusions 167
9. Terms and Definitions 169
   References 175

Chapter 5

Ecology of Terrestrial Fungal Entomopathogens

Ann E. Hajek

1. Introduction	193
2. Life Histories of Insect/Fungus Systems	195
3. Survival of Entomopathogenic Fungi during Unfavorable Periods	199
3.1. Inactive Resting Stages	200
3.2. Persistence in Soil and Facultative Saprophytic Growth	204
4. Activity of Entomopathogenic Fungi during Favorable Periods	206
4.1. Conidial Adhesion	206
4.2. Conidial Germination	208
4.3. Infection and Growth *in Vivo*	209
4.4. Host Death and Production of Spores for Dispersal	215
4.5. Fungal Dispersal	218
4.6. Conidial Survival and Transmission after Dispersal	222
5. Epizootiology	225
5.1. Epizootic Determinants	226
5.2. Community-Level Interactions	228
6. Relevance to Biological Control	234
References	236

Chapter 6

Mahoney Lake: A Case Study of the Ecological Significance of Phototrophic Sulfur Bacteria

Jörg Overmann

1. Introduction	251
2. The Habitat	252
3. Species Composition of the Bacterial Plate	256
4. Factors Controlling Growth of *Amoebobacter purpureus*	260
5. What Are the Mechanisms of the Accumulation and Disappearance of *A. purpureus*?	266
6. Bacterial Interactions in the Chemocline: Anoxygenic Phototrophs and Sulfate Reducers	269
7. Bacterial Interactions in the Chemocline: Anoxygenic Phototrophs and Sulfur Reducers	272

8. The Coupling between Phototrophic and Sulfate-Reducing Bacteria: General Implications 274
9. Significance of Purple Sulfur Bacteria for Oxic Water Layers 278
10. Paleomicrobiology of Mahoney Lake 280
    References 284

Chapter 7

Ecology and Biogeochemistry of *in Situ* Groundwater Bioremediation

Robert T. Anderson and Derek R. Lovley

1. Introduction 289
2. Diversity and Distribution of Microorganisms in Pristine Aquifers 290
   2.1. Oligotrophic Nature of the Subsurface 290
   2.2. Types of Microorganisms in the Subsurface 291
   2.3. Numbers and Distribution of Microorganisms 291
3. Microbial Ecology and Biogeochemistry of Contaminated Aquifers 294
   3.1. Terminal Electron-Accepting Processes in Contaminated Aquifers 301
   3.2. *In Situ* Bioremediation of Specific Contaminants 306
4. Conclusions 330
   References 331

Index 351

1

Effects of UV Radiation on Phytoplankton

DONAT-P. HÄDER

1. Stratospheric Ozone and Solar UV Radiation

1.1. The Ozone Layer

Ozone is distributed throughout the atmosphere with its highest concentration in the stratosphere between about 15–40 km. Both its production and breakdown are powered by solar ultraviolet radiation, though at different wavelengths. The cycle of generation and destruction keeps the concentration constant, but on an extremely low level: If compressed under atmospheric pressure, the total ozone layer would be on the order of a few millimeters thick. Therefore, gaseous pollutants such as chlorinated fluorocarbons (CFCs) can effectively decrease ozone density, even though they are emitted in relatively small amounts (about 1 million tons per year).

It has been proven beyond reasonable doubt that, indeed, pollutants of anthropogenic origin are responsible for stratospheric ozone destruction by catalytic action: Chlorine atoms are cleaved from the CFC molecule by ultraviolet radiation and abstract an oxygen atom from ozone, leaving an oxygen molecule behind. The reactive chlorine is regenerated and enters the cycle again until it is removed from the stratosphere; the half-life of most CFCs is on the order of 100 years (Vaida *et al.*, 1989; Rowland, 1989).

The Antarctic ozone hole was the first observed incidence of stratospheric ozone depletion. It was first documented in the early 1980s and has increased in size and depth ever since (Kerr, 1990; Wei, 1991; Schoeberl and Hartmann, 1991) and reached a new record low in 1995, dropping to below 100 Dobson units from its normal 300–350 Dobson units (1 Dobson unit = 0.01 mm path

---

DONAT-P. HÄDER • Institut für Botanik und Pharmazeutische Biologie, Friedrich-Alexander-Universität, D-91058 Erlangen, Germany.
Advances in Microbial Ecology, Volume 15, edited by Jones. Plenum Press, New York, 1997.

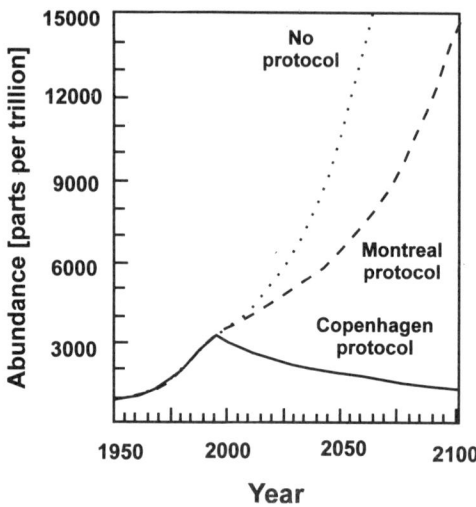

Figure 1. Predicted global ozone values for the coming decades.

length). The Antarctic ozone hole covers an area of the size of the continental U.S. (Bidigare, 1989; Lubin et al., 1989; Karentz and Lutze, 1990). It affects South America and Australia by drawing ozone from lower latitudes, documented from as far away as Santiago, Chile (Atkinson et al., 1989; Proffitt et al., 1989).

Measurements from balloon and aircraft have shown increased levels of chlorine in the stratosphere (Hofmann and Deshler, 1991; Brune et al., 1991; Schnell et al., 1991; Anderson et al., 1991). During the flight of a research aircraft in the stratosphere from South America toward the Antarctic, an increase of reactive chlorine was measured that was closely inversely correlated with decreasing ozone concentrations. In recent years, beginning in 1993, abnormal high chlorine concentrations and concomitant decreases in ozone layer have been found over the Northern hemisphere (Madronich et al., 1994).

In addition to the polar holes, decreases in stratospheric ozone are observed over most latitudes, except for the tropics (Heath, 1988; Hough and Derwent, 1990; Brasseur, 1989). Due to the long lifetime of CFCs and the slow phaseout of production and emission of these substances, the current predictions envision a peak of ozone depletion around the year 2000, and decreased ozone values will prevail well into the next century (Fig. 1).

1.2. Solar UV Radiation

Ozone is an effective filter for solar short-wave ultraviolet radiation (UV-B, 280–315 nm, CIE definition). Therefore, ozone depletion results in increased levels of UV-B radiation that has been measured under the ozone hole(s) (Grant, 1988; Frederick et al., 1989). Recently, increased UV-B levels have also been detected over industrialized areas of the Northern hemisphere (Webb, 1991; Frederick, 1990; Frederick et al., 1991; Blumthaler and Ambach, 1990).

1.3. Penetration of Solar UV-B Radiation into the Water Column

Based on the use of insensitive instruments, earlier researchers assumed that short wavelength UV radiation penetrates only a short distance into the water column. More recent instruments allow us to measure the spectral distribution of solar radiation in dependence of depth in the water column. The penetration not only depends on the wavelength (Jerlov, 1970) but also on the transmission, which can vary substantially between freshwater eutrophic water and clear open oceanic waters. In costal waters with high turbidity and large concentrations of gelbstoff UV-B penetrates only a few decimeters or meters into the column (Smith and Baker, 1978; Siebeck and Böhm, 1987; Piazena and Häder, 1994) while in clear oceanic waters UV-B has been measured to penetrate to depths of dozens of meters (Baker and Smith, 1982).

Because of the step drop of solar radiation over several orders of magnitude in the UV-B range, accurate spectral measurements can only be performed with a double monochromator spectroradiometer (Piazena and Häder, 1994). An additional problem in measurements in the water column is the fact that especially short wavelength radiation undergoes multiple (Rayleigh) scattering. Even a short distance from the surface a considerable proportion of the radiation is upwelling (comes from below), even if there is no reflection from the bottom. This problem was solved by designing a spherical detector that receives radiation from a 4 π geometry. Record penetration was measured in Antarctic waters (Gieskes and Kraay, 1990; Smith et al., 1992) where UV-B radiation could be detected at a depth of several tens of meters (Fig. 2) using a new instrument (LUVSS) that has a 0.2 nm resolution from 250–350 nm and a 0.8 nm resolution from 350–700 nm (Smith et al., 1992). The spectroradiometer is installed on a remotely operated vehicle that dives to a certain depth and measures data while moving upward in the water column. During a recent cruise in Antarctic waters, measurements clearly allowed researchers to determine when the ozone hole vortex was over the measurement site.

2. Phytoplankton

This chapter includes effects of ultraviolet radiation on freshwater microorganisms and ecosystems, but the most far-reaching effects are expected in marine habitats because of their immense size, which equal terrestrial habitats in biomass productivity.

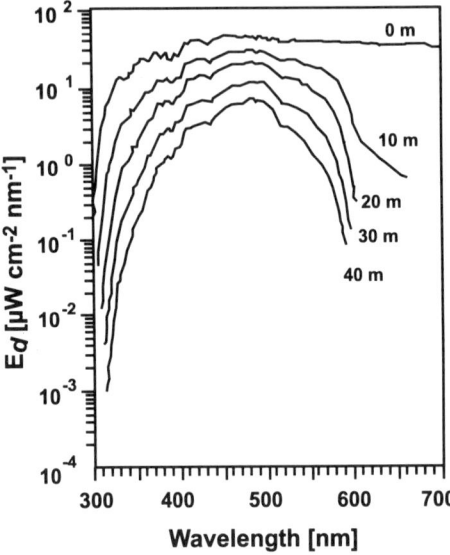

Figure 2. Penetration of solar radiation into clear oceanic waters in the Antarctic (modified from Smith et al., 1992).

2.1. Phytoplankton and Global Carbon Cycles

Compared to terrestrial ecosystems, the standing crop of phytoplankton is rather small but the photosynthetic productivity equals that of terrestrial ecosystems (Fig. 3). It is estimated that the aquatic ecosystems incorporate about 10^{11} tons of carbon annually into organic material (Houghton and Woodwell, 1989). This is balanced by a similar amount of carbon taken up from the atmosphere by terrestrial plants. In both cases, most of this is released after the decay of the organic material. This delicate balance of uptake and release is disturbed by the release of about 5 Gt of carbon from fossil fuel burning and another 2 Gt from (mostly tropical) deforestation. However, only 3 of the total of extra 7 Gt have been found to pile up in the atmosphere and contribute to the greenhouse effect. The reminder of 4 Gt is thought to be removed from the cycle by what is called the biological pump in the oceans: Organic and inorganic carbon falls out of the upper layers of the water column in the form of oceanic snow and is deposited in the deep sea.

2.2. Global and Vertical Distribution of Phytoplankton

Most microorganisms are unicellular and lack an epidermal layer, thus they are easily affected by short wavelength irradiation. There are two indications that

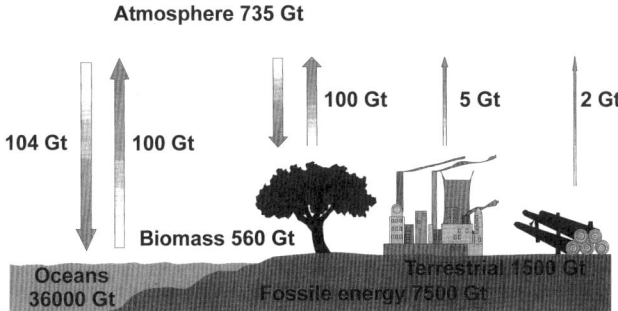

Figure 3. Global carbon fluxes (in Gt per year) and sizes of major reservoirs (in Gt) (modified from Houghton and Woodwell, 1989).

phytoplankton is under considerable UV-B stress even at ambient levels of solar radiation: Phytoplankton is not uniformly distributed in the oceans as shown by satellite pseudocolor images (Viollier et al., 1980; Lohrenz et al., 1988). Surface chlorophyll concentrations are highest in the circumpolar regions, while in equatorial waters the concentrations are 100 to 1000 times smaller. This observation corresponds with the observation that UV-B levels are much higher in the tropics and subtropics than at higher latitudes. Also the ratio of UV-B:PAR is much higher in the tropics than in any other ocean (Smith, 1989). Other factors, including nutrient concentration, temperature, and salinity also control phytoplankton density. The only exceptions of this global distribution are the large concentrations of phytoplankton in the upwelling areas along the continental shelves where high turbidity and gelbstoff concentrations are found accompanied by high nutrient concentrations.

The second indication for a high UV-B sensitivity of phytoplankton is that the large blooms in mid-latitudes occur in spring, when water temperatures have risen sufficiently, but disappear when the level of solar ultraviolet irradiation increases into the summer. There is often a second smaller algal bloom in autumn, permissive temperatures and sufficient nutrients provided.

Phytoplankton depend on the availability of solar energy for their growth and metabolism and are therefore not equally distributed throughout the water column (Häder, 1991b; Häder and Worrest, 1991; Häder et al., 1995). Being photosynthetic organisms, phytoplankton occupies the top of the water column to obtain sufficient light. On the other hand, these organisms do not tolerate the unfiltered solar radiation at the surface. As a consequence they tend to move to a specific depth, which results in a typical vertical distribution (Cabrera and Montecino, 1987). Depending on the transparency of the water and the incident irradiance, this depth varies between a few decimeters (in turbid coastal waters and eutrophic freshwater lakes) to tens of meters (in clear open oceanic waters).

To understand where the organisms are, a useful biological concept is "the euphotic zone," which is defined as the depth of the water column that allows 0.1 % of photosynthetically active radiation (PAR) to penetrate. This corresponds to the irradiance where photosynthesis is balanced by respiration. This typical vertical distribution pattern of phytoplankton is disturbed by passive mixing due to high wind and waves (Smith, 1989; Ignatites, 1990). The consumers follow the primary producers and are thus exposed to the same radiation regime as the phytoplankton (Brown and Cochrane, 1991).

2.3. The Biological Food Web

Phytoplanktonic organisms are the primary biomass producers in both freshwater and marine ecosystems. The primary consumers (zooplankton) include unicellular and multicellular organisms and feed on the primary producers. The following levels in the food web are free swimming organisms (nekton), including krill, mollusk, and fish and crab larvae, feeding in turn small fish, mollusks, and crustaceans. The final consumers are large fish, birds, and mammals, including humans. During each transition from one trophic level to the next the amount of biomass is reduced by a factor of about ten. In addition to the direct effects of solar UV-B radiation, the consumers are affected indirectly when the productivity of the primary producers decreases.

3. UV-B Effects on Phytoplankton

In contrast to higher plants, phytoplankton are not protected by an epidermal layer, and both marine and freshwater phytoplankton are affected by solar UV radiation as indicated by numerous studies over the last decade (El Sayed, 1988a,b; Bidigare, 1989; Cullen and Lesser, 1991; Raven, 1991; Karentz and Lutze, 1990; Karentz, 1991; Helbing *et al.*, 1991; Ekelund, 1991; Karentz *et al.*, 1991c; Häder, 1994). This results in the fact that many aquatic ecosystems are under considerable UV-B stress even at current radiation levels (Worrest *et al.*, 1980, 1981a,b; Worrest, 1982; Lorenzen, 1979; Calkins and Thordardottir, 1980; Smith *et al.*, 1980; Maske, 1984).

Many of the first measurements were carried out using artificial UV sources, the emission spectrum of which strongly deviated from that of solar radiation. The action spectra of inhibition responses in phytoplankton differ between species and even between individual physiological responses. Therefore, the results from exposure to different artificial radiation sources are difficult to interpret. One way out of this dilemma are so-called exclusion studies: Phytoplankton are exposed to ambient solar radiation, and increasing amounts of shorter wavelengths are removed by using appropriate cut-off filters (e.g., Schott WG filters) or filter foils. The alternative is to increase the irradiance of ambient radiation by moving the experimental site from mid-latitudes toward the equator or up onto a higher mountain. At both locations higher ambient UV-B radiation is found (Häder and Häder, 1988a,b; Gerber and Häder, 1993).

Recent research has concentrated on two separate issues. One is to study the mechanisms of inhibition by UV-B radiation; this can best be done in the laboratory under constant conditions. The other is to investigate the results of exposure to solar radiation under natural conditions. For this purpose ecologically relevant organisms and taxonomic groups should be selected that constitute the important biomass producers such as dinoflagellates, cryptophyceae, and diatoms.

3.1. UV-B Effects on Motility, Orientation, and Vertical Distribution

As discussed above, phytoplankton display a distinct vertical distribution in the water column to optimize their light input for growth and survival. Some organisms rely on active propelling mechanisms based on flagella or cilia. Others are capable of modifying their buoyancy by producing gas vacuoles (Walsby *et al.*, 1992) or oil droplets (Gosink *et al.*, 1993). The vertical migrations of the organisms are controlled by endogenous tidal or circadian rhythms (Lindholm, 1992). Many phytoplankton move toward the surface before sunrise, to be ready for the utilization of solar energy when the sun rises, and move into deeper waters at night. Some dinoflagellates are known to migrate up to 15 m up and down in the water column (Burns and Rosa, 1980). Another important influence controlling vertical migrations are external physical and chemical factors such as the magnetic field lines of the earth or thermal and chemical gradients (Macnab, 1985; Poff, 1985; Esquivel and de Barros, 1986); light and gravity may be the most important stimuli for the orientation in the water column (Nultsch and Häder, 1988; Häder, 1988a,b, 1991a, b; Bean, 1985). Fig. 4 shows the patterns of vertical distribution over a 38-h period (Eggersdorfer and Häder, 1991a,b). The cells start moving toward the surface before sunrise and accumulate at or near the surface before noon. At times of excessive irradiation the cells move deeper down into the water column to come back later in the afternoon. During the night they again move deeper down in the water column.

The most effective orientation with respect to light is phototaxis, which can be positive (toward the light source), negative (away from the light source) or perpendicular to the light direction (diaphototaxis) (Fig. 5). Many organisms use positive phototaxis at low irradiances to move toward the surface and negative phototaxis at high irradiances to escape excessive exposure near the water surface (Häder *et al.*, 1981). The upward movement is augmented by negative gravitaxis, which guides the cells toward the surface even in turbid water or before sunrise (Häder, 1987).

Earlier studies have shown that the motility of many phytoplankton organisms is affected by exposure to solar radiation (Häder, 1985, 1986a,b; Häder and

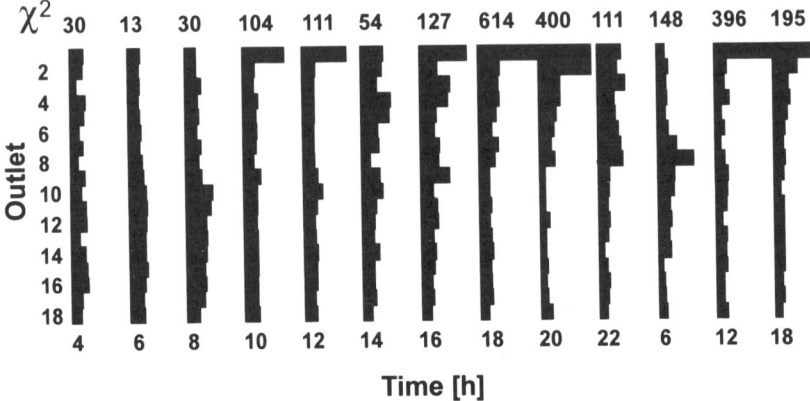

Figure 4. Vertical distribution of the marine dinoflagellates *Peridinium faeroense* in a 3-m water column (modified after Eggersdorfer and Häder, 1991b).

Häder, 1988a,b,c, 1989a,b, 1991a, ; Häder *et al.*, 1990a,b). Both the percentage of motile cells and the swimming velocity of the still motile cells decrease dramatically when exposed to unfiltered radiation (Fig. 6). Even though UV-A and visible radiation also are involved in this inhibition, UV-B has a strong effect that is far higher than predicted from the low proportion it has in solar energy. This is also supported by the fact that artificial UV radiation (without the visible band) also impairs motility in phytoplankton.

In addition to the inhibition of motility, the precision of orientation is affected both in freshwater and marine flagellates. Even short exposure times to solar radiation strongly decrease the degree of gravitaxis in most organisms

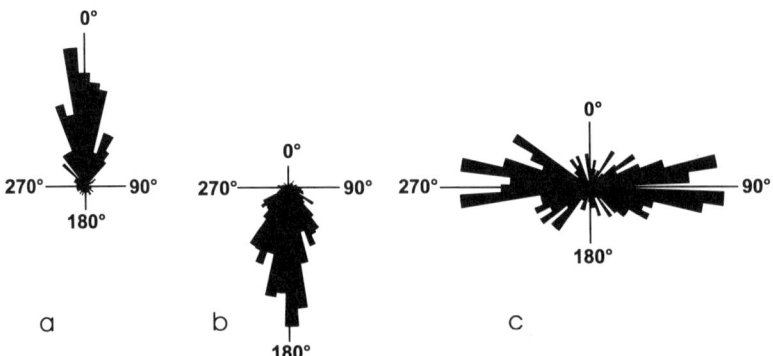

Figure 5. Circular histograms for positive (a, *Peridinium gatunense*), negative (b, *Euglena gracilis*) and diaphototaxis (c, *Peridinium gatunense*).

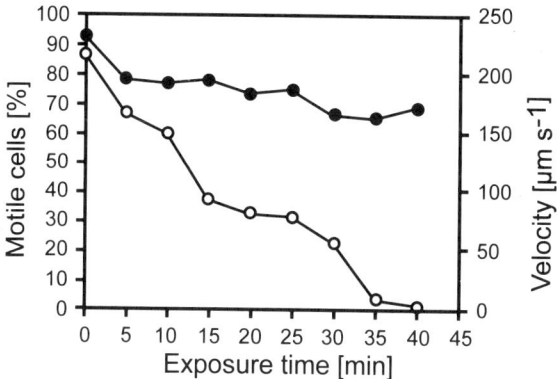

Figure 6. Inhibition of motility (percentage of motile cells, open circles, and swimming velocity, closed circles) in *Gymnodinium* sp. (Y100) by solar radiation.

investigated so far (Häder and Liu, 1990a,b). However, one interesting effect was found in several dinoflagellates that stop moving after excessive UV irradiation and, since the cells are heavier than water, sediment in the water column to a greater depth. An unidentified *Gymnodinium* (Y100) even reversed the direction of gravitaxis and actively moved downward when exposed to excessive solar radiation (Tirlapur *et al.*, 1993). This behavior may be an effective escape mechanism to avoid stress situations such as excessive solar radiation (Häder and Liu, 1990a; Eggersdorfer and Häder, 1991). Also, phototaxis is affected by solar radiation (Häder *et al.*, 1990a,b, Häder, 1985, 1986b) (Fig. 7). As with motility, orientation with respect to light and gravity also is affected by artificial UV radiation (without the visible component), indicating the inhibitory role of short wavelength radiation, even though effects by UV-A and visible radiation cannot be excluded.

Biochemical work in a model flagellate, *Euglena gracilis*, shows that UV-B radiation selectively destroys the proteins of the photoreceptor organelle, the paraflagellar body (PFB, Häder and Brodhun, 1991). The PFBs with the flagella still attached were isolated from the cells (Brodhun and Häder, 1990) and the proteins solubilized and separated by 2D gel electrophoresis. The comparison of the protein patterns from UV-B exposed cells with those from control cells clearly indicate the loss of specific PFB proteins (Häder and Brodhun, 1991) (Fig. 8). In addition, the chromophoric groups, pterins and flavins (Galland *et al.*, 1990), are bleached by excessive radiation. Since the position in the water column is of considerable ecological significance, any inhibition of motility and orientation with respect to external stimuli by increased UV-B radiation would diminish the chances for growth and survival. It is interesting to note that the organisms have no capability to respond to increased solar UV-B radiation, but

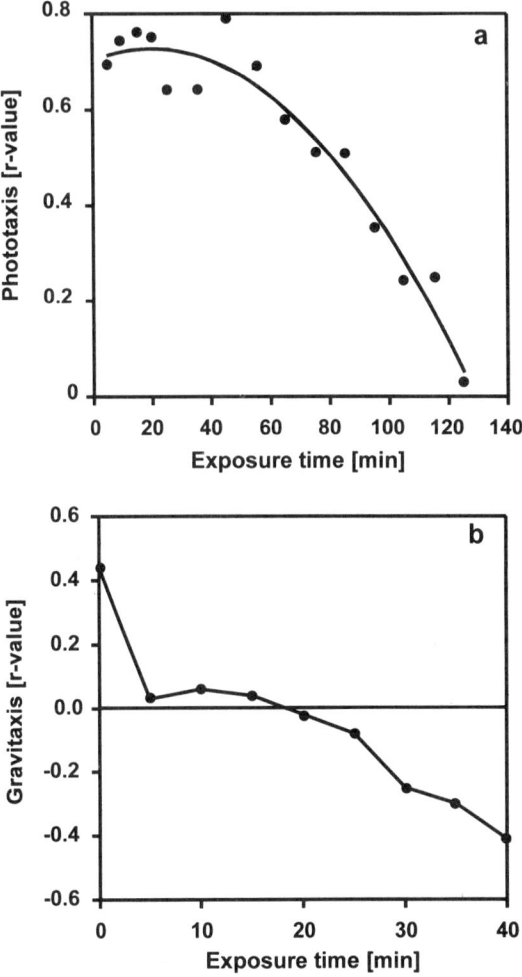

Figure 7. Inhibition of phototaxis (a) and gravitaxis (b) by solar radiation. Ordinate: r-value, a statistical measure for the precision of orientation.

rather use UV-A and visible radiation to orient in their environment, and consequently cannot move to lower levels within the water column with lower UV-B doses to protect themselves from the UV stress.

3.2. Cyanobacteria

Cyanobacteria are believed to have developed during a time when neither oxygen nor a protecting ozone layer was present in the atmosphere. Despite the

Figure 8. Loss of proteins from the PFB, the photoreceptor for phototaxis in *Euglena gracilis* after UV-B radiation. 3-D representations of 2-D electrophoresis gels (a, control; b, after UV-B exposure).

considerable levels of solar ultraviolet radiation believed to impinge on the earth surface at that epoch, most cyanobacteria are very sensitive to UV-B. This apparent contradiction can be explained by assuming that these organisms have developed at considerable water depth, which absorbs most of the short wavelength radiation. In accordance with this hypothesis is the fact that many cyanobacteria are adapted to very low PAR levels, corresponding to a few percent of unfiltered solar radiation.

Due to their capacity to utilize atmospheric nitrogen, a characteristic which they share with other prokaryotes, cyanobacteria occupy a central position in nutrient cycling. They use the enzyme nitrogenase to convert N_2 directly into ammonium (NH_4), a form through which nitrogen enters the food chain. It has been calculated that cyanobacteria fix over 35 million tons of nitrogen annually (Häder et al., 1989), which is thus available for use by higher plants. The role of nitrogen-fixing cyanobacteria as a natural biofertilizer, increasing the fertility of soils, especially rice paddy fields, has been demonstrated by several investigators (Stewart, 1980; Kumar and Kumar, 1988; Sinha and Kumar, 1992; Huang and Chow, 1988). The biotechnological use of cyanobacteria as a natural substitute for artificial nitrogen fertilizer has been discussed (Venkataraman, 1981; Padhy, 1985). Up to 20–30 kg of biologically fixed nitrogen could be provided per hectare of rice crop per season. The relevance of this technology is significant taking into account that rice crops cover about 40 million ha.

Cyanobacteria also optimize the niche in their habitat by light-controlled movement responses. Filamentous cyanobacteria are especially capable of gliding motility, and they use photophobic responses (reversal of movement at light/dark boundaries) to prevent entering too bright or too dark areas (Häder and Hoiczyk, 1992). Thus, the ecological consequences of decreased motility due to solar UV are similar to those in other motile organisms. Upon exposure to solar or artificial UV radiation the percentage of motile filaments and their linear velocity have been found to decrease (Häder et al., 1986; Häder and Häder, 1990a; Donkor and Häder, 1991). Exposure experiments in Ghana have shown that cyanobacteria exposed to tropical solar radiation were affected within minutes; motility and the orientation mechanisms were impaired (Donkor et al., 1993a,b). The damage encountered by the cells can only partially be repaired and only after short-term exposure as shown by recovery experiments.

The effects of artificial UV-B irradiation on growth, survival, pigmentation, nitrate reductase (NR), glutamine synthetase (GS), and total protein profile have been studied in a number of N_2-fixing cyanobacterial strains isolated from rice paddy fields in India (Sinha et al., 1995a). *In vivo* NR activity was found to increase, while *in vivo* GS activity decreased following exposure to UV-B for different durations in all tested organisms. SDS-PAGE analysis of the total protein profile of the cells treated with UV-B shows a linear decrease in the protein content with increasing UV-B exposure time. Complete elimination of most of

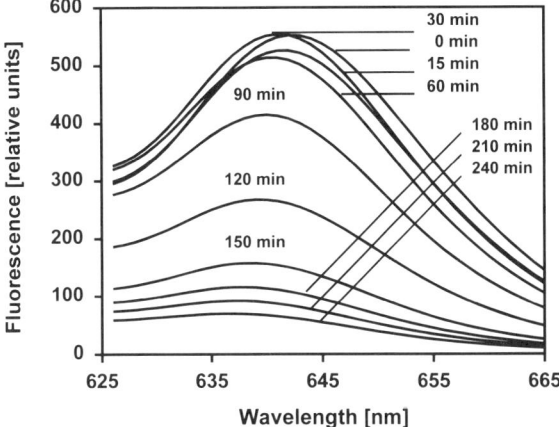

Figure 9. Fluorescence emission spectra (excited at 620 nm) of the lowest fraction from a sucrose gradient obtained after solubilization of cytoplasm from the cyanobacterium *Anabaena* sp. before and after increasing exposure times to artificial UV radiation from a transilluminator (modified from Sinha *et al.*, 1995b).

the protein bands occurred after about 2 h of UV-B exposure in *Nostoc carmium* and *Anabaena* sp., whereas total protein destruction occurred only after 150 min of UV-B treatment in *Nostoc commune* and *Scytonema* sp. Also, different organisms showed different effects in terms of growth and survival. Complete killing of *Anabaena* sp. and *Nostoc carmium* occurred after 2 h of UV-B exposure, whereas in the case of *Nostoc commune* and *Scytonema* sp. this happened only after longer exposure times. Cyanobacteria are also responsible for making available a significant share of the nitrogen consumed by eukaryotic phytoplankton algae in the oceans (Paerl and Bebout, 1988; Carpenter and Romans, 1991) and freshwater habitats (Storch *et al.*, 1990).

Pigment content, particularly phycocyanin, severely decreased following UV-B irradiation in all strains tested so far. Spectroscopic analysis of the photosynthetic pigments indicated that the phycobilins are bleached first, followed by the carotenoids, and finally chlorophyll *a* is affected. Detailed studies of the kinetics of phycobilin destruction shows that in the initial phase of UV exposure the fluorescence yield increases, indicating that the excitation energy cannot be passed to the reaction centers and consequently is lost in the form of fluorescence (Fig. 9). Later during exposure the fluorescence yield decreases again, indicating a gradual loss in biliproteins. In addition, a shift to shorter wavelengths is encountered, which has been interpreted to reflect a disassembly of the phycobilisomes first into hexamers, then trimers, and finally (α,β) monomers.

3.3. Metabolism and Development

Solar UV-B radiation affects many physiological and biochemical reactions in microorganisms. At lower doses the radiation inhibits growth and disturbs the endogenous rhythms in many microorganisms (Worrest, 1982). Recent investigations in Antarctic waters have indicated that carbon dioxide incorporation can be reduced by as much as 6–12 % in the top 10–20 m due to increased UV-B radiation (Smith et al., 1992). However, the UV-A component caused a quantitatively similar inhibition as did UV-B (Smith et al., 1980; Mitchell, 1990). The ATP content in Antarctic phytoplankton also decreased measurably under the ozone hole (Vosjan et al., 1990; Karentz et al., 1991a).

Short wavelength radiation bleaches the photosynthetic pigments (Nultsch and Agel, 1986; Häder et al., 1988). In contrast to higher plants, most phytoplankton do not tolerate excessive solar radiation. The bleaching was quantified by measuring absorption spectra in cells of the marine *Cryptomonas maculata* immobilized in solid agar when exposed to solar radiation. Difference absorption spectra were calculated after increasing exposure times (Fig. 10).

Biochemical analysis of protein extracts from cells before and after UV-B exposure using SDS-PAGE shows a drastic loss of several proteins (Zündorf and Häder, 1991; Gerber and Häder, 1992). Further separation of the proteins on an FPLC anion-exchange column indicates the loss of chlorophyll protein complexes. As a consequence of the loss of proteins and pigments from the photosynthetic apparatus, oxygen production decreases drastically (Smith et al., 1980; Zündorf and Häder, 1991; Häberlein and Häder, 1992). In addition to excessive visible radiation, UV-B plays a strong role in the inhibition, as indicated by exclusion studies using UV cut-off filters (Fig. 11). In contrast, respiration does not seem to be affected as drastically as photosynthetic oxygen production (Schäfer et al., 1993).

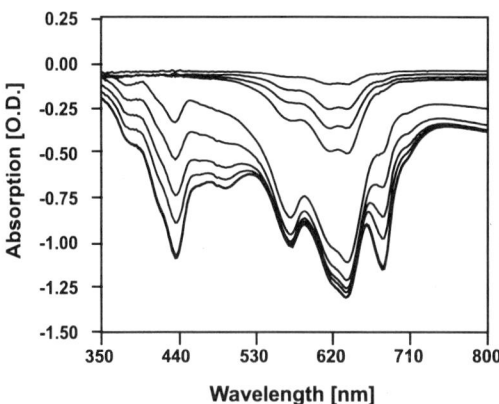

Figure 10. Absorption difference spectra of the marine flagellate *Cryptomonas maculata* after increasing exposure times to artificial UV radiation from a transilluminator.

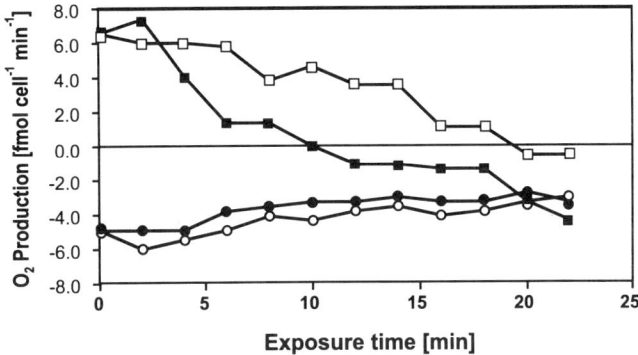

Figure 11. Inhibition of photosynthetic (squares) and respiratory (circles) oxygen exchange in the dinoflagellate *Gymnodinium* (Y100) before and during exposure to unfiltered solar radiation (closed symbols) and radiation devoid of the UV component (filtered through a GG 400 cut-off filter, closed symbols).

While most instrumentation requires transfer of the biological material into the laboratory, a novel hardware device was developed that allows the determination of photosynthetic and respiratory oxygen uptake and release in organisms in their natural habitat, even using solar radiation as an actinic light source. The chamber can be used above water or can be lowered into the water column. The data of oxygen concentration, irradiance, and temperature are constantly monitored by a laptop computer and stored in disk files (Häder and Schäfer, 1992a,b). The experimental data indicate that for many organisms only a limited irradiance window allows net photosynthetic oxygen production under natural conditions; at too low irradiances respiration exceeds photosynthesis and at too high irradiances photosynthesis is shut down by photoinhibition, at least in species not adapted to unattenuated solar radiation.

Another new tool for the determination of the status of the photosynthetic apparatus is pulse amplitude modulation (PAM) fluorescence (Schreiber *et al.*, 1986). Induced chlorophyll fluorescence can be successfully used to obtain valuable information on the physiological parameters that control the photosynthetic apparatus (Renger and Schreiber, 1986; Krause and Weis, 1991). The basic assumption of fluorescence quenching analysis is that two different processes with different time kinetics can reduce the maximal fluorescence yield, F_m. The faster of the two is photochemical quenching, which is effectively suppressed by application of a short saturating pulse that closes all the reaction centers of PS II. A slower process is non-photochemical quenching, which is thought to be mainly based on the energization of the thylakoid membrane (Schreiber *et al.*, 1995). Genty *et al.* (1989) and Weis and Berry (1987) developed empirical expressions for the quantum yield based on the fluorescence parameters measured during quenching analysis. The validity of these expressions has been supported by concomitant gas exchange measurements (Schreiber et al., 1994).

While PAM fluorescence techniques were first used in higher plants, the techniques were later successfully applied to measurements in unicellular algae, even when in dilute suspensions. However, several algal groups show a qualitatively different behavior than higher plants, which has been interpreted to be based on different regulatory mechanisms in various algal taxonomic groups (Büchel and Wilhelm, 1993; Ting and Owens, 1992).

3.4. Targets of UV-B Radiation

Action spectra have been measured for a number of UV-B effects in many microorganisms and for very different responses (Häder and Liu, 1990b; Cullen et al., 1992; Häder et al., 1991, 1994). Fig. 12 shows the action spectrum for the inhibition of photosynthesis in the cyanobacterium *Nodularia spumigena*, which forms large algal blooms in the Baltic Sea. The DNA is certainly one of the targets for damaging UV-B radiation (Peak et al., 1985). The most common mechanism by which UV-B radiation affects DNA is the formation of pyrimidine dimers, a mechanism which has also been found in animal tissues (Yasuhira et al., 1992). Most organisms possess the enzyme photolyase that repairs damage in the DNA caused by UV-B radiation (photorepair) through the removal and replacement of thymine dimers; this enzyme is induced by UV-A and blue light

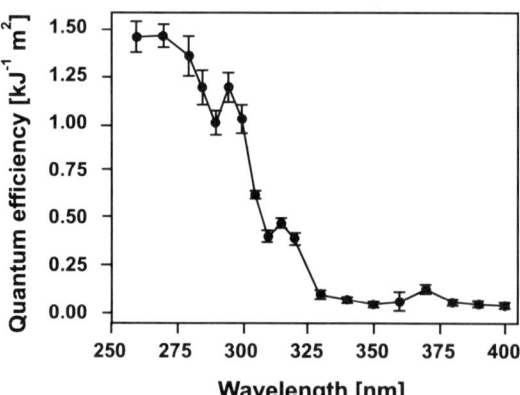

Figure 12. Action spectrum for the inhibition of photosynthesis the cyanobacterium *Nodularia spumigena* based on fluence rate response curves.

(Yamamoto et al., 1983; Hirosawa and Miyachi, 1983). However, a number of UV-B responses do not depend on DNA damage (Häder and Häder, 1988a) since the response occurs within 10 min, which is too short to account for protein resynthesis mediated by DNA. In addition, photoreactivation could not be observed in these cases (Häder and Häder, 1988a; Häder et al., 1986).

Photodynamic reactions are another potential mechanism by which ultraviolet radiation affects living cells (Ito, 1983). When a chromophore molecule absorbs a high energy photon, the excess excitation energy can lead to the formation of singlet oxygen or free radicals. Both routes result in aggressive molecular species that are known to destroy membranes and other cellular components. Photodynamic reactions have been found to play a role in some UV-B-induced types of damage, e.g., in the ciliate *Stentor coeruleus* (Häder and Häder, 1991b), but not in the UV-B-induced inhibition of motility in cyanobacteria or green flagellates (Häder et al., 1986).

Proteins are also targets of UV-B radiation; aromatic amino acids in particular strongly absorb short wavelength radiation. While at sublethal doses these may be regenerated, higher doses eventually cause irreversible damage. This result is supported by biochemical analysis of the proteins involved in photoperception, motility, and also the photosynthetic apparatus.

In the photosynthetic apparatus several targets of UV-B radiation have been identified, including the D1/D2 protein complex associated with photosystem II. Destruction of this protein complex decreases the noncyclic photosynthetic electron transport (Renger et al., 1989). Other targets are the water splitting site of the photosynthetic apparatus and the reaction center of photosystem II (Bhattacharjee et al., 1987; Bhattacharjee and David, 1987). Furthermore, the integrity of the membranes is affected, caused by a decrease in the lipid content.

3.5. Protective Strategies

Higher plants are known to produce UV-B-absorbing substances that are synthesized and stored in the epidermal layer when induced by short wavelength solar radiation (Beggs et al., 1986; Tevini et al., 1983; Murali and Teramura, 1985). Similarly, a group of UV-absorbing substances has been found in phytoplankton, which could be identified as mycosporine-like amino acids (Carretto et al., 1990; Karentz et al., 1991b; Raven, 1991). These substances are even passed to the next trophic level in the food web by predation. Cyanobacteria have been found to produce a UV-B-induced shielding substance, called scytonemin, which is incorporated into the slime sheaths of the organisms. The production of this substance prevents most of the UV-B-dependent bleaching of chlorophyll (Garcia-Pichel and Castenholz, 1991). Other cyanobacteria were found to produce shock proteins in response to UV irradiation (Shibata et al., 1991).

4. Consequences of UV-B Damage in Aquatic Ecosystems

One of the most important consequences of solar UV-B-induced damage is the loss of biomass production in aquatic ecosystems. In contrast to terrestrial ecosystems, aquatic ecosystems have a rather small standing crop, but the productivity is comparative to that of terrestrial ecosystems (Houghton and Woodwell, 1989). Because of the enormous size of the marine ecosystems, even a small loss in biomass productivity has noticeable adverse effects (Häder et al., 1995). However, decreases related to increased solar UV-B radiation are difficult to assess since no reliable data are available for the era before ozone depletion; thus, our present knowledge is too limited to predict exact losses and damages on a global basis. Any sizable reductions in primary productivity is bound to result in significant reduction in fisheries catch (Hardy and Gucinski, 1989).

The annual uptake of atmospheric CO_2 by aquatic ecosystems has been estimated to amount to about 100 Gt (calculated as carbon) annually, which equals that incorporated by all terrestrial ecosystems taken together. Thus, the phytoplankton in the oceans is a major biological sink for atmospheric CO_2 (Gaundry et al., 1987), and any decrease in the phytoplankton populations will result in an increase in the atmospheric CO_2 concentration, augmenting the greenhouse effect and the consequent sea level rise (Schneider, 1989).

Sensitivity of phytoplankton to solar UV-B radiation varies between species and during their developmental cycle; therefore, changes in the species composition are predicted as a consequence of ozone depletion (McLeod and McLachlan, 1959; Worrest et al., 1978, 1981a,b). This effect will be relayed through the subsequent trophic levels in the food web. Changes in the species composition also bear the risk of the development of toxic algal blooms (such as dinoflagellates or cyanobacteria), which are known to cause poisoning in fish, mussels, crabs, and other primary and secondary consumers, including humans.

5. Conclusions

Based on intensive research during the past few years there is ample evidence that aquatic habitats are under considerable UV-B stress even at ambient ultraviolet radiation levels. Any substantial depletion of the ozone layer will have detrimental effects on the aquatic ecosystems. Due to the enormous size of the systems, even a small loss in productivity could have substantial impacts on a global scale. Ecosystems have a limited buffer capacity to tolerate moderate stresses; when this capacity is exceeded, the system responds with disproportionally large changes. Our current knowledge is too limited to identify this threshold. While experimental data indicate substantial consequences of enhanced levels of solar UV-B radiation, extrapolating laboratory findings to the natural habitat is difficult and uncertain, and long-term monitoring on a global basis, e.g., by satellite imaging is necessary.

ACKNOWLEDGMENTS. This work was supported by financial aid from the Bundesminister für Forschung und Technologie (project KBF 57), the European Community, and the State of Bavaria (BayForKlim).

References

Anderson, J. G., Toohey, D. W., and Brune, W. H., 1991, Free radicals within the Antarctic vortex: The role of CFCs in Antarctic ozone loss, *Science* **251**:39-46.

Atkinson, R. J., Matthews, W. A., Newman, P. A., and Plumb, R. A., 1989, Evidence of the mid-latitude impact of Antarctic ozone depletion, *Nature* **340**:290-294.

Baker, K. S., and Smith, R. C., 1982, Spectral irradiance penetration in natural waters, in: *The Role of Solar Ultraviolet Radiation in Marine Ecosystems* (J. Calkins, ed.), Plenum Press, New York, pp. 233-246.

Bean, B., 1985, Microbial geotaxis, in: *Membranes and Sensory Transduction* (G. Colombetti and F. Lenci, eds.), Plenum Press, New York, pp. 163-198.

Beggs, C. J., Schneider-Ziebert, U., and Wellmann, E., 1986, UV-B radiation and adaptive mechanisms in plants, in: *Stratospheric Ozone Reduction. Solar Ultraviolet Radiation and Plant Life* (R. C. Worrest and M. M. Caldwell, eds.) NATO ASI Series, Vol. G8, Springer, Heidelberg, pp. 235-250.

Bhattacharjee, S. K., and David, K. A. V., 1987, UV-sensitivity of cyanobacterium *Anacystis nidulans*: Part II—a model involving photosystem (PSII) reaction centre as lethal target and herbicide binding high turnover B protein as regulator of dark repair, *Indian J. Exp. Biol.* **25**:837-842.

Bhattacharjee, S. K., Mathur, M., Rane, S. S., and David, K. A. V., 1987, UV-sensitivity of cyanobacterium *Anacystis nidulans*: Part I—evidence for photosystem II (PSII) as a lethal target and constitutive nature of a dark-repair system against damage to PSII, *Indian J. Exp. Biol.* **25**:832-836.

Bidigare, R. R., 1989, Potential effects of UV-B radiation on marine organisms of the Southern Ocean: Distributions of phytoplankton and krill during Austral spring, *Photochem. Photobiol.* **50**:469-477.

Blumthaler, M., and Ambach, W., 1990, Indication of increasing solar ultraviolet-B radiation flux in alpine regions, *Science* **248**:206-208.

Brasseur, G. P., 1989, A dent outside the hole?, *Nature* **342**:225-226.

Brodhun, B., and Häder, D.-P., 1990, Photoreceptor proteins and pigments in the paraflagellar body of the flagellate *Euglena gracilis*, *Photochem. Photobiol.* **52**:865-871.

Brown, P. C., and Cochrane, K. L., 1991, Chlorophyll *a* distribution in the southern Benguela: Possible effects of global warming on phytoplankton and its implication for pelagic fish, *Suid-Afrikaanse Tydskrif vir Wetenskap* **87**:233-242.

Brune, W. H., Anderson, J. G., Toohey, D. W., Fahey, D. W., Kawa, S. R., Jones, R. L., McKenna, D. S., and Poole, L. R., 1991, The potential for ozone depletion in the Arctic polar stratosphere, *Science* **252**:1260-1266.

Büchel, C., and Wilhelm, C., 1993, In vivo analysis of slow chlorophyll fluorescence induction kinetics in algae: progress, problems and perspective, *Photochem. Photobiol.* **58**:137-148.

Burns, N. M., and Rosa, F., 1980, *In situ* measurements of the settling velocity of organic carbon particles and ten species of phytoplankton, *Limnol. Oceanogr.* **2**:855-864.

Cabrera, S., and Montecino, V., 1987, Productividad primaria en ecosistemas limnicos, *Arch. Biol. Med. Exp.* **20**:105–116.

Calkins, J., and Thordardottir, T., 1980, The ecological significance of solar UV-B radiations on aquatic organisms, *Nature* **283**:563–566.

Carpenter, E. J., and Romans, K., 1991, Major role of the cyanobacterium *Trichodesmium* in nutrient cycling in the North Atlantic Ocean, *Science* **254**:1356–1358.

Carreto, J. J., Carignana, M. O., Daleo, G., and de Marco, S. G., 1990, Occurrence of mycosporine-like amino acids in the red tide dinoflagellate *Alexandrium excavatum:* UV-photoprotective compounds, *J. Plankton Res.* **12**:909–921.

Cullen, J. J., and Lesser, M. P., 1991, Inhibition of photosynthesis by ultraviolet radiation as a function of dose and dosage rate: results for a marine diatom, *Marine Biol.* **111**:183–190.

Cullen, J. C., Neale P. J., and Lesser, M. P., 1992, Biological weighting function for the inhibition of phytoplankton photosynthesis by ultraviolet radiation, *Science* **258**: 646–650.

Donkor, V., and Häder, D.-P., 1991, Effects of solar and ultraviolet radiation on motility, photomovement and pigmentation in filamentous, gliding cyanobacteria, *FEMS Microbiol. Ecol.* **86**:159–168.

Donkor, V. A., Amewowor, D. H. A. K., and Häder, D.-P.,1993a, Effects of tropical solar radiation on the motility of filamentous cyanobacteria, *FEMS Microbiol. Ecol.* **12**:143–148.

Donkor, V.A., Amewowor, D. H. A. K., and Häder, D.-P.,1993b, Effects of tropical solar radiation on the velocity and photophobic behavior of filamentous gliding cyanobacteria, *Acta Protozool.* **32**:67–72.

Eggersdorfer, B., and Häder, D.-P., 1991a, Phototaxis, gravitaxis and vertical migrations in the marine dinoflagellate, *Prorocentrum micans*, *Eur. J. Biophys.* **85**:319–326.

Eggersdorfer, B., and Häder, D.-P., 1991b, Phototaxis, gravitaxis and vertical migrations in the marine dinoflagellates, *Peridinium faeroense* and *Amphidinium caterii*, *Acta Protozool.* **30**:63–71.

Ekelund, N. G. A., 1991, The effect of UV-B radiation on dinoflagellates, *J. Plant Physiol.* **138**:274–278.

El Sayed, S. Z., 1988a, Fragile life under the ozone hole, *Natural History* **97**:73–80.

El Sayed, S. Z., 1988b, Productivity of the Southern Ocean: A closer look, *Comp. Biochem. Physiol.* **90B**:589–498.

Esquivel, D. M. S., and de Barros, H. G. P. L., 1986, Motion of magnetotactic microorganisms, *J. Exp. Biol.* **121**:153–163.

Frederick, J. E., 1990, Trends in atmospheric ozone and ultraviolet radiation: Mechanisms and observations for the northern hemisphere, *Photochem. Photobiol.* **51**:757–763.

Frederick, J. E., Snell, H. E., and Haywood, E. K., 1989, Solar ultraviolet radiation at the earth's surface, *Photochem. Photobiol.* **50**:443–450.

Frederick, J. E., Weatherhead, E. C., and Haywood, E. K., 1991, Long-term variations in ultraviolet sunlight reaching the biosphere: Calculations for the past three decades, *Photochem. Photobiol.* **54**:781–788.

Galland, P., Keiner, P., Dörnemann, D., Senger, H., Brodhun, B., and Häder, D.-P., 1990, Pterin- and flavin-like fluorescence associated with isolated flagella of *Euglena gracilis*, *Photochem. Photobiol.* **51**:675–680.

Garcia-Pichel, F., and Castenholz, R. W., 1991, Characterization and biological implications of scytonemin, a cyanobacterial sheath pigment, *J. Phycol.* **27**:395–409.

Gaundry, A., Monfray, P., Polian, G., and Lanabert, G., 1987, The 1982–1983 El Nino: a 6 billion ton CO_2 release, *Tellus* **39B**:209–213.

Genty, B. E., Briantais, J. M., and Baker, N. R., 1989, Relative quantum efficiencies of the two photosystems of leaves in photorespiratory and non-photorespiratory conditions, *Plant Physiol. Biochem.* **28**:1–10.

Gerber, S., and Häder, D.-P., 1992, UV effects on photosynthesis, proteins and pigmentation in the flagellate *Euglena gracilis*: biochemical and spectroscopic observations, *Biochem. System. Ecol.* **20**:485–492.

Gerber, S., and Häder, D.-P., 1993, Effects of solar irradiation on motility and pigmentation of three species of phytoplankton, *Env. Exp. Biol.* **33**:515–521.

Gieskes, W. C., and Kraay, G. W., 1990, Transmission of ultraviolet light in the Weddell Sea. Report on the first measurements made in Antarctic, *Biomass Newsletter* **12**:12–14.

Gosink, J. J., Irgens, R. L., and Staley, J. T., 1993, Vertical distribution of bacteria in Arctic sea ice, *FEMS Microbiol. Ecol.* **102**:85–90.

Grant, W. B., 1988, Global stratospheric ozone and UVB radiation, *Science* **242**:1111.

Häberlein, A., and Häder, D.-P., 1992, UV effects on photosynthetic oxygen production and chromoprotein composition in the freshwater flagellate *Cryptomonas* S2, *Acta Protozool.* **31**: 85–92.

Häder, D.-P., 1985, Effects of UV-B on motility and photobehavior in the green flagellate, *Euglena gracilis*, *Arch. Microbiol.* **141**:159–163.

Häder, D.-P., 1986a, Effects of solar and artificial UV irradiation on motility and phototaxis in the flagellate, *Euglena gracilis*, *Photochem. Photobiol.* **44**:651–656.

Häder, D.-P., 1986b, The effect of enhanced solar UV-B radiation on motile microorganisms, in: *Stratospheric Ozone Reduction, Solar Ultraviolet Radiation and Plant Life*, (R.C. Worrest, and M.M. Caldwell, eds.), Springer Verlag, New York, pp. 223–233.

Häder, D.-P., 1987, Polarotaxis, gravitaxis and vertical phototaxis in the green flagellate, *Euglena gracilis*, *Arch. Microbiol.* **147**:179–183.

Häder, D.-P., 1988a, Ecological consequences of photomovement in microorganisms, *J. Photochem. Photobiol. B: Biol.* **1**:385–414.

Häder, D.-P., 1988b, Signal perception and amplification in photoresponses of cyanobacteria, *Biophys. Chem.* **29**:155–159.

Häder, D.-P., 1991a, Effects of enhanced solar ultraviolet radiation on aquatic ecosystems, in: *Biophysics of Photoreceptors and Photomovements in Microorganisms* (F. Lenci, F. Ghetti, G. Colombetti, D.-P. Häder, and P.-S. Song, eds.), Plenum Press, New York, pp. 157–172.

Häder, D.-P., 1991b, Phototaxis and gravitaxis in *Euglena gracilis*, in: *Biophysics of Photoreceptors and Photomovements in Microorganisms*, (F. Lenci, F. Ghetti, G. Colombetti, D.-P. Häder, and P.-S. Song, eds.), Plenum Press, New York, pp. 203–221.

Häder, D.-P., and Brodhun, B., 1991, Effects of ultraviolet radiation on the photoreceptor proteins and pigments in the paraflagellar body of the flagellate, *Euglena gracilis*, *J. Plant Phys.* **137**:641–646.

Häder, D.-P., and Häder, M., 1988a, Inhibition of motility and phototaxis in the green flagellate, *Euglena gracilis*, by UV-B radiation, *Arch. Microbiol.* **150**:20–25.

Häder, D.-P., and Häder, M., 1988b, Ultraviolet-B inhibition of motility in green and dark bleached *Euglena gracilis*, *Current Microbiol.* **17**:215–220.

Häder, D.-P., and Häder, M.A., 1989c, Effects of solar UV-B irradiation on photomovement and motility in photosynthetic and colorless flagellates, *Environ. Exp. Bot.* **29**:273–282.

Häder, D.-P., and Häder, M., 1989a, Effects of solar radiation on photoorientation, motility and pigmentation in a freshwater *Cryptomonas*, *Botanica Acta* **102**:236–240.

Häder, D.-P., and Häder, M. A., 1989b, Effects of solar and artificial radiation on motility and pigmentation in *Cyanophora paradoxa*, *Arch. Microbiol.* **152**:453–457.

Häder, D.-P., and Häder, M., 1990a, Effects of solar radiation on motility, photomovement and pigmentation in two strains of the cyanobacterium, *Phormidium uncinatum*, *Acta Protozool.* **29**:291–303.

Häder, D.-P., and Häder, M., 1990b, Effects of UV radiation on motility, photo-orientation and pigmentation in a freshwater *Cryptomonas*, *J. Photochem. Photobiol. B: Biol.* **5**:105–114.

Häder, D.-P., and Häder, M., 1991a, Effects of solar and artificial UV radiation on motility and pigmentation in the marine *Cryptomonas maculata*, *Env. Exp. Bot.* **31**:33–41.

Häder, D.-P., and Häder, M. A., 1991b, Effects of solar radiation on motility in *Stentor coeruleus*, *Photochem. Photobiol.* **54**:423–428.

Häder, D.-P., and Hoiczyk, E., 1992, Gliding motility, in: Algal Cell Motility (M. Melkonian, ed.), Chapman and Hall, New York, pp. 1–38.

Häder, D.-P., and Liu, S.-L., 1990a, Effects of artificial and solar UV-B radiation on the gravitactic orientation of the dinoflagellate, *Peridinium gatunense*, *FEMS Microbiol. Ecol.* **73**:331–338.

Häder, D.-P., and Liu, S.-M., 1990b, Motility and gravitactic orientation of the flagellate, *Euglena gracilis*, impaired by artificial and solar UV radiation, *Curr. Microbiol.* **21**:161–168.

Häder, D.-P., and Schäfer, J., 1994a, In-situ measurement of photosynthetic oxygen production in the water column, *Environm. Monit. Assessm.* **32**:259–268.

Häder, D.-P., and Schäfer, J., 1994b, Photosynthetic oxygen production in macroalgae and phytoplankton under solar irradiation, *J. Plant Physiol.* **144**:293–299.

Häder, D.-P., and Worrest, R. C., 1991, Effects of enhanced solar ultraviolet radiation on aquatic ecosystems, *Photochem. Photobiol.* **53**:717–725.

Häder, D.-P., Colombetti, G., Lenci, F., and Quaglia, M., 1981, Phototaxis in the flagellates, *Euglena gracilis* and *Ochromonas danica*, *Arch. Microbiol.* **130**:78–82.

Häder, D.-P., Watanabe, M., and Furuya, M., 1986, Inhibition of motility in the cyanobacterium, *Phormidium uncinatum*, by solar and monochromatic UV irradiation, *Plant Cell Physiol.* **27**:887–894.

Häder, D.-P., Rhiel, E., and Wehrmeyer, W., 1988, Ecological consequences of photomovement and photobleaching in the marine flagellate *Cryptomonas maculata*, *FEMS Microbiol. Ecol.* **53**:9–18.

Häder, D.-P., Worrest, R. C., and Kumar, H. D., 1989, Aquatic ecosystems, *UNEP Environmental Effects Panel Report*, pp. 39–48.

Häder, D.-P., Häder, M., Liu, S.-M., and Ullrich, W., 1990a, Effects of solar radiation on photoorientation, motility and pigmentation in a freshwater *Peridinium*, *BioSystems* **23**:335–343.

Häder, D.-P., Liu, S.-M., Häder, M., and Ullrich, W., 1990b, Photoorientation, motility and pigmentation in a freshwater *Peridinium* affected by ultraviolet radiation, *Gen. Physiol. Biophys.* **9**:361–371.

Häder, D.-P., Worrest, R. C., and Kumar, H. D., 1991, Aquatic ecosystems, *UNEP Environmental Effects Panel Report*, pp. 33–40.

Häder, D.-P., Worrest, R. C., Kumar, H. D., and Smith, R. C., 1994, Effects of increased solar ultraviolet radiation on aquatic ecosystems, *UNEP Environmental Effects Panel Report*, pp. 65–77.

Häder, D.-P., Worrest, R. C., Kumar, H. D., and Smith, R. C., 1995, Effects of increased solar ultraviolet radiation on aquatic ecosystems, *AMBIO* **24**:174–180.

Hardy, J., and Gucinski, H., 1989, Stratospheric ozone depletion: implications for marine ecosystems, *Oceanogr. Mag.* **2**:18–21.

Heath, D. F., 1988, Non-seasonal changes in total column ozone from satellite observations, 1970–86, *Nature* **332**:219–227.

Helbing, E. W., Villafañe, V., Ferrario, M., and Holm-Hansen, O., 1991, Impact of natural ultraviolet radiation on rates of photosynthesis and on specific marine phytoplankton species, *Marine Ecology Progress Series* **80**:89–100.

Hirosawa, T., and Miyachi, S., 1983, Inactivation of Hill reaction by long-wavelength ultraviolet radiation (UV-A) and its photoreactivation by visible light in the cyanobacterium, *Anacystis nidulans*, *Arch. Microbiol.* **135**:98–102.

Hofmann, D. J., and Deshler, T., 1991, Evidence from balloon measurements for chemical depletion of stratospheric ozone in the Arctic winter of 1989–90, *Nature* **349**:300–305.

Hough, A. M., and Derwent, R. G., 1990, Changes in the global concentration of tropospheric ozone due to human activities, *Nature* **344**:645–648.
Houghton, R. A., and Woodwell, G. M., 1989, Global climatic change, *Sci. Amer.* **260**:18–26.
Huang, T.-C., and Chow, T.-J., 1988, Comparative studies of some nitrogen-fixing unicellular cyanobacteria isolated from rice fields, *J. Gen. Microbiol.* **134**:3089–3097.
Ignatiades, L., 1990, Photosynthetic capacity of the surface microlayer during the mixing period, *J. Plankton Res.* **12**:851–860.
Ito, T., 1983, Photodynamic agents as tools for cell biology, in: *Photochemical and Photobiological Reviews* (K.C. Smith, ed.), Volume 7, Plenum Press, New York, pp. 141–186.
Jerlov, N. G., 1970, Light–general introduction, in: *Marine Ecology* (O. Kinne, ed.), Volume 1, Wiley, New York, pp. 95–102.
Karentz, D., 1991, Ecological considerations of Antarctic ozone depletion, *Antarctic Sci.* **3**:3–11.
Karentz, D., and Lutze, L. H., 1990, Evaluation of biologically harmful ultraviolet radiation in Antarctica with a biological dosimeter designed for aquatic environments, *Limnol. Oceanogr.* **35**:549–561.
Karentz, D., Cleaver, J. E., and Mitchell, D. L., 1991a, DNA damage in the Antarctic, *Nature* **28**:350.
Karentz, D., Cleaver, J. E., and Mitchell, D. L., 1991b, Cell survival characteristics and molecular responses of Antarctic phytoplankton to ultraviolet-B radiation, *J. Phycol.* **27**:326–341.
Karentz, D., Mc Euen, F. S., Land, M.C., and Dunlap, W. C., 1991c, Survey of mycosporine-like amino acid compounds in Antarctic marine organisms: Potential protection from ultraviolet exposure, *Marine Biology* **108**:157–166.
Kerr, R. A., 1990, Another deep Antarctic ozone hole, *Science* **250**:370.
Kumar, A., and Kumar, H. D., 1988, Nitrogen fixation by blue-green algae, in: *Plant Physiology Research* (S.P. Seu, ed.), Society for Plant Physiology and Biochemistry, First International Congress of Plant Physiology, New Delhi, pp. 15–22.
Lindholm, T., 1992, Ecological role of depth maxima of phytoplankton, *Arch. Hydrobiol. Beih. Ergebn. Limnol.* **35**:33–45.
Lohrenz, S. E., Arnone, R. A., Wiesenburg, D. A., and DePalma, I. P., 1988, Satellite detection of transient enhanced primary production in the western Mediterranean Sea, *Nature* **335**:245–247.
Lorenzen, C. J., 1979, UV radiation and phytoplankton photosynthesis, *Limnol. Oceanogr.* **24**:1117–1120.
Lubin, D., Frederick J. E., Booth, C. R., Lucas, T., and Neuschuler, D., 1989, Measurements of enhanced springtime ultraviolet radiation at Palmer Station, Antarctica, *Geophys. Res. Lett.* **16**:783–785.
MacNab, R. M., 1985, Biochemistry of sensory transduction in bacteria, in: *Sensory Perception and Transduction in Aneural Organisms* (G. Colombetti, F. Lenci, and P.-S. Song, eds.), Plenum Press, New York, pp. 31–46.
Madronich, S., McKenzie, R. L., Caldwell, M. M. and Björn, L. O., 1994, Changes in ultraviolet radiation reaching the Earth's surface, *Environmental Effects Panel Report,* United Nations Environmental Program, pp. 1–13.
Maske, H., 1984, Daylight ultraviolet radiation and the photoinhibition of phytoplankton carbon uptake, *J. Plankton Res.* **6**:351–357.
McLeod, G. C. and McLachlan, J., 1959, The sensitivity of several algae to ultraviolet radiation of 2537 Å. *Physiol. Plant.* **12**:306–309.
Mitchell, B. G., 1990, Action spectra of ultraviolet photoinhibition of Antarctic phytoplankton and a model of spectral diffuse attenuation coefficients, in: *Proceedings of Workshop on Response of Marine Phytoplankton to Natural Variations in UV-B Flux,* Scripps Institution of Oceanography, La Jolla, Appendix H.
Murali, N. S., and Teramura, A. H., 1985, Effects of ultraviolet-B irradiance on soybean. VI.

Influence of phosphorus nutrition on growth and flavonoid content, *Physiol. Plant.* **63**:413–416.

Nultsch, W., and Agel, G., 1986, Fluence rate and wavelength dependence of photobleaching in the cyanobacterium *Anabaena variabilis*, *Arch. Microbiol.* **144**:268–271.

Nultsch, W., and Häder, D.-P., 1988, Photomovement in motile microorganisms II, *Photochem. Photobiol.* **47**:837–86.

Padhy, R. N., 1985, Cyanobacteria employed as fertilizers and waste disposers, *Nature* **317**:475–476.

Paerl, H. W., and Bebout, B. M., 1988, Direct measurement of O_2-depleted microzones in marine *Oscillatoria*: relation to N_2 fixation, *Science* **241**:441–445.

Peak, J. G., Peak, M. J., Sikorski, R. S., and Jones, C. A., 1985, Induction of DNA–protein crosslinks in human cells by ultraviolet and visible radiations: action spectrum, *Photochem. Photobiol.* **41**:295–302.

Piazena, H., and Häder, D.-P., 1994, Penetration of solar UV irradiation in coastal lagoons of the Southern Baltic Sea and its effect on phytoplankton communities, *Photochem. Photobiol.* **60**:463–469.

Poff, K. L., 1985, Temperature sensing in microorganisms, in: *Sensory Perception and Transduction in Aneural Organisms* (G. Colombetti, F. Lenci, and P.-S.Song, eds.), Plenum Press, New York, pp. 299–307.

Proffitt, M. H., Fahey, D. W., Kelly, K. K., and Tuck, A. F., 1989, High-latitude ozone loss outside the Antarctic ozone hole, *Nature* **342**:233–237.

Raven, J. A., 1991, Responses of aquatic photosynthetic organisms to increased solar UVB, *J. Photochem. Photobiol., B: Biol.* **9**:239–244.

Renger, G., and Schreiber, U. 1986, Practical applications of flourometric methods to algae and higher plant research, in: *Light Emission by Plants and Bacteria*, Volume 47 (Govindjee, J. Amesz, and D.C. Fork, eds.), Academic Press, New York, pp. 587–619.

Renger, G., Völker, M., Eckert, H. J., Fromme, R., Hohm-Veit, S., and Gräber, P., 1989, On the mechanisms of photosystem II deterioration by UV-B irradiation, *Photochem. Photobiol.* **49**:97–105.

Rowland, F. S., 1989, Chlorofluorocarbons and the depletion of stratospheric ozone, *Am. Scientist* **77**:36–46.

Schäfer, J., Sebastian, C., and Häder, D.-P., 1993, Effects of solar radiation on motility, orientation, pigmentation and photosynthesis in a green dinoflagellate *Gymnodinium*, *Acta Protozool.* **33**:59–65.

Schneider, S. H., 1989, The changing climate, *Sci. Am.* **261**:38–47.

Schnell, R. C., Liu, S. C., Oltmans, S. J., Stone, R. S., Hofmann, D. J., Dutton, E. G., Deshler, T., Sturges, W. T., Harder, J. W., Sewell, S. D., Trainer, M., and Harris, J. M., 1991, Decrease of summer tropospheric ozone concentrations in Antarctica, *Nature* **351**:726–729.

Schoeberl, M. R., and Hartmann, D. L., 1991, The dynamics of the stratospheric polar vortex and its relation to springtime ozone depletions, *Science* **251**:46–52.

Schreiber, U., Schliwa, U., and Bilger, W., 1986, Continuous recording of photochemical and non-photochemical chlorophyll fluorescence quenching with a new type of modulation fluorometer, *Photosynth. Res.* **10**:51–62.

Schreiber, U., Bilger, W., and Neubauer, C., 1994, Chlorophyll fluorescence as a nonintrusive indicator for rapid assessment of in vivo photosynthesis, in: *Ecophysiology of Photosynthesis. Ecological Studies*, Volume 100 (E. D. Schulze, and M. M. Caldwell, eds.), Springer Verlag, Berlin, pp. 49–70.

Schreiber, U., Endo, T., Mi, H., and Asada, K., 1995, Quenching analysis of chlorophyll fluorescence by the saturation pulse method: particular aspects relating to the study of eukaryotic algae and cyanobacteria, *Plant Cell Physiol.* **36**:873–882.

Shibata, H., Baba, K., and Ochiai, H., 1991, Near-UV irradiation induces shock proteins in *Anacystis nidulans* R-2; possible role of active oxygen, *Plant Cell Physiol.* **32**:771–776.
Siebeck, O., and Böhm, U., 1987, *Untersuchungen zur Wirkung der UV-B-Strahlung auf kleine Wassertiere*, BPT Bericht, Gesellschaft für Strahlen- und Umweltforschung, Munich, pp. 84.
Sinha, R. P, and Kumar, A., 1992, Screening of blue-green algae for biofertilizer, in: *Proceedings of the National Seminar on Organic Farming* (P. L. Patil, ed.), Pune, India, 95–97.
Sinha, R. P., Kumar, H. D., Kumar, A., and Häder, D.-P., 1995a, Effects of UV-B irradiation on growth, survival, pigmentation and nitrogen metabolism enzymes in cyanobacteria, *Acta Protozool.* **34**:187–192.
Sinha, R. P., Lebert, M., Kumar, A., Kumar, H. D., and Häder, D.-P.,1995b, Spectroscopic and biochemical analyses of UV effects of phycobiliproteins of *Anabaena* sp. and *Nostoc carmium*, *Bot. Acta* **108**:87–92.
Smith, R., 1989, Ozone, middle ultraviolet radiation and the aquatic environment, *Photochem. Photobiol.* **50**:459–468.
Smith, R. C., and Baker, K. S., 1978, Penetration of UV-B and biologically effective dose-rates in natural waters, *Photochem Photobiol.* **29**:311 323.
Smith, R. C., Baker, K. S., Holm-Hansen, O., and Olson, R., 1980, Photoinhibition of photosynthesis in natural waters, *Photochem. Photobiol.* **31**:585–592.
Smith, R. C., Prezelin, B. B., Baker, K. S., Bidigare, R. R., Boucher, N. P., Coley, T., Karentz, D., MacIntyre, S., Matlick, H. A., Menzies, D., Ondrusek, M., Wan, Z., and Waters, K. J., 1992, Ozone depletion: Ultraviolet radiation and phytoplankton biology in Antarctic waters, *Science* **255**:952–959.
Stewart, W. D. P., 1980, Some aspects of structure and function in N_2-fixing cyanobacteria, *Annu. Rev. Microbiol.* **34**:497–536
Storch, T. A., Saunders, G. W., and Ostrofsky, M. L., 1990, Diel nitrogen fixation by cyanobacterial surface blooms in Sanctuary Lake, Pennsylvania, *Appl. Environm. Microbiol.* **56**:466–471.
Tevini, M., Thoma, U., and Iwanzik, W., 1983, Effects of enhanced UV-B radiation on germination, seedling growth, leaf anatomy and pigments of some crop plants, *Z. Pflanzenphysiol.* **109**:435–448.
Ting, C. S., and Owens, T. G., 1992, Limitations of the pulse-modulated technique for measuring the fluorescence characteristics of algae, *Plant Physiol.* **100**:367–373.
Tirlapur, U., Scheuerlein, R., and Häder, D.-P., 1993, Motility and orientation of a dinoflagellate, *Gymnodinium*, impaired by solar and ultraviolet radiation, *FEMS Microbiol. Ecol.* **102**:167–174.
Vaida, V., Solomon, S., Richard, E. C., Rühl, E., and Jefferson, A., 1989, Photoisomerization of OCIO: a possible mechanism for polar ozone depletion, *Nature* **342**:405–408.
Venkataraman, G. S., 1981, Blue-green algae: a possible remedy to nitrogen scarcity, *Curr. Sci.* **50**:253–256.
Viollier, M., Tanré, D., and Deschampes, P. Y., 1980, An algorithm for remote sensing of water color from space, *Boundary-Layer Meteorol.* **18**:247–267.
Vosjan, J. H., Döhler, G., and Nieuwland, G., 1990, Effect of UV-B irradiance on the ATP content of microorganisms of the Weddell Sea Antarctica, *Neth. J. Sea Res.* **25**:391–394.
Walsby, A. E., Kinsman, R., and George, K. I., 1992, The measurement of gas volume and buoyant density in planktonic bacteria, *J. Microbiol. Meth.* **15**:293–309.
Webb, A. R., 1991, Solar ultraviolet radiation in Southeast England: the case for spectral measurements, *Photochem. Photobiol.* **54**:789–794.
Wei, D.-W., 1991, On the formation of the Antarctic ozone hole and its trend predictions, *Science in China B* **34**:95–103.
Weis, E., and Berry, J., 1987, Quantum efficiency of photosystem II in relation to energy-dependent quenching of chlorophyll fluorescence, *Biochim. Biophys. Acta* **894**:198–208.

Worrest, R. C., 1982, Review of literature concerning the impact of UV-B radiation upon marine organisms, in: *The Role of Solar Ultraviolet Radiation in Marine Ecosystems* (J. Calkins, ed.), Plenum Press, New York, pp. 429–457.

Worrest R. C., van Dyke, H., and Thomson, D., 1978, Impact of enhanced simulated solar ultraviolet radiation upon a marine community, *Photochem. Photobiol.* **27**:471–478.

Worrest, R. C., Brooker, D. L., and van Dyke, H., 1980, Results of a primary productivity study as affected by the type of glass in the culture bottle, *Limnol. Oceanogr.* **25**:360–364.

Worrest, R. C., Thompson, B. E., and van Dyke, H., 1981a, Impact of UV-B radiation upon estuarine microcosms, *Photochem. Photobiol.* **33**:861–867.

Worrest, R. C., Wolniakowski, K. U., Scott, J. D., Brooker, D. L., Thompson B. E., and van Dyke, H., 1981b, Sensitivity of marine phytoplankton to UV-B radiation: Impact upon a model ecosystem, *Photochem. Photobiol.* **33**:223–227.

Yamamoto, K. M., Satake, M., Shinagawa, H., and Fujiwara, Y., 1983, Amelioration of the ultraviolet sensitivity of an *Escherichia coli* recA mutant in the dark by photoreactivating enzyme, *Mol. Gen. Genet.* **190**:511–515.

Yasuhira, S., Mitani, H., and Shima, A., 1992, Enhancement of photorepair of ultraviolet-induced pyrimidine dimers by preillumination with fluorescent light in the goldfish cell line. The relationship between survival and yield of pyrimidine dimers, *Photochem. Photobiol.* **55**:97–101.

Zündorf, I., and Häder, D.-P., 1991, Biochemical and spectroscopic analysis of UV effects in the marine flagellate *Cryptomonas maculata*, *Arch. Microbiol.* **156**:405–411.

2

CASE: Complex Adaptive Systems Ecology

JAN MOLIN and SØREN MOLIN

1. Microbial Ecology

1.1. *In Situ* studies

Microbial ecology is special in the sense that it is at the same time microscopic and macroscopic. Obviously, studies of microorganisms in general are connected with analyses of microcommunities, and there is an increasing tendency to perform detailed studies of microbial ecosystems, which has become possible through an amazing development of microscopic *in situ* techniques. It is, however, the macroscopic part of microbial ecology, that most evidently has the greatest impact on life on the planet, and therefore the majority of attention has been devoted to the development of this field of research (Atlas and Bartha, 1993; Brock and Madigan, 1988). This part relates to the geochemical cycles of important elements like carbon, oxygen, sulphur, nitrogen, and hydrogen. Bacteria are main actors in the conversion of these elements through their different oxidation states, and the global balances between these states have a fundamental significance for the understanding of the composition and maintenance of the atmosphere and the biosphere on earth.

There is also a traditional focus on studies of symbiotic relationships in microbial ecology. The close and mutually dependent interaction pattern between two or more organisms are model examples of cooperative activities in nature, and the interactions developed between symbiotic partners are also elements of adaptive and developmental traits of significance in biological evolution. In some cases symbiosis may attain an irreversible state in which two organisms form an integrated life form (Amann *et al.*, 1991; Bryant *et al.*, 1967). Other variants of

---

JAN MOLIN • Institute of Organization and Industrial Sociology, Copenhagen Business School, DK-2200N Blaagaardsgade 23B, Denmark.   **SØREN MOLIN** • Department of Microbiology, Technical University of Denmark, DK-2800 Lyngby, Denmark.
Advances in Microbial Ecology, Volume 15, edited by Jones. Plenum Press, New York, 1997.

symbiotic relationships are different forms of parasitism that, for example, result in infectious diseases, and such cases of course also attract a lot of interest.

In contrast to the fields of general microbiology and molecular microbiology, which most often refer to simple laboratory model systems (test tubes), microbial ecology is directed towards studies of natural ecosystems. Such systems may be terrestrial or aquatic, and they may be real-world natural (a lake, a marine location, a field, etc.), or they may be natural models established in the laboratory in form of various types of microcosms. In all such cases the overall metabolic processes (conversion of C, O, N, etc.) may be monitored chemically. Such studies do not necessarily depend on knowledge about the involved microorganisms, but often it will be logical to identify the players. In this way taxonomy has its natural role in microbial ecology, and through many comparative studies of different ecosystems showing the same types of metabolic activities, it has been possible to predict the presence of specific groups of bacteria (Atlas and Bartha, 1993; Brock and Madigan, 1988).

In conclusion, the major classical themes of microbial ecology are concerned with fundamental metabolic processes and the participation of the organisms responsible for these processes as they take place in natural environments. The accumulated knowledge about these aspects has resulted in our present view on the global elemental cycles, and at the same time we see increasingly detailed information concerning specific transformations of matter, in particular ecosystems composed of different combinations of microorganisms.

The recent development of a number of specific and very sensitive tools and methods in the field of molecular biology has also attracted attention among microbial ecologists. Above all, the possibility of identification of both well-known and unknown organisms in very complex samples has resulted in a vast increase in the information about composition of microbial communities (Stahl *et al.*, 1988; Amann *et al.*, 1990; Ward *et al.*, 1990). Microelectrodes and tracer techniques have made detailed analyses of local environments possible (Revsbech and Jorgensen, 1988), and the powerful scanning confocal laser microscopes have revolutionized the visualization of the microbial world (Caldwell *et al.*, 1992).

This means that old questions may be approached with new methods and tools, as has most certainly happened. The entire field of bacterial taxonomy has been radically influenced by the introduction of rRNA hybridization (Woese, 1987), non-culturable organisms may be detected *in situ* (Amann *et al.*, 1991; Ward *et al.*, 1992), bacterial activity is recorded at infinitely small levels (Rodrigues *et al.*, 1992), etc. However, the modern biological techniques have also given rise to new questions, which either were not addressable previously or which have been introduced to the field by scientists from other disciplines, who have caught interest in microbial ecology simply because such questions may now be answered (Marshall, 1994).

We certainly do not attempt to present a full review of the field of microbial ecology; neither do we intend to review the current trends in microbial ecology. However, we do wish to relate our models and ideas to basically all corners of microbial ecology, and therefore we directly address the microbial ecologists, asking them to come forward with critique and comments to the picture we lay out.

The microbial ecologists often make a point of distinguishing sharply between microbial physiologists and ecologists; the former group usually work with monocultures in laboratory media, whereas the true ecologists only as an exception deal with single organisms, and certainly would not easily accept the laboratory as an ecologically relevant environment.

We do not share this view fully, but do agree that test tubes containing one strain are hardly ecologically relevant. Nevertheless, we want to argue that some laboratory-based systems may possess ecologically significant features to the same extent as any natural system, and of course the experimental designs for such laboratory-based systems are often much easier to construct and analyze than those directed towards natural systems.

The basic features we find necessary and satisfactory for microbial ecosystems in the present context are contained in the formulation of CASE (complex adaptive systems ecology):

- a microbial community in a certain environment
- the community normally consisting of more than one organism
- the system maintaining some basic features over time (stable) and remaining relatively insensitive to environmental changes (homeostasis)
- the components of the system being organized

To the extent that such microbial community systems can be established and maintained in the laboratory they have significant relevance as model systems for studies of ecologically significant traits (Slater and Hartman, 1982). Therefore, when exploring systemic features and attempting to challenge models for ecosystems it may turn out to be quite productive to create controllable systems as a first step. However, it is equally important to stress that eventually conclusions from such model systems must be transferred to natural communities and ecosystems.

2. CASE

2.1. A Paradigmatic Platform

The analysis of microbial community systems will not be operational within the confinement of conventional laboratory designs. Investigations of complex patterns of microbial ecosystems demand a fundamental reorientation concerning methodological issues. This is, however, a consequential step to take. When we choose to abandon single-organism-based test tube studies in order to comprehend the characteristics of living systems *in situ* we need a paradigmatic reformulation. The conventional experimental designs grew out of a scientific tradition, and as such, conventional analyses performed in the laboratory today work as a consistent way of operationalizing the underlying assumptions of the dominating paradigm. To radically transcend this methodology we need an alternative paradigm (Kuhn, 1962). Without an explicit paradigm we have no platform for the development of a new strategic methodology.

We use the term paradigm to indicate the set of axioms, assumptions, or fundamentals that enable us to create a 'meaningful' order. It is very much like a map of reality. A paradigm is not reality itself, but the directions we use to find our way. Thus, the term indicates on one hand the experiments, or set of procedures, that every member of the scientific discipline learns to appreciate as a necessary methodology to sustain the quality of scientific research; on the other hand, the term "paradigm" also has the broader meaning, as indicated above, associated with a fundamental belief system or map of reality: the lenses through which one sees everything (Schwartz and Ogilvy, 1979).

We wish to describe an emergent paradigm in both senses of the term. We wish to introduce a methodology and research designs that transcend our conventional laboratory-based techniques and to describe a guiding paradigm, a consistent framework that will allow us to model and analyze the characteristics of complex, adaptive ecological systems.

Such conventional paradigms grow out of scientific tradition, and generally the paradigm, despite its implicit nature, will be dominant at the cutting edge of international research. This will be true even more, as long as the underlying scientific paradigm serves as a platform for smooth communication, exchange of results, and professional validation (Molin and Molin, 1988).

We will argue that there is a widely shared conventional paradigm resting on the assumption that there is one, tangible reality 'out there,' and that science one day will be able to build a convergent body of knowledge about this reality (Morgan, 1987). This ontological position allows one to design projects from a series of implicit assumptions, according to which phenomena are (Lincoln, 1985):

- simple and probabilistic
- hierarchically ordered
- of a mechanistic or machinelike nature
- predictable in a determinate universe
- to be explained through laws of direct causality
- assembly systems constructed from series of simple units
- objectively accessible

According to a Stanford Research Institute study across disciplines such as physics, chemistry, mathematics, biology, political theory, and linguistics,

Schwartz and Ogilvy (1979) demonstrated the way these paradigmatic assumptions are widely shared; furthermore, they identify the contours of an emerging alternative. Despite the fact that the positivist perception of a single measurable reality seems to dominate both the natural and the social sciences, theories and models from both mathematics and physics have demonstrated a need to formulate a new paradigm based on the ontological stance that reality is constructed (. . . or created).

The emerging paradigm of multiple realities matches the social constructivist position that plays an increasingly stronger role in modern sociology. The multiple realities gives the emergent paradigm a set of new assumptions to work from, where phenomena are:

- complex and diverse
- heterarchically ordered
- of a holographic nature
- unpredictable in an indeterminate universe
- to be understood through analyses of mutual causality
- complex structures rising from morphogenesis
- contextually tied constructions reflecting changing perspectives.

2.1.1. Naturalistic Inquiry

With the multiple realities follows the conception of an uncontrollable universe that leaves the researcher very little certainty in his way of knowing (Rorty, 1980). Leaving the conventional paradigm and its "naive realism," scientific studies arise through processes of explicit choice. Knowledge is no longer the result of a search for—or an exposure of—some inherent and definite truth. Knowledge is a construction bounded by time and space reflecting the assumptions and values of the researcher who built the design. The significance of the assumptions is that the researcher is forced to choose. Science becomes a question of choices: choice of focus, choice of system/environment boundary, choice of measures, choice of analytic tools, choice of theoretical constructs, etc. (Morgan, 1987).

Our position in the present text is close to this emergent paradigm. With the term "naturalistic inquiry" (Lincoln and Guba, 1985) we want to stress a paradigmatic platform that may allow us to merge experience, methods, and theory from different disciplines. We find that a naturalistic approach may contribute strongly to the field of microbial ecology as we have observed the growing need in this field to handle issues of complexity, diversity, and indeterminacy. However, the paradigmatic discussion has consequences. Introducing a naturalistic inquiry as a paradigmatic platform for the analysis of microbial communities has an impact on all the choices that take the researcher into a position of analyzing ecological issues. If we adopt the ontological stance upon which the naturalistic inquiry rests—the notion of multiple, socially constructed realities—it is, and should be, fundamental to the choices we make concerning our future research designs. Every step in the process of operationalizing research issues into concrete study should reflect the new paradigmatic criteria.

These naturalistic criteria for the design of scientific study, according to Lincoln and Guba (1985), may be described in fourteen operational characteristics:

1. a natural setting: the construction of *in situ* studies
2. the human instrument: the role/impact of the researcher
3. tacit knowledge: the use of intuitive, informal data
4. qualitative methods: the utilization of "soft" measures
5. purposive sampling: the explicit choice of focus
6. inductive data analysis: the appreciation of multiplicity
7. grounded theory: the theory emerging from data and context
8. emergent design: the iterative process of research
9. negotiated outcome: the conscious use of peer reviews
10. case study: the description and analysis of local particulars
11. idiographic interpretation: the concern for the particulars of the case
12. tentative application: the shift from generalization to possible transferability
13. focus-determined boundaries: the conscious choices of the observed system
14. trustworthiness: the call for credibility, transferability, dependability, and confirmability.

Let us argue briefly for the different characteristics, and do so by grouping them into 5 categories.

2.1.1.a. Subjectivity. Naturalistic inquiry rests on two characteristics concerning the role of the researcher. The call for a *human instrument* (2) and for the explicit use of *tacit knowledge* (3) highlights the researcher's profile as one that consciously involves himself in studies of natural phenomena, and conducts his inquiries as a kind of participant observation. It may be a microcosmos that is investigated, but the subjective analysis of a natural community (however micro it may be) is very much affected by the way that the researcher reads himself with his tacit knowledge into the context (e.g., by his choice of focus, the questions he asks, or the frames of analysis he applies).

The human instrument points out that whatever mechanistic conception we employ as an intermediary instrument in the observation of natural phenomena, the focus, interpretation and general impact on the observed context is still a human factor. The analysis of complex adaptive systems demands a human touch, in as much as no non-human instrument will be able on its own to adjust meaningfully to the changings conditions of a living system under observation.

Even in microbial studies we need to acknowledge the inquirer/respondent relationship (Guba, 1985). In naturalistic inquiry the researcher enters a "dialogue" with the field. However obscure the size of community and the disproportion between observer and object, still the relationship is interdependent and parallel to the transformation of information that we normally call dialogue.

Utilization of tacit knowledge points out that the analysis of complex and diverse phenomena in various systemic set-ups takes the researcher into an interpretative mode, where only part of his data derive from direct observation (counts, identifications, and description of specific functions). In the analysis of adaptive systems such formal data only become meaningful when integrated with non-linguistic and intuitive impressions of local contingencies. The subjective assessment of systemic properties plays a central part in the way that hard data are ascribed meaning. Any researcher applies, explicitly or in most cases implicitly, his particular ideas of natural order as he projects himself unto the system in focus (Toulmin, 1961).

2.1.1.b. Focus. Our paradigmatic platform gives the subjective researcher an obligation to construct his scientific field—the target for his inquiry. There are no pre-given objects to analyze; instead, there are multiple realities for the researcher to transform into researchable units. Three characteristics (1, 5, and 13) underline this fundamental point of departure. The researcher is free to define the system he wants to address, but he is obliged to confine his design to a natural setting. As the multiple realities spring from specific contexts, it follows that a subjective researcher must confront the system in its natural condition (or under conditions that may be considered to match this natural setting). Within this natural setting he is free to draw the line between what is the system to be studied and what is outside this system. It is his choice of focus that leads him to define the boundaries of the natural system that shall be his principal frame for inquiry. On this background a wide range of purposive samplings may be employed. No one sampling method (i.e., concerned with objectivity or representativity) will be dominant.

In this way the natural setting points out that microbial ecology is based on *in situ* experiments that may be carried out in or out of laboratory conditions, as long as the conditions of natural context are preserved. Context is not just a stimulus environment but a nested arrangement of structures and processes. As such, the natural setting is the only possible way of getting a glimpse of "reality in flight"; to see how processes are both constrained by contexts at the same time as they shape contexts (Pettigrew, 1995).

Focus-determined boundaries point to the increasingly important role for designing scientific studies guided by the researchers appreciation of systemic problems and issues. Naturalistic case studies more and more address the dilemmas and ambiguities expressed by the complex systems that are accessible to analysis. Over time it becomes less interesting to let our preconceptions of "important phenomena" control the research agenda. Instead, careful exploration of microbial communities with in-depth knowledge of specific systemic characteristics may allow us to assess what appear to be the points of imbalance, contingent problems, or cases of outright malfunction. It is this dialogue between the naturalist and the observed system that allows us to define the focus for intervention. Focus and boundaries therefore derive from the conscious choices made by an intelligent observer, who allows himself to communicate constructively with the complex adaptive system in front of him.

Purposive sampling points to the pragmatic need to choose the system for analysis that will allow you to apply your research focus most constructively. Outside a positivist framework, random or representative sampling lose meaning. As we address a multiplicity of realities the idea of purposive sampling becomes a call for explicit choices. The alternative to random or representative sampling is far from "anything goes" (Feuerabend, 1975). Purposive sampling covers a series of different sampling criteria, which each may give you access to fruitful analyses, from sampling extreme or deviant cases to simple convenience sampling (Patton, 1980).

2.1.1.c. Methodology. The characteristics of subjectivity and focus have methodological implications. There are four criteria (4, 6, 8, and 10) that the subjective researcher must address. On a basis of multiple realities there is a need for coherent, purposively sampled case studies that explicitly probe the system in the specific context of time and place. This will involve not only conventional quantitative measures and counts, but it will also demand an interpretative, qualitative approach that allows the researcher to enter a dialogue with his focused system. The human instrument and the role of tacit knowledge play a significant part of any dialogue between researcher and system. It is the conscious qualitative interpretation that allows us to benefit directly from these subjective characteristics that are always part of a constructed reality. The qualitative interpretations rest on inductive processes, where the contextual understanding of the system grows out of the subjective confrontation between researcher and system. There is no pre-given objective structure or theory to be deduced from our case studies. Instead, we have cases that may enable us to construct meaningful interpretations of systemic properties in given contexts. The inductive data generation becomes an iterative process that will force the researcher to adjust his study underway as a kind of emergent design.

Case study points out that we are no longer in the business of compiling general data in scientific reports. Data are not absolute facts, and they are not general in the sense that they may be later accumulated and convergently brought together. Data constitute one possible description of contextual relationships. Data are tied to the specific contingencies of the observed system, and as such they are confined in time and space. Instead of extracting data and generating conventional scientific reports (be it research protocols, articles, papers, or lectures) we should construct case descriptions. To appreciate the data one needs to understand the relationship to the time and context that spawned, harbored, and supported it (Lincoln and Guba, 1985).

Qualitative methods point to the utilization of a wider methodological repertoire that may be applied in confrontation with the multiple realities inherent in case studies. It is important to emphasize that a naturalistic inquiry is not anti-quantitative. There are numerous examples of highly useful data derived from quantitative measurements, and there are as many future opportunities for the naturalistic inquirer to go on designing quantitative analyses. There is, however, also a need to transcend from the quantitative toolbox, and realize the obvious fit between systemic complexity and diversity and qualitative designs. To monitor and comprehend the local contingencies in a microbial case study you certainly need more than quantitative data. It is the qualitative appreciation of multiplicity that allows us to shift from the question of whether reality exists to the more promising question of what we can make of it (Kelly, 1955).

Inductive data analysis points to an approach where the careful, focused monitoring of patterns, elements, and relations gradually allows you to construct descriptions of systemic characteristics. It is a qualitative search that is based on exploration and dialogue with the case study samples. Instead of forecasting the presence of systemic structures to be empirically verified, the inductive data analysis sees systemic modelling as an end (or an intermediate result). It is the analysis of contextual activities that may lead us to the construction (understanding) of local systemic structures.

Emergent design points to the perhaps somewhat trivial notion that a qualitative and inductive approach involves you in a feedback chain with the observed system (dialogue), through which the focus and attention of observation and analysis may be self-correcting. As there is no *a priori* theory or model to control the research process, it becomes an iterative process to evaluate and adjust the focus and attention of the study. It is the prerogative of the researcher to constantly choose if and how the design should change (Van de Ven, 1992).

2.1.1.d. Analysis. How do we interpret the data that we construct in the qualitative and inductive case study. The naturalistic approach offers three characteristics of the analytical contingencies (7, 9 and 11). First, it is self-evident that any nomothetic (i.e., search for general patterns and laws) ambition is out of place. With multiple realities depicted in focused case studies, and the way that these may allow us to construct knowledge about the system-in-context, we have to abandon the dream of finding universal truths, and confine ourselves to the conscientious idiographic endeavours of constructing local knowledge. The models and theories that we bring forth are the negotiated pictures of a reality that we may construct in a meaningful discussion with our peers. Results are stipulated outcomes that we were able to get agreement on, after having spent time generating idiographic data from a specified context. Thus, theory is grounded in the context of our research, and it represents the negotiated outcomes of scientific discussion.

In this way, *idiographic interpretation* points to the logical consequences of the contextual quality of our casuistic approach. Naturalistic inquiry confines itself to the thorough analyses of contexts. There is no illusion of lawlike generalization (nomothetic research), only the explicit appreciation of the particulars of the case in point. On the basis of quantitative and qualitative data, interwoven with the tacit knowledge of the individual scientist, the idiographic approach may be used to investigate the configurational patterns constituted by the parts of phenomena under study, and the way in which these parts fit within the wider context (Tsoukas, 1989).

Grounded theory points directly to the consequences of an inductive analysis. As the naturalistic approach works from the observation of systemic characteristics towards the formulation of multiple constructed realities, theory formation will always be based on data (Glaser and Straus, 1967). No *a priori* theory could possibly encompass the complexity of adaptive systems in general—no model could possibly explain the multiplicity of realities that we may encounter. Through the careful analysis of local contexts we may, however, be able to construct theories about the systemic conditions that we have observed to be present in the focus of our attention (Turner, 1983).

Negotiated outcomes describe the conditions for theory and modeling that may be the most difficult to accept. Accustomed to a positivist (or post-positivist) research agenda, where the purpose of study is the documented presentation of an objective, true reality, it is not easy to accept the idea that results (theory about a specified systemic context) are based on negotiation. The naturalistic inquiry into multiple realities deprives us of the possibility of letting facts speak for themselves. Thus, the value and importance of scientific results may no longer be inferred from the data themselves. What may be a substantial result in the analysis of a particular systemic setup, may be of only marginal interest in studies of different contexts (Burrell and Morgan, 1979). With multiple realities the value and perspective of concrete outcomes have no absolute scale to be measured against—no convergent body of knowledge to fit into. If results are no longer just pieces in the "grand puzzle," then we need other means of assessing the content and quality of our results. This calls for, first, a capacity for subjective interpretation developed by the individual researcher and second, the ongoing exchange of ideas, data, and results with colleagues. Probably, these two types of assessment will be mutually interdependent.

2.1.1.e. Scientific Control. A very apparent aspect of a paradigmatic platform is the way that it stages the public discourse of science (12, 14). Today the communicative rules of this discourse demonstrate the appreciation of truth and results. The way professionals conduct their discussions in journals and at conferences, as well as the way that modern media convey scientific endeavours to the public, is a unidimensional tribute to generalizability. The cornerstone in today's dominant paradigm is knowledge that is considered objective, true, and generalizable. Only on this basis may scientific results be assessed, only on this basis may findings be applied in the interest of society. Within the naturalistic frame of reference, generalizability is obsolete. It is no surprise that the outcome of idiographic and qualitative case studies may in no way be considered to be generalizable. Instead, we are left with the possibility of tentative application (if we can identify comparable contexts), and a responsibility to demonstrate trustworthiness (we do not see "anything goes" as a desirable alternative). In light of the paradigmatic concept of negotiated outcome, the need to be trustworthy becomes an integral part of future scientific control.

Tentative application points to consequenses of idiographic case studies. With contextual data and in the absence of generalizable knowledge, it is not viable to construct results that are applicable to multitudes of different contexts. The problem of application springs from the ontological position with multiple realities, advising us to be more hesitant than desired to forward theories and outcomes that may apply to systemic conditions in general. In the absence of generalizable laws, the studies may only allow us to propose experiments that test the transferability of results between specified contexts; under such circumstances application becomes a question of fittingness. In different contexts that appear to suit the requirements central to our observed system, we may assume that our results may be meaningful. But we can not be sure.

Criteria for trustworthiness points to the most controversial consequences of naturalistic inquiry. Without generalizability—an idiographic approach based on the subjectively chosen focus and boundaries—it becomes self-contradictory to uphold any desire to meet fundamental scientific criteria, such as internal and external validity, reliability, and objectivity. To many researchers this probably means the end of science. Still, the alternative is not anarchic. We need future efforts to replace them with socially negotiated criteria such as credibility, transferability, dependability, and confirmability (Guba, 1981). Credibility is produced when we focus our research on issues that makes sense to our peers; and credibility comes from the results that are accepted by our peers and the larger audience of different interest groups. Transferability means the degree of similarity between different empirically accessible contexts known to the original investigator. Dependability is a complex assessment of the system–context relation to take into account both factors of instability and factors of phenomenal- or design-induced change. Finally, confirmability means a definition of qualitative objectivity, where the emphasis is placed on the data itself, in as much as observation should be reliably factual and confirmable in a formal audit process.

The fourteen characteristics described in groupings of subjectivity, focus, methodology, analysis, and scientific control strip the researcher of all his accustomed security, and they challenge his professional identity. If it is no longer possible to produce generalizability, what then is the purpose of science. If there is no longer a firm monolithic body of knowledge about the real world, what then can we make of our present experience and accumulated data. If our focus is the study of contextual particulars in relative cases, what then will we ever know about life's big issues (Whitehead, 1967).

The answers are really quite simple. Instead of being a pawn in the game (generating tiny bits of knowledge to a huge puzzle), the naturalistic inquiry liberates the researcher to pursue a forceful role as constructivist (Molin and Molin, 1988). There is no overarching rationality (Rorty, 1980); there are no grand narratives (Lyotard, 1984). There are similarities across changing realities; there are tangible standards across different contexts; but no matter how much we may learn about these regularities, it will not bring us closer to the dream of knowing it all. Insight is a holographic feature. It is through the inductive and tacit appreciation of the diversity and uniqueness found in idiographic analyses that we get insight. In the context of particulars, when we have purposively sampled our systems in focus, we have a platform for the construction of a local reality. The morphogenesis of this local reality is itself a living metaphor for the way that complex adaptive systems may be captured and explained as "images of organization," rephrasing Morgan (1986).

The conventional paradigm holds a promise of letting us "read" the real world. To many scientists today this is recognized to be an illusion. Science has traded the drive for adventure, exploration, and limitless curiosity for security, structure, and coherence. It may make it easier to conduct our research, easier to develop designs and employ our refined instruments. It is, however, an open question as to what the value is of the accumulated knowledge generated through such endeavours (Weick, 1989).

The paradigmatic platform that we have described is not one we have invented in our backyard. It is greatly indebted to the work of Schwartz and Ogilvy, and especially to the unique contribution from Lincoln and Guba. Our efforts have been focused on a wish to take naturalistic inquiry (developed within disciplines that traditionally address higher order living systems: individuals, groups, organizations, and societies), and reconstruct it to be a paradigmatic platform for studies in microbial ecology. As we shall demonstrate in the following section, this platform fits into an open systemic perspective.

3. SCIO—Parameters for Description

3.1. Development of an Analytical Model

In order to investigate ecological communities in all their complexities we need a scientific framework. In the field of microbial ecology, research designs focus on isolated aspects of the total community, e.g., patterns of interaction, behavior of specific organisms, or composition of specific layers in a structured environment. The systems theory framework takes the fundamental view that the systemic properties may only be understood if we investigate the interrelated impacts of these varying aspects. Our insight into complex adaptive systems rests on our ability to go beyond the scientific tradition of isolating units, processes, or patterns for detailed monitoring.

Bohm (1983) makes a distinction between manifest and non-manifest orders. The manifest orders of elements is what we have traditionally observed under laboratory-based test-tube conditions. In today's microbial ecology we still see how research designs attempt to disclose manifest order by translating known designs and techniques to fit *in vivo* observation conditions. The non-manifest order—the fundamental network of interconnections—is, however, rarely addressed.

A naturalistic approach will attempt to combine knowledge of the elements, their activities, and positions in a community (manifest order) with analyses of patterns of mutual interdependencies that characterize the observed community and its immediate environment (non-manifest order). As known from open system theory (Buckley, 1967), elements of a system are only the visible tip of a very complex, dynamic domain of interdependency patterns.

The essential characteristic of the open system is its organization, which refers to the structural set-up that is controlled by information and fueled by energy (Wilden, 1972). Today we have access to a substantial scientific groundwork of theories and models that conceptualize the characteristics of living organisms as cases of open systems; this also implies models of how to make such cases of open systems empirically accessable (Bateson, 1972; von Bertalanffy, 1968). The modern systems theory (MST) that produced this systemic groundwork since the late 1960s took the analyses from their origin in biology and physics, and made an application onto living systems in general, including highly developed social systems (Buckley, 1967; Maruyama, 1963; Wilden, 1972; Rapoport, 1986). Today there is a wide application of systemic concepts and models in sociology and organization theory (Scott, 1992; Morgan, 1986; Weick, 1979). At the same time, the modern systems theory approach has been reintegrated into biology as part of a general description of autopoietic structures and self-regulation (Maturana and Varela, 1980; Morin, 1992; Berkowitz and Tschirgi, 1988; Ulrich and Probst, 1984).

The CASE perspective rests on modern systems theory. Our point of departure is the idea that every atom, cell, or crystal, as well as any organism, community, or culture constitutes an organization. With enormous amounts of data about entities and units in biological systems we still have no viable theory or model of their organization (von Bertalanffy, 1968). The open systemic approach investigates the organization of complex systems. It tries to describe the way that organizational characteristics are constituted by the dynamic interrelationships between the organization in focus and its immediate environment. There is no evidence that nature works with fixed types of organizational forms, but there is reason to believe that the organization of a living system reflects a dynamic balance arising from the mututal interdependency between system and environment (Boulding, 1953; Glassman, 1973; Hall and Fagan, 1969).

A systemic approach is contextual in the sense that the boundary between system and environment is arbitrary and due to the choices made by the researcher. Where to draw the line is an epistemological question. The lines drawn between system and environment by our conventional models of reality are such lines. These lines may be part of the conventional wisdom, but nevertheless they are of a methodological nature; they are not real (Wilden, 1972). It is the design of the empirical study that defines the particular universe to be investigated as a possible configuration of objects (Hall and Fagan, 1969). According to convention and/or tradition, we have classes of such objects waiting to be analyzed, which does not change the fundamental point about the constructivist perspective. Constructivism points to the idea that the systemic analysis starts with a conscious definition of the boundary between system and environment (Berger and Luckmann, 1966).

Contextuality, thus, implies that the systemic analyses of any organized community must involve consideration of the interdependency between system and environment in a chosen period of time. Contextuality points to the information that is carried by the complex, adaptive system as a repertoire based on previous experience and learning processes. The contextual preconditions in this way trigger specific systemic responses that reflect parts of the overall repertoire. It is the context that brings out the expressivity of the system in focus.

This leaves us with a very ambiguous and chaotic research field. There are no givens, no standards, no fundamentals. There are only objects and phenomena that we may wish to define as systems. This is not new. It is simply an aspect of our history of research, which we tend to forget or ignore (Whitehead, 1929). A cell is still a cell, and that is a natural, given object; or is it? Modern systems theory does not make other types of research obsolete. It argues that the knowledge we gain may be unidimensional or insufficient. The coherent analysis of complex systems demands investigations that are not confined to isolation of single cells in laboratories. The systems approach involves the analysis of the cell in the presence of other cells of the same and/or a different kind. It takes into account the role that the cell plays in the total organization of cells in the investigated system or community. It tries to explain the distribution of functions and regulation of activities within each cell, as well as accross the different cells in the system (Hall and Fagan, 1969).

To conduct a naturalistic inquiry we need a model of the open systemic structures that will allow us to combine data on both the manifest and the nonmanifest order. It is the basic assumption in this article that we may monitor

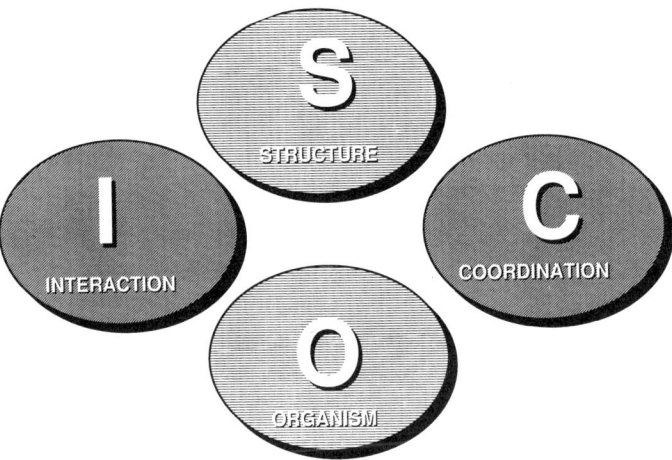

Figure 1. The SCIO model. Presentation of the four systemic parameters useful in characterization of microbial communities.

an open system based on four parameters that we shall further describe according to the SCIO model, a general model of complex adaptive systems (Fig. 1). It is an analytical framework based on four equivalent parameters that are all empirically accessible. The SCIO model contains a complementary set of parameters each of which describes unique classes of systemic properties (Jaques et al., 1978):

- S is a structural parameter
- C is a coordination parameter
- I is an interaction parameter
- O is an organism parameter

The characteristics of each of the four parameters may be described as follows:

S, for structure, refers to the positions of the component elements as single units or subpopulations in a spatial (three-dimensional) arrangement. It is the positioning of organisms, the identification of microcolonies, aggregates, substructures, etc., and the constitution of stable configurations in time and space. Structure describes a level of analysis where elements contribute to a systemic capacity to generate new internal solutions to changing environmental conditions. Structure springs from the chosen demarcation of the researchable system, whereby the scientist draws a line between what is inside and what is outside the object of analysis.

C, for coordination, refers to the presence of mechanisms and processes that bind together activities shared by parts of or all of the population. It is the communication links between the component elements. Coordination is the result of feedback loops affecting the concrete behavior of individuals (or groups) through information processing and any later consequential adjustments. Coordination describes a level of analysis where elements contribute to a systemic capacity to develop an internal repertoire for handling changing interrelationships. Coordination represents the script or the blueprint of the interrelated processes, whereby organisms form local, semi-stable structural arrangements.

I, for interaction, refers to the concrete and directly observable activity of organisms reacting to a shared set of signals and responses. It is the reflexual character of the interrelationsship between some or all of the component elements that share local substructural conditions or pathways. Interaction describes a level of analysis where elements contribute to a systemic capacity for standardized activities.

O, for organism, refers to the types, characteristics, and relative numbers of the component elements. It is a taxonomic classification of individuals (normally derived from existing data about the population), and it is a phenotypical identification of the individual charateristics as it comes across in the present context. Organism describes a level of analysis where elements contribute to a systemic capacity for basic metabolism. The organisms express a phenotypical profile that is generated by the structurally nested and coordinated activities of interactions with other organisms in the system.

3.2. *In Situ* Parameter Analysis

The system boundaries are defined by choices made by the investigator, whether the object is a naturally existing biosystem or an artificially created system. Lake sediments may be defined vertically, i.e., the various layers of organisms related to the depth in the water column, whereas the horizontal dimensions may be considered less significant. In this way the boundaries are defined as the top of the water column and the deepest position where measurements are performed; thus, only two dimensions are considered. A plant rhizosphere carries a different boundary problem: There is no objective way of distinguishing between the plant root environment and the external soil environment. Instead, the rhizosphere may be defined mechanically as whatever comes off the roots when these are shaken for a certain period of time—a truly pragmatic boundary definition, but also operational and in many cases reproducible. A sewage plant and a biofilm in the lab have physical boundaries (walls or edges of the slide supporting the biofilm), which may be chosen, but they are not necessarily useful due to the complexity and size of these systems. Smaller parts of such systems are therefore often defined by the investigator based on more or less arbitrary choices.

We thus have to face the fact that no matter how natural or unnatural the ecosystem we study may be, the precise line drawn around the object is the choice of the investigator, and such boundaries are rarely or never related to the inherent properties of the systems. We do not consider this a problem for the analysis of systems such as those defined here (mainly because it is inevitable), but we certainly wish to emphasize this intervention of the investigator as a factor to be aware of in relation to the interpretations to be made of the results obtained.

Having discussed the problems of defining system boundaries in all biological communities (especially the invisible ones as the microbial communities are), it of course becomes equally important to maintain that it is possible to characterize the most important external parameters in the environment surrounding the biosystem: The lake water above the sediments and the light intensity and oxygen content; the humidity and content of organic compounds of the soil around the plant roots, like the composition of the plant exudates reaching the microbes from within the roots; the flow-through of waste water in the sewage plant and its contents of degradable and non-degradable compounds; and the physico-chemical properties of the substrate flowing over the biofilm and the mixing efficiency. The important aspect is to reach a point where key boundary spanning interactions between the physically and functionally defined system and the surrounding conditions can be addressed and characterized.

With an emphasis on system parameters like structure and interactions it becomes very important to develop experimental approaches that are based on non-destructive measurements. In many cases, however, destructive methods may provide very useful information about the community composition (cell types, content of carbon, nitrogen, and other nutrients, etc.), and such data represent important information for subsequent interpretations of the dynamics of the system (behavior and organization).

The increasing demand for non-destructive methods makes analyses of microbial communities very difficult and very much dependent on technological developments. Changes in the environment surrounding the community may have very significant impacts on the community structure and interactions, and simple parameters like the tension of oxygen or pH should be monitored with precision around and within the community. The development of microelectrodes meets such demands for mapping important environmental parameters as well as reporting on microbial activities in the community (Revsbech and Jorgensen, 1988). Future designs of microelectrodes for other types of metabolically important compounds will contribute significantly to the mapping of the environmental and systemic extracellular parameters.

Microscopic methods have also been developed to the point where they may serve the purpose of providing non-destructive measurements of an increasing number of parameters. Specific staining of bacterial species with various molecular probes combined with microscopy allows population descriptions that go far beyond simple recordings of the present flora; the precise mapping of cellular positions results in descriptions of community configurations (Amann et al., 1991; Poulsen et al., 1994). With the recent development of scanning microscopy (the laser confocal microscope) the complete survey in all three dimensions of a community has become possible (Lawrence et al., 1991).

The four parameters, SCIO, are assumed to be empirically accessible using methods from microbiology, biochemistry, molecular biology, and microbial ecology. In the following we will describe the methods and the methodological approaches available for characterization of microbial communities based on the four parameters. In order to make the connection to practical studies more obvious, the parameters are discussed in the order OISC.

3.2.1. Organisms

Natural microbial communities most often consist of complex mixtures of microbial species whose relative frequency may vary a great deal. The classical approach towards an analysis of such populations is to cultivate bacteria on laboratory substrates as pure cell lines, followed by identification of the individuals. This is practically an impossible task due to the complexity of the community, and also because the majority of the bacteria are not culturable. However, it must be emphasized that population analysis without cultivation and identification of at least the major community constituents is problematic. It is therefore crucial that new methods for cultivation and characterization of single species are developed.

An alternative to traditional microbial identification is population analysis based on DNA reassociation kinetics (Torsvik et al., 1990). With this approach there is no direct information available about the identity of the individual organisms in the community; instead, the analysis provides a complexity profile, which at any given time and after any given change of environmental conditions reflects the status and development of the organism composition.

The most powerful technique at present for organism identification in environmental samples is that of rRNA identification (Pace et al., 1986; Paster et al., 1994). Bacterial taxonomy is changing rapidly these years due to the systematic analysis of rRNA sequences that are almost ideal indicators of evolutionary relationships (Woese, 1987). Direct analysis of rRNA sequences present in complex samples are possible without any step of cultivation using the PCR method to create libraries of probes (Ward et al., 1992; Giovannoni, 1991). The simplicity and speed of the PCR based methods make detailed analyses of microbial communities possible. The information obtained relates to the complexity of the community, but also provides specific data on identification at a number of different taxonomic levels (Amann et al., 1995).

The method of rRNA hybridization is furthermore useful in the context of *in situ* identification of single cells (Amann *et al.*, 1995). Specific rRNA probes labelled with fluorochromes are precise and highly sensitive tools for single-cell identification in the fluorescence microscope, and when used in combination with the flow cytometer they may also provide quantitative information such as the relative proportion of a certain organism (Moller *et al.*, 1995; Jepras *et al.*, 1995). To some extent similar information can be obtained by fluorescence microscopy. The limitation of the method is that only a few species can be investigated at one time.

Finally, molecular tagging of a distinct organism is an interesting possibility in cases where the purpose is to specifically follow one particular organism in a complex community (Chalfie *et al.*, 1994; Pickup, 1991). Introduction of such tags requires growth and manipulation of the organism, followed by introduction of the tagged cells to the community. This scenario is often not possible in natural ecosystems, but in laboratory model systems it represents an excellent choice of specific organism analysis.

In conclusion, the O parameter (organisms) is directly accessible for analysis through a combination of traditional and modern molecular approaches. The choice of method of course depends on the complexity of the system, the need for specific information, and the available base of instrumentation. Community analysis has always been important in the context of microbial ecology, and there is no doubt that method improvement will progress continuously. It should be emphasized once more that there is also a strong need for improved methods of organism cultivation in order to develop a better understanding of the organisms and their different repertoires of performance in different environmental contexts.

3.2.2. Interactions

There is a tradition in microbial ecology to write up the mass/energy balance of an ecosystem (Hojbjerg and Sorensen, 1993; Nybroe, 1995). The mass balance determined as the net flow of energy source (and of elements in general) through the system and the increase in biomass provides the same type of information about interactions as the population profile based on, e.g., DNA reassociation kinetics provides about organisms. The obtained information is valuable for the overall (macroscopic) picture of the composition and processes of the system, respectively, and the consequences of environmental changes on the population profile and the biochemical reactions can be recorded through such investigations.

For a more detailed investigation of systemic interactions there is a need for other types of analysis. Microorganisms respond to their surrounding environment by mobilizing different parts of their genetic repertoire (Neidhardt *et al.*, 1990). Only a limited number of environmental factors are specifically registered and responded to by the organisms, but other factors may have general impacts on the behavior of the individuals. Thus, the overall chemico-physical conditions of the environment such as temperature, pH, ionic strength, availabilty of nutrients, presence of oxygen, etc. may have severe effects on the performance of the organisms; these are, however, distinct from the very specific interactions that exist between particular environmental components and specific receptor systems in the cells.

The interactions between organisms and the external environment constitute what may be called the biochemistry of the system (Wolfaardt *et al.*, 1994b). Signals from the environment are converted to cellular responses through the complicated networks of gene expression and gene control. In communities where different organisms coexist there is the obvious possibility that two or more species together may exploit certain compounds present in the environment by combining their metabolic capacities. We will refer to such cases as community metabolic pathways, and they are important examples of interactions taking place at the community level.

In order to monitor interactions and understand their connections to the population and the environment, it is necessary to build up a strong base of information about the physiology of the participating organisms derived from laboratory-based experiments with pure cultures. The cellular repertoire can only be studied in precise terms from very simple experimental scenarios. On the other hand, many physiological properties in bacteria are more or less universal, which means that common tools and analytical strategies may be applied to a broad spectrum of species.

The molecular methods offer approaches to studies of specific interactions in specific locations. The presence of essential nutrients in the external environment or locally in the immediate neighborhood of a group of cells can be revealed by using molecular probes for the physiological state of the cells. The ribosome content estimated by quantitative *in situ* rRNA hybridization reflects the growth physiology of the cells (the more ribosomes, the more physiologically active are the cells), and in the absence of nutrients the concentration of ribosomes is expected to be low (Maaloe and Kjeldgaard, 1966; Moller *et al.*, 1995; Poulsen *et al.*, 1995). Similarly, cell size and DNA content will reflect the physiological state and thus the nutritional situation of the organisms (Maaloe and Kjeldgaard, 1966; Moller *et al.*, 1995). Under extreme conditions the cells may be physiologically inactive (dormant), and hence have no interactions with the environment or with other cells in the population.

A number of reporter genes have been designed that, depending on their inherent features, may report about the environmental conditions and the cellular responses to these. Thus, constitutive expression of bacterial luciferase requires oxygen and metabolic activity of the cell, and, consequently, the turn-off of light indicates either lack of oxygen or lack of ATP in the cell. Fusion of reporters to genes activated by certain stress signals (starvation, draught, high osmotic pressure, etc.) can also be applied to indicate local environmental conditions and, therefore, also provide information about specific interactions between surroundings and cells (Kragelund et al., 1995). In all cases, the molecular tools in several cases supply information both about the state of the cells and about the conditions of the environment, and together this knowledge is an excellent background for reading the interactions going on in the particular location. In addition, the level of resolution of the interactions between cells and surroundings is the single cell and its microenvironment.

The methods and arguments presented above for studying interactions between organisms and environment can be applied to studies of interactions between different organisms. The different approaches available (tracers (Robarts and Zohar, 1993), microelectrodes, molecular tools) are useful at different levels of resolution, and for improved understanding of microbial ecosystems it is imperative that there is a continuous development of all the methods, since none of them offer high-level resolution and global analysis at the same time.

3.2.3. Structure

Community structure in microbial ecology is often presented as a list of identified organisms present in a given environmental sample or system (Wagner et al., 1993). We will refer to structure in a different way: the three-dimensional distribution of organisms (positioning) (Poulsen et al., 1994), and the occurrence of specific compartments with functional or constructional significance (extracellular matrix, channels, pores, etc.) (Massol-Dey et al., 1995). For microbial communities this definition assumes the use of microscopic equipment, although there are some very well-studied examples from natural environments that present their structural features without the need for microscopic examination. For example, some aquatic sediments have structural properties that can be described as alternating layers of groups of bacteria reflecting the nutritional conditions down through the sediment core (Brock and Madigan, 1988). However, this only provides a rough overview of the positioning of the organisms; the bacteria are mostly bound to sediment particles, and their actual location on these can only be revealed by microscopy.

The scanning electron microscope has in the past offered impressive views of the structural organization of microbial communities (Dempsey, 1981; Robinson et al., 1984), but a problem always was that the preparation methods for electron microscopy might have serious effects on the structures to be examined (and in any case they are, of course, destructive). Looking at microbial communities with the light microscope also is not ideal; the pictures are two-dimensional and the noise from unfocused parts of the viewing field is often totally damaging to the final picture. However, the use of light microscopy offers a feature not carried by electron microscopy: Simultaneous observations of organism identity (using specific hybridization probes) and of overall positioning of the cells; and the communities may be looked at directly without any seriously interfering preparation steps.

The most ideal instrument for structure analysis at the microscopic level should therefore combine the scanning possibility of the electron microscope with the repertoire of the light microscope. The scanning confocal laser microscope (SCLM) offers such an ideal analytical potential. The collection of information as discrete points and removal of information from locations that are not in focus (in the point), coupled to the scanning of plane after plane, and finally the passing of all images through image analysis leads to a detailed and well-resolved illustration in three dimensions of the observed object. The images thus obtained constitute the most obvious possibilities for discovering specific structures in the community. SCLM is compatible with many of the molecular probing methods (involving fluorescence), and it is therefore one of the most important new methods for analysis of complex community structures (White *et al.*, 1987; Brakenhoff *et al.*, 1988; Caldwell *et al.*, 1992, 1993).

One of the immediately apparent observations that is made when studying bacterial surface communities is the uneven distribution of cells and the existance of substructures such as microcolonies, channels, and extracellular polymer areas. Such structural features may be observed in many different types of microbial communities (Stams *et al.*, 1989), and after analysis of several systems it should be possible to categorize the various structural features and elements such that their presence or absence can be determined as any other system parameter.

3.2.4. Coordination

It is possible to view a microbial community as an aggregation of individuals trying to get the most out of the community at the expense of the nearest competitors. Every cell lives in its own world without realizing (or appreciating) the presence of the others. Alternatively, the community can be described as a multicellular organism (Shapiro, 1988; Wolfaardt *et al.*, 1994a) consisting of a large number of differentiated cells that are all connected in a functional and structural network. In this view, every cell is communicating with the rest. It is a major assumption in the present analysis of microbial communities that at least some parts of the community features (structurally and functionally) are coordinated in such a way that struggling for existence is based on coexistence, rather than on survival of the fittest individual (Leigh, 1983). However, this does not rule out that individual organisms under certain conditions compete for substrate or try to elimate each other by toxin production. Thus, it is not a question of either competition or collaboration, but rather a combination of the two.

Communication between individuals and between sectors of individuals is an essential prerequisite for coordinated activities. The language of this type of communication must be of a chemical nature, but in contrast to simple metabolic intermediates, which are just substrates for further biochemical transformations, signals for coordinated activities are targeted at cellular control loops. This means that production of extracellular signals from some cells may influence other cells at the level of complex phenotypic traits (involving many different genes connected by joint control factors). An excellent example of this type of communication is the quorum-sensing system based on the family of acylated homoserine lactones (Dunlap and Greenberg, 1991; Greenberg *et al.*, 1979; Kaiser and Losick, 1993). Many different bacteria produce these compounds and also possess receptors for transduction of the signals to the final destination, promoter regions upstream of certain genes. It has recently been shown that bacterial swarming—a highly coordinated pattern of movements of thousands of cells—is in part controlled by homoserine lactone derivatives (Eberl *et al.*, 1996). Other similar cellular activities have been shown to be regulated or influenced by these extracellular signal molecules (Bainton *et al.*, 1992; Passador *et al.*, 1993). The presence of such signals in a complex microbial community indicates coordinated activities, and most likely other examples of communication will be discovered in the future.

Coordinated activities in a community may also be documented by the presence of ordered structures and activities (Shapiro, 1988). The finding of channels and pores, through which import of nutrients and export of waste products may take place (Massol-Dey *et al.*, 1995), could indicate that many individuals have been involved in creating such specific structural elements within the community, in other words, they have worked together rather than against each other. This type of conclusion is indirect in the sense that observations are made that can most easily be understood if it is assumed that coordination is exerted at some level in the system. Likewise, coordinated behavior can be indicated by the systemic responses to perturbations: If one group of organisms is influenced by a perturbation from outside, and the impact is significant in other parts of the community resulting in a change of community performance, then it is strongly indicated that the community performance is the result of coordinated activities.

3.3. SCIO in General

We have discussed how the individual systemic parameters, S, C, I and O, can be analyzed in microbial communities. They are all subject to experimental monitoring using a mixture of microbiological, biochemical, and molecular methods, and they provide relevant information about important features of the community under study. The microscope (and in particular the SCLM) offers structural information (S); the isotope tracers, microelectrodes and molecular reporters offer interactive information (I); *in situ* rRNA hybridization probes offer information about organisms (O); and, finally, a mixture of chemical analysis and indirect observations of systemic response patterns offer information about coordination (C). An important conclusion from this is that microbial communities can be analyzed very thoroughly with methods that are commonly used by the experimental microbiologist. The SCIO model at this point offers a framework for organizing the measurements and the obtained data. It must be emphasized, however, that the type of parameter investigation presented above does not provide much understanding of how the communities perform. The data collection secures an important base of information; understanding comes out of integrated studies of these parameters in the context of the dynamics of the system.

4. Behavior and Organization

4.1. Levels of Analysis

Based on the definitions of the four parameters, the following presentation of the SCIO model points to two separate levels of analysis and their interrelation: 1) The organism and the interaction parameters constitute what may be called a behavioral level of analysis; and 2) the coordination and the structural parameters constitute what we may call an organizational level of analysis. What we suggest is to distinguish between the different levels of analysis (keep parameters apart), and still combine data from the four different parameters in the model (generate coherence) (Fig. 2).

The distinction between the two empirical levels of analysis is important for understanding the difference between descriptions according to the performance pattern of specific organisms, and descriptions according to the organizational impact from structure and coordination.

By observing and carefully monitoring the distinct activities of the systemic elements (organisms), we may learn about the systemic functionality—we may get a picture of what goes on. By analyzing the structural distribution and coordination of activities, on the other hand, we may learn about why the systemic functionality works.

We need an understanding of the way individual elements and their monitorable activities in the system is conditioned by the structural set-up, and of the way our observation of patterns of interaction is conditioned by the coordination processes. Any element taken out of a natural setting becomes a different element that may (or may not) express a different phenotypical pattern of activities in response to new and changing structural conditions. Any pattern of interactions placed under different systemic conditions may (or may not) change according to

Complex Adaptive Systems Ecology

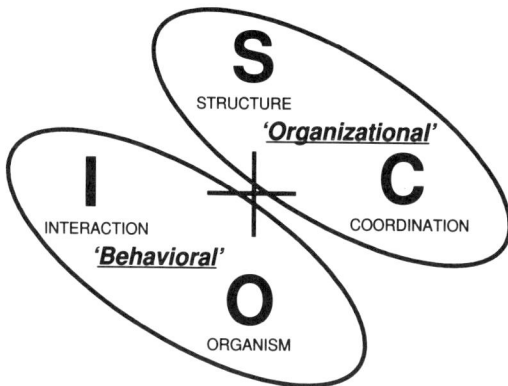

Figure 2. Levels of analysis. Community analysis may be performed at two separate levels defined by specific combinations of the system parameters. The behavioral level represents the performance of the community defined by the I and O parameters (the system phenotype); the organizational level represents the configuration and activity distribution of the community defined by the S and C parameters (the system genotype).

different coordinative constraints. Parallel to this, we need to monitor the concrete activities, as performed by specific organisms, to fully comprehend how coordinative patterns are enacted, and why structural set-ups prevail.

Principally, our argument is that it is not sufficient to give taxonomic and/or phenotypical accounts of the involved organisms in a microbial community. It is not sufficient to map three-dimensionally the numbers and positions of the participating organisms. Only by combining the two sets of information can one gain insight of how the system has organized itself. The combination gives you an accurate picture of what types of organisms (and how many) are located where in the structure. It determines whether two or more specific types of organisms appear to be located together in different parts of the microbial structure, and it enables one to see whether specific organisms are located randomly as single units, or whether they form substructures at given positions in the total configuration. Still, this provides little information as to the way that the system responds to its immediate environment.

In order to recognize and understand the patterns of these systemic responses, data on the distribution and regulation of functionalities is needed. In other words, we need to know how information and energy are transformed and controlled, i.e., the behavioral energy transformation patterns, and how these relate to regulatory loops of information. No matter how precisely we may monitor the interactions (the metabolism and general stimulus/response activities carried out by specific organisms on specific locations in the structure) we will only have an indication of "the what" as a description of systemic behavior. Regardless of how much information we receive about the coordination (the regulatory communication loops across organizational units and substructures) we will at best be able to obtain indications of "the why" as a description of the systemic self-regulation.

If we bring together the data on interactions and coordination, if we analyze the what and the why, we get a description of the systemic functionality. Interaction data allow us to monitor the specific characteristics of stimulus/response patterns, i.e., the way the distributed functionality of the system is carried out by specific types of organisms (or sub-groupings of organisms). Coordination data allow us to monitor how the distributed functionalities are regulated, i.e., the way the specific parts of the system uphold a specified kind of interaction, and how this pattern of behavior is adjusted according to changing conditions in the system. The functional analysis produces knowledge about the systemic capacity and its inherent interdependencies. As such it tells us nothing about how these capacities are constituted (the mutual interdependencies between the four parameters). To answer the question of how, we need an analysis of structure and organisms.

This type of analysis provides us with an illustration of the configuration of the microbial community. It allows us to draw a three-dimensional model of its composition, its mix, and the distribution of all its component elements (organisms). As such, we have a recognizable system that may be perturbed, stimulated, etc., and monitored in a real-time study in which systemic patterns of behavior (not to be mistaken by organism behavior) may be recorded. In this way we are able to describe systemic activity by indication of what parts of the system are active under what kinds of stimulation. We may also be able to describe what types of organisms are responsible for the systemic response.

In this way we recognize how any empirical entry via one of the parameters calls for subsequent analyses of one or more of the other three parameters. We have to distinguish between data from the four parameters as well as between the two qualitatively different levels of analysis; however, at the same time these intermediate findings and descriptions have to be merged to create a picture of the specific system context based understanding of the object of our study.

If we take a closer look at the first level analyses we get a picture of what has dominated research design for the last decades. Typically we see taxonomic studies of single cells under laboratory conditions, and/or descriptions of global cycles of elements (carbon, nitrogen, etc). These studies furnish a worthwhile and necessary knowledge about the organisms and the way simple transformations are conducted. Often information deduced from these different studies is not combined, but is treated as unique sets of data. Knowledge about interactions and organisms arise from direct observations and interpretations. No matter how much information we may produce from analyses of these two parameters we will never be able to understand the complex systemic features and how these influence the patterns of interactions and the conditions for the organisms. Still, they are integral parts of the systems approach, and with their behavioral point of departure they represent empirical entries that are relatively easy to specify, isolate, and control.

If we take a closer look at the second level of analysis we get a picture of complex systemic properties. Coordination and structure are almost totally missing from the microbial ecology research agenda. Often coordination is mistaken for interaction, and is reduced to simple flows of biochemical reactions; and structure is mistaken for sediments, or it is reduced to quantitative counts of the organisms present. Both parameters imply direct observations and interpretations, although these require much more elaborate longitudinal studies than are needed for the analyses at the first level. Both parameters are integral parts of the systemic understanding, and combined they furnish a description of the system's organizational characteristics.

The two levels of analysis, the behavioral and the organizational, combine well to describe the system at large. In general, the four parameters are unique empirical entries—each in its own right. Still, we have to admit that in practice the interpretation of data from the two organizational parameters may easily overlap or presuppose the presence of one or more of the others. The parameters are empirically based constructs, and as such the element of interpretation signifies a level of knowledge about one or more of the other three. Contrary to the tradition in studies of organisms or interaction, we find that this interrelatedness between the four parameters also concerns these first level studies. To fully understand the behavioral characteristics of organisms and their interaction we need to know something about the regulatory patterns and the structural configuration of the system. If we do not confront the first level analyses with this kind of information we get information that is specific to an artificial context. This gives data that are ultimately irrelevant, unless we can substantiate that the artificial context, in principle, resembles true *in vivo* conditions. Similarly, any account of the regulatory processes and how these fit into a structural arrangement presupposes intimate knowledge about the organisms present and their habitual patterns of interaction.

The two-level investigation, comprising a behavioral analysis and an organizational analysis, is based on studies of the dynamics of the communities; the key words are change and time. Through the choice of parameters made by the investigator, the analysis will be directed towards one or the other level. Monitoring of the community activities must be performed under conditions of serial sampling of data, such as on-line monitoring or time-lapse monitoring, implying that non-destructive methods are employed. The optimal approach will be one that exclusively involves constant surveillance and measurements, but in several cases that is not possible (see below). Alternatively, sampling from the community for analysis of specific community parameters (e.g., organism composition) without interfering significantly with the community itself is a reasonable approach. In the laboratory it may be possible to establish communities in very reproducible forms that may allow parallel studies of a series of 'identical' communities. These may be analyzed one by one at different times in their development.

Methods for non-destructive and real-time analysis may be applied to natural ecosystems as well as to laboratory systems (Sjollema et al., 1989), at least as long as the monitoring activity is directed against organisms and interactions (behavior). This conclusion seems reasonable since the necessary methods (discussed in connection with the SCIO parameter investigations), with their sophistication as far as specificity and sensitivity are concerned, can be applied to nearly any complex biosystem. The parameters relevant for the description of individual cells, general metabolic activities (like oxygen consumption), signal–response interactions at the level of gene activity, and even positions of different types of organisms also can be monitored for systems in natural environments. We may therefore learn a lot about the life processes of specific organisms under conditions that are either models of natural environments or the natural environments themselves, and the new molecular tools available for such studies have greatly expanded this potential. However, as soon as we approach questions concerning community organization, monitoring of organisms and interactions does not provide the answers.

The system parameters S (structure) and C (coordination) are key parameters in the context of system organization, although the two other parameters are, of course, also important. If direct monitoring methods of the types described above are to become the basis for descriptions of system organization and to a large extent also behavior, it is crucial that knowledge about complex, adaptive microbial communities is built up from studies of systems whose environmental conditions are controllable. Characterization of coordinating activities will depend on the possibility of controlling the challenges to the biosystems. Surface communities of microorganisms (biofilms), designed in the laboratory and developing under controllable conditions, represent biosystems that are complex and adaptive without being uncontrollably natural (Caldwell, 1993, 1995; Caldwell and Lawrence, 1988; Characklis et al., 1990; Costerton et al., 1987, 1994; Marshall, 1989). They seem to us to constitute useful model systems from which behavioral and organizational patterns may be derived, which subsequently can be tested in more natural systems. The possibility of freely combining various manipulations of the conditions with monitoring methods should provide unique opportunities for studying community parameters that have not been accessible before. Let us introduce the most important experimental designs, which may be applied in our analyses.

4.2. Interference as an Experimental Approach

In the following, three different environmental conditions representing different analytical approaches to biofilm studies are presented as increasing levels of direct interference, and in all cases the interferences are carefully planned as means of challenging the system to expose its reaction repertoire. In other cases (such as naturally located systems) such interferences may happen without being planned at all.

"Steady state" in biology has various meanings: For microorganisms it may mean resting cells that no longer grow, but that also do not die (constant cell density, no apparent changes of the population). However, it may also mean cells growing exponentially in suspension (constant external conditions, identical growth constants for all measurable cell parameters). In the present context of surface based microbial communities we define steady state as a design in which the environmental conditions are kept constant (flow of nutrients, etc.) (Moller et al., 1995). Under such conditions investigations of the community behavior may be monitored without disturbing influences from a changing environment, and among other interesting types of information one may learn about the system's self-organization and self-development. The basic experimental approach will be on-line monitoring of specific parameters related to individual, subpopulation, or systemic behavioral activities.

The full control of the environmental conditions relative to laboratory-based biofilm communities makes it possible, in principle, to vary any one environmental parameter while the rest are kept constant. If such changes are imposed on already established communities, the methods of analysis described above for steady-state designs should be directly applicable (on-line measurements). In classical physiology (Maaloe and Kjeldgaard, 1966; Neidhardt et al., 1990) performed on test tube cultures, microbial performance in a number of environments (media) with one parameter being varied could be studied quite accurately due to the reproducibility of such test tube experiments. Even though complex systems are not reproducible to the same extent, it is most likely possible in many cases to obtain information about the effect of the changes at the level of individual cells, subpopulations, and even the entire community.

More dramatic interferences than adjusting nutrient concentration or availability may expose the response repertoire at many different levels of the community. If the perturbation is very severe, it may even be possible to register in very direct ways the added value of societal life compared with that of an individual (better chance of coping when present in complex community than when alone) (Leigh, 1983). Once more, a steady-state situation is the experimental basis on which these interferences are assessed. The perturbations should be of the kind that will be very hard to cope with for the individual organisms under conditions of monoculture systems or suspended systems, and may cover interferences such as local community destruction, introduction of foreign elements (other organisms or physical items), changes of the population (addition or removal), and physical shocking. This type of analysis challenges the ecosystem stability and buffer capacity, and it also provides means of investigating effects on natural ecosystems and their balances.

In principle, it is conceivable that the cells present in a complex community will react in a broad variety of ways to variations of the physical and chemical properties of the external environment. These include:

- no reaction at all
- stochastic behavior
- standardized reactions (same response independent of introduced change)
- inconsistent reactions (different reactions in different positions)

With the monitoring techniques available today, it is possible to obtain very precise and concrete observations and measurements of organisms and their interactions with the surroundings (O/I parameter). What is very interesting is that additional information concerning contextual activities and performances of organisms and their local surroundings is produced at the same time with indications about the S/C parameters of the community. The accumulated data, from the first basic investigations of monoculture biofilms developing under constant external conditions, to the performance of complex systems going through relatively simple variations of the surroundings, will already represent a huge pool of inputs for the categorization of the system's behavior and organization.

In the following section we describe in more detail various experimental approaches to interferences from investigator induced changes on microbial community systems. The experimental methods relevant for the analyses are basically the same as those decribed for the fundamental parameter analysis of surface communites under constant conditions. When the communities become more complex (composition) and the conditions are changing, it will in many cases be advantageous to combine two or more methods of direct observation.

4.2.1. Changing Surroundings

The system discussed is what we had previously termed biofilm, meaning bacteria growing on surfaces (substrata) and surrounded by a liquid substrate passing the surface as a flow (Caldwell, 1995). The composition of the microbial community may be artificially created by the investigator, or it may be a natural mixture from a source in the environment. The surroundings in this type of set-up are represented by the liquid substrate. We will discuss changes as either introduction of physical/chemical variations of the surroundings, which do not have severe local or total effects on the community, or introduction of perturbations that more directly interfere with the community and its constituents. Changes of conditions are imposed on microbial communities that have already been established and monitored for some time under constant conditions (steady state).

4.2.1.a. Effects Generated by Variations of Physical Parameters. The primary parameter considered here is temperature. All bacteria react to temperature changes, and in a mixed community the reaction pattern may be quite complex if the different species have different reaction profiles to variations in temperature. For a specific organism variations of temperature within a certain range may not change the overall composition of the cells, but only affect the rate constants of the enzymatic reactions in the cell (Neidhardt *et al.*, 1990). This type of reaction may not create any significant changes of the physiological states of the cells, although growth rates may change dramatically. However, outside such temperature ranges the cellular response may be of a totally different type: Heat shock and cold shock are reactions to what may be experienced as extreme temperatures by the particular organism, and they normally create very different physiological states of the cells, at least transiently (Bukau, 1993; Hegarty and Weeks, 1940; Jones and Inouye, 1994).

Direct observations of the effects of changing physical conditions of the surroundings may be carried out at the level of single cells or at the levels of sectors of the community. All monitorable cellular features may be studied during the transition from one set of conditions to another, and to the extent that such measurements are species specific, the possibility exists that reaction patterns of several types of organisms with different types of temperature responses may be recorded and analyzed in various parts of the community. In this way internal environmental differences may be revealed as variations in species-specific reactions, depending on the position in the community. It is known from studies of monocultures of many bacteria in test tubes that reactions to changes in temperature to some extent depends on the physiological state of the cells: Fast-growing cells are often much more sensitive to extreme temperatures than cells that have been otherwise stressed, e.g., by starvation for nutrients for some time (Givskov *et al.*, 1994a, 1994b; Nystrom *et al.*, 1992). Such information about the heat/cold responses of the monitored species in the community is of course very valuable for the interpretation of the community reactions to changing temperatures.

4.2.1.b. Effects Generated by Variations of Chemical Parameters. The flow of nutrients over the surface community is the target for changing the chemical environment. There are several ways of introducing changes in the chemical composition of the surroundings (Marshall *et al.*, 1989), but as far as the nutrients (C-, N-, P-, S-sources, etc.) are concerned, the simplest is to adjust the flow rate. By changing the flow rate the community will be confronted with different degrees of nutrient limitation, and as described above for temperature changes, different bacteria will have different uptake capacities for different nutrients. Therefore, flow rate changes will affect the community in a complex fashion. Species-specific reactions may be monitored directly using the molecular probes available for measuring growth and metabolic activity.

Important information about the potential physiological reactions of specific organisms to nutrient limitation may be obtained from monoculture chemostat experiments (Nystrom *et al.*, 1990; Delaquis *et al.*, 1989; Eberl *et al.*, 1996). It is to be expected, however, that the cellular reations to changed flow rates will depend on the internal environment in the particular position of the community. Fine tuning of the degree of nutrient limitation through adjustments of the substrate flow rate will at the same time test the organism's interactions with the external environment and provide information about the dominance of the signals received by the organisms in a particular location. In other words, the monitoring of the SCIO parameters under conditions of changing substrate flow rates gives rise to a contextual understanding of the particular community.

Other alterations of the chemical constitution of the environment may be obtained by addition of compounds with known regulatory properties (inhibitors, catalysts), and obviously there are unlimited possibilities of addition of new nutrients. Chemical changes of the surroundings will expose the adaptive repertoire of the community at the levels of the single organism, a sector of the community, and the entire system in much the same way as we discussed for temperature changes. However, for some organisms a lot of information is available about cellular responses to nutrient limitations (Kjelleberg, 1993), and it is therefore possible to define much more precise expectations in connection with specific flow-rate changes. With this type of background the actual reactions of the organism in various parts of the community may yield a similarly more precise understanding of the local environment in the immediate proximity of the observed cells. Likewise, if more than one organism, whose test tube reactions have been well characterized, are present in the community, their actual reactions in the complex, associated community may bring quite a few surprises relative to the expected reactions.

4.2.2. System Perturbations

The major distinction between the environmental changes described in section 4.2.1. and the perturbations to be described in the following is that the latter interference is directed against the internal elements of the system, whereas the external surroundings are kept constant. (In some cases this distinction is not totally clear, as for example in the case of surface changes, but since the substratum in the conventional definition of a biofilm is part of the biofilm, we have chosen to consider this element an internal component). Thus, these types of manipulations affect the system in relation to the self-generating changes that take place internally as the system develops. Such interferences comprise all sorts of gross changes of important sectors of the system, and they tend to have more dramatic consequences for the system than those caused by changing the surroundings. We have chosen to discuss four different types of perturbations, but the list is obviously not exhaustive:

- local destruction
- introduction of new substrata (focal or total)
- introduction of new elements
- population changes (selective additions or eliminations)

4.2.2.a. Effects Generated by Introduction of Local Destructions. Destruction of part of a microbial community may take place spontaneously as a consequence of the flow of substrate, which from time to time will loosen some of the biofilm material (sloughing) (Delaquis *et al.*, 1989). This is frequently observed, but the timing is of course random, and the position likewise. Similar removals of sectors of the community may be obtained by direct physical interference through which well-defined areas are cut out of the biofilm. Micromanipulations under the microscope or laser-light burning of defined positions are possible methods. It is important that these destructive manipulations do not interfere seriously with the external surroundings, and one parameter to be concerned with, which is very sensitive to surface alterations of this kind, is the flow pattern of substrate passing the community. With these precautions in mind it is possible to make such precise intrasystemic manipulations.

An interesting and obvious target for observation is the recovery of the community in the area of destruction. It is clear that free space for recolonization is created, and the first question is which organisms move in and take over. Although direct observation may provide information about specific organisms and their potential for expansive colonization, the interpretation of the behavioral pattern after local destruction is made very difficult because factors like the internal environmental conditions at the destroyed locus at the time of destruction, and the specific interactions between the organisms in that locus and their coordination, will be nearly impossible to determine and yet be very important for the reaction pattern. On the other hand, as soon as some information is available about the S/C parameters and the specific structural functions of loci in the community, specified destructions of such positions may allow more rigorous interpretations.

4.2.2.b. Effects Generated by Introduction of New Substrata. There is one aspect of particular interest concerning the substratum, which distinguishes it from other internal parts of the biofilm: In contrast to the developing internal environments of the biofilm community, which cannot be controlled or manipulated directly, the substratum represents an important inner environment, whose properties may be subject to manipulations, and which can be specifically created by the investigator (Fletcher and Loeb, 1979; Dalton *et al.*, 1994). In the present context of perturbations, the technological challenge is to design surfaces, the properties of which can be modified after a biofilm has developed (e.g., change of hydrophilicity). If the entire substratum is suddenly changed in this way, it would offer excellent possibilities for studying reactions to a local environmental change that occurs without any changes taking place in the surroundings. The most obvious part of the community from which to observe reactions is the inner cell layers (relative to the surface substratum), and it may therefore be relevant to compare reaction patterns from this type of changed conditions with those induced by changing the surrounding substrate flow, which first of all will affect the outer cell layers. In particular, measurements of the distribution of reactions through the community as a consequence of the two types of interference may yield important information about key systemic features.

4.2.2.c. Effects Generated by Introduction of New Elements. In simple clean biofilm scenarios from the laboratory, the introduction of new elements is perhaps difficult to imagine if these elements are not new species of bacteria (we will discuss this perturbation separately). However, in natural environments non-related elements may easily appear as invaders of the community, and since the artificial biofilms of the laboratory eventually should be seen as models of nature, we find it relevant to go through a brief discussion of this type of perturbation. The elements may be physical objects or biological materials (including living organisms or parts of living organisms), but here we will only mention biological agents.

This kind of perturbation gives rise to the same monitoring possibilities that have already been discussed above (resulting, of course, in an extra opportunity for comparisons of systemic reactions to different perturbation scenarios). But the specific introduction of a new element also introduces a new aspect of the systemic reaction repertoire to be exposed: The introduction, establishment, and stable integration of a new element.

Predator microorganisms (ciliates, amoebae, etc.) are supposed to be important for maintaining low numbers of bacteria in natural environments. They are fairly easy to work with in the laboratory, and it would therefore be a simple thing to introduce a predating organism in small number in a biofilm community (Caron, 1987). The first question is whether grazing on the surface of the community is possible (if the substrate flow is too strong the predator will never get a chance to hold on); the next question is whether it will have a possibility to get access to deeper parts of the community. We can thus imagine many levels of establishment ranging from quick removal of the predator over transient grazing of the surface to permanent integration in the community resulting in either total elimination of the bacteria or establishment of a balance between predator and prey.

From the community point of view the relevant observations of predator-invaded biofilms are concerned with the reactions of the community members that are not eaten (Moller *et al.*, 1997). It is also of interest to monitor the fate of those species that are known to be liked by the predator, and their possible survival potential relative to their activities and positions in the community. In this way, we will gain information about systemic properties that are not easily reduced to individual features of the single species. Thus, once more we have an opening available for directing questions with relevance for the S/C parameters of the system.

4.2.2.d. Effects Generated by Introduction of Population Changes. The population profile of a community may be changed in a number of different ways. First we have to distinguish between selective *introduction* of organisms and selective *elimination* of organisms, i.e., either adding a number of cells of one or a few species to the system, or removing a number of cells of one or a few from the system. Elimination in the present context does not involve physical destruction of sectors of the community (discussed above), but rather implies species specific killing from outside or similar types of interference.

A second characteristic of relevance to this type of perturbation is connected to the types of organisms to be introduced. Either the organisms are of a type already present in the community, in which case the change in the population is quantitative, or they belong to a new class, in which case the change is qualitative.

The introduction of new organisms provides possibilities for analysis of two principally different targets: 1) the incoming organisms and their reactions to the new environment (i.e. the biofilm community) and 2) the reactions of the recipient community after the new organisms have been introduced. The first type of analysis deals exclusively with parameters of the I/O domain, in which all questions are concerned with the specific cells and their interactions with the resident population and its environmetal conditions. The only systemic aspect to be studied from the invader's point of view is to what extent it becomes integrated in the community and in which pattern this integration takes place. The introduced organism may show new functionalities when confronted with the biofilm community, but the information still resides in the I/O domain. Basically, the second type of analysis is completely analogous with the one described for introduction of new elements. However, there is one aspect that was not relevant for the previous discussion of alien components, but which may have very special implications in cases where a specific subpopulation is suddenly boosted, or where better fit organisms appear: internal competition. The so-called steady state of the community, which is the experimental system to which the new organisms are added, represents a specific balance between the population of the community and the external conditions (homeostasis). As we have previously argued, this balance is internally created, it is constantly developing, and the steady-state characteristic only refers to the external surrounding conditions. A sudden change of the population by an introduction of a significant change of the population profile has the potential to change this balance, and hence this type of perturbation carries a possibility to study the buffer effect of the systemic balance. In this analysis of the resident population and its reaction pattern, competition for nutrients, space, etc. is one of many interesting issues to study.

This type of direct attack on the inner systemic balance also becomes the source of indicators for the S/C parameters of the system. We can imagine that other types of perturbations affecting the systemic balance may eventually create an indirect image of the regulatory routes of the system. In a different biological scenario these very same questions are seen as fundamental for society: risk assessments in connection with release of genetically engineered microorganisms.

The elimination of specific organisms from the community without interfering directly with others may be accomplished by addition of specific compounds (antibiotics) to the community (e.g., via the substrate flow) that have very specific effects on one or a few species (Korber *et al.*, 1994). Alternatively, one species of the community may have been engineered with a suicide system that can be activated from outside by addition of an otherwise neutral chemical agent to the substrate (Molin *et al.*, 1993). This type of approach will eliminate the relevant (sensitive) cells from the entire community independent of their location, and thereby represent a removal of a set of activities with no possibility of their recreation. This contrasts with the local destruction perturbation, which could also be preferentially directed towards a specific subpopulation, but which in no way could guarantee the total elimination of that species, and therefore also would imply the possibility of a recreation of the very same subpopulation at a later time.

The interesting observations after this kind of perturbation are concerned with the pattern of reactions of the remaining community. Once more we obtain data about the inner environments and their significance for the local population, but as before we do not have any specific knowledge about the particular features of these inner environments that are actually sensed by the organisms. We can only register the gross and the more specific reactive responses of the constituents, and use the accumulated information later when we have more understanding of the S/C parameters and the role in these of the eliminated organism. However, the direct observations of the community reactions after removal of member species may as so many times before be reinterpreted in the framework of systemic properties. How does the system reconfigure? Which new processes are induced to compensate for the loss of others? How does the balance of the remaining species develop? Are the elimination effects distributed to all parts of the community? With these and other similar questions we once again approach the systemic properties of the community on the basis of experiments designed to investigate very directly the behavior of specific organisms and their specific interactions with the surroundings.

5. Identity

5.1. Contextual Understanding

The application of the SCIO framework on microbial communities involves two lines of empirical inquiry: a behavioral analysis that monitors the organisms and their activities, including transformation of energy and information in the systemic set-up; and an organizational analysis that describes the configuration of the open system and the coordination between its component elements in time and space. The open system may be further characterized as to where it fits into a larger ecological domain (Boulding, 1953). The open system demonstrates a dependency on environmental aspects of variation, fluctuation, and competition. The degree of stringency in the systemic responses to such changing environmental conditions plays a central role in the overall assessment of the open system's capacity for adjustment and expression of systemic specificity (Riedl, 1984).

It may be self-evident, but still not insignificant, to notice that the open system is a non-trivial machine: The open system is never analytically determinable. The complexity of its organization and the contextuality of its functioning tell us that the systemic responses once observed for a given stimulus may not be reproduced when the same stimulus is given at any later time (von Foerster, 1984). Thus, on one hand, we see how the open system shows a dependency on the environmental contingencies, and, on the other hand, the system is surprisingly independent of momentary environmental changes (Glassmann, 1968). The interdependent relationship between environment and system in this way demonstrates a dialectical precondition for systemic adaptation and development. The system is open internally as well as externally in the sense that the interchanges may result in significant changes in the nature of systemic elements themselves (Buckley, 1969). At the same time the system becomes an autonomous, dynamic entity that is both resilient and coherent as it compensates for the continuous flux of perturbations (Christensen and Molin, 1995).

The systemic control mechanisms serve as a machinery for effecting stabilization (Boulding, 1953). By means of feedback processes enabling a preservation of networks despite replacement of some of the component elements, the open system's most characteristic feature may be described as constancy amid change. It is a direct implication that it is the network of relations, not the nature of the elements, which defines the characteristics of a system (Rapoport, 1986). These analytical points rest on the assumption that any open, complex system forms a dynamic relationship with the environment. Some (epistemological) boundary is necessary to draw as conditions for communication, but it is always arbitrary in the sense of being methodological rather than real.

It is the relationships (no matter how the boundary is defined) that signify the systemic identity. Identity becomes synonymous with the maintenance of the dynamic system, reflecting both the flux of external stimuli and the form and structure of the system itself (Berkowitz and Tschirgi, 1988). Despite the range of possible changes (internally as well as in the dynamic relations to the environment) the open system is by nature conservative, i.e., the effects of small input changes will be negated and the steady state maintained by reversible adjustment (Bateson, 1972). According to our open system perspective we may argue that the systemic responses to environmental impacts only amount to internal, homeostatic modifications, preserving the fundamental systemic identity under changing conditions. The open system is autonomous in the sense that it does not orient itself towards the environment as such; all it does serves the fundamental purpose of preserving the systemic identity (Rapoport, 1986).

When we introduce the term "identity" we do not refer to the traditional connotation in biology. The identity of an open system does not signify a taxonomic identification. As indicated in the discussion just above, systemic identity expresses a dialectical relationship between collective features and unique characteristics. This intricate dialectic is the basis of a qualitative description of the systems identity (Mead, 1962). The open system identity comes out of a contextual balance between ignoring external impacts (to preserve unique characteristics) and relying on the environmental exchanges (to develop collective features). This involves a series of antagonistic processes (paradoxes), by which the relationship between system and environment may be described. For types of paradoxes, we refer to processes of passive imitation and expressive specificity, processes of adjustment and adaptation, and processes of reflexual standardization and learning.

Any open system will demonstrate the capacity to administer these paradoxical processes, and doing so the open system develops its identity. Each system employs its specific configuration to handle the paradoxes. Each system develops its identity playing a part in an overall ecological context (Hejl, 1984).

Intuitively we feel that some systems are more robust than others (Rapoport, 1986). It appears that they demonstrate a totality of organizational and behavioral relations that are tighter in structure and less flexible in performance. In contrast, other systems show a remarkable regenerative potentiality due to the characteristics of their internal coordination loops (Bateson, 1980). Finally, we have learned to appreciate how some systems show a remarkable capacity to swing back from medium disturbancies into stable states (Riedl, 1984). Such examples are characteristic of the identity formation in open systems.

5.2. Coupling and Regulation

We think that the identity concept is paramount to the understanding of the system–context interrelatedness. What may appear as paradoxes makes perfectly good sense as processes of organization if we consider the relationship between system and environment to be mutually contingent (Giddens, 1984)

To generate contextual understanding of a complex, adaptive system, we therefore need an analytical framework that may allow us to capture the dynamics of systemic identity. To do so we introduce the four parameters of the SCIO model as axes in a two dimensional model shown in Fig. 3.

The vertical dimension focuses on the organism/structure relationship extending from a loose to a tight systemic coupling. This coupling dimension may

Complex Adaptive Systems Ecology

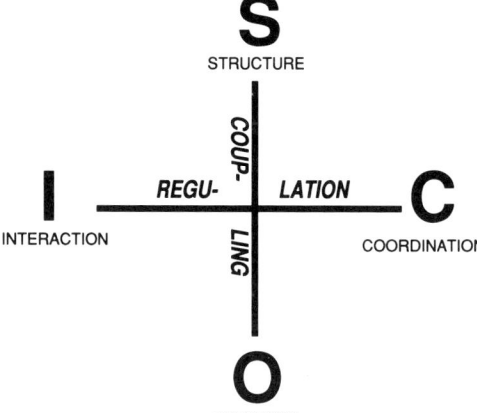

Figure 3. Contextual understanding. Systemic traits may be exposed through interferences that are aimed at identifying specific couplings in the community (S and O parameters) and specific regulatory links (I and C parameters).

be used to classify the open system according to how closed to outside forces (coupling produces stability), and how open to impacts (looseness gives flexibility), the system is (Orton and Weick, 1990). The coupling axis describes the relationship between the individual community members and the particular structures of the community, defining a structural perspective. In order to assess the degree of coupling for a community it is necessary to determine which organisms are there (identification methods), to determine where they are (microscopic position analysis), and, finally, to analyze how they react to changes (challenging the positioning of the individuals and the observed structures by introducing changed external conditions, or even perturbations). If structural patterns decay over time after more or less drastic changes, the degree of coupling is high; if the introduced changes only affect small and random parts of the community, the degree of coupling is low. The methods employed to reveal these systemic reactions are those already described for parameter analysis combined appropriately.

The horizontal dimension focuses on the interaction/coordination relationship, extending from a low to a high systemic regulation. The regulation dimension may be used to classify the open system according to how restricted the repertoire of the component elements is institutionalized to be (regulation produces effectiveness), and how autonomously the units interact (low control induces local adjustments) (DiMaggio, 1988). The regulation axis describes the relationship between the specific activities in the community and the coordination of specific functions, defining a process perspective. Metabolic activities in the community will produce intermediate compounds that directly or indirectly may have severe regulatory effects on community members and their performances. In order to expose such loops it is necessary to interfere either with the chemical compounds directly or with the organisms responsible for making them, and over time investigate the consequences of such interferences at the level of processes and functions. If interference at one particular step in a metabolic activity elicits a drastic change in functional performance in other parts of the system, the degree of regulation is high; if the interference only affects a particular pathway, the degree of regulation is low.

The investigation designs required for describing systemic coupling and regulation are combinations of parameter determinations and real-time studies after environmental changes. The target for the investigations is the response pattern of the systems; the monitoring methods are those already known to the microbiologist and discussed previously. These experiments should not only address the systemic responses to external stimuli at a given time. Systemic identity is not a simple resultant of external forces, but develops from the structuration processes through which system and environment are mutually dependent. The system enacts its immediate environment, and addresses external stimuli that are compatible with its organizational set up. As such the system also affects its environment.

The two axes representing degrees of coupling and regulation, respectively, depict an analysis of system traits that may be recorded through the use of the methods and tools already described for the analysis of the SCIO parameters. The most important distinction from the parameter analysis, however, is the absolute need for investigations over time. The structure of a system can be described on the basis of a still-picture, and the organism composition likewise. In contrast, the degree of coupling or regulation can only be disclosed as system behavior over time. Moreover, the SCIO parameters (at least SIO) may be determined in systems under steady-state conditions, whereas coupling and regulation can only be analyzed as consequences to changes of conditions. The experimental scenarios described in section 4 produce data and observations relevant to the present discussion.

5.3. The Four Analytical Windows for Identification

The two dimensions create four analytical "windows" through which we may define four principally different systemic types (Fig. 4).

Until this point most of the analyses according to SCIO have shown an intrasystemic focus (although it has constantly been underlined how these analyses were confined to the specific characteristics in time and space of the systemic environmental conditions). Through the introduction of the coupling/regulation dimensions we now focus directly on the system/context relationship. Each quadrant in the two-dimensional model (Fig. 4) describes basic characteristics of the relationship between system and environment (Giddens, 1984).

The contextuality will be described in the following conceptualizations of

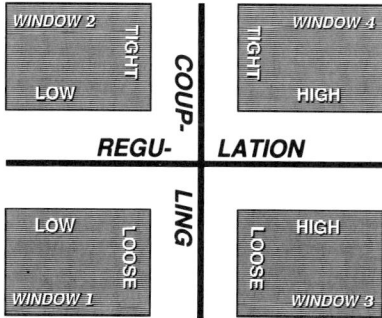

Figure 4. Systemic identity. Combining the coupling and the regulation traits exposed through intervention experiments allows a classification of communities defined by the two axes (coordinate system). The windows are equivalent to the conventional quadrants of coordinate systems.

the four windows as combinations of responsiveness (how easily the system is affected by environmental impacts) and distinctiveness (with how much precision is the system capable of reacting to the environmental impacts) (Orton and Weick, 1990).

Each window should be understood as a principle description of how a complex system relates to its immediate environment. As we have pointed out above, the context and the observed system are constructions of the researcher, and hence the model will provide no generalizable classification of complex adaptive systems. The model requires that the researcher in each case specify the nature of the contextual conditions and the systemic set-up, allowing him or her to perform studies of patterns of responsiveness/distinctiveness. Within the confinement of the research design, the researcher may in this way be able to describe the systemic identity according to the principle types of the four windows. He or she will, however, have no basis for claiming this identification as being generalizable across different contexts.

From the proposed analysis of system identity we obtain a typology that may allow us to handle more systematically longitudinal studies of the behavior and organization of known systems *in situ* as described previously. This framing of our observations (which of the four window types are we dealing with) enables us to categorize systemic performance under given conditions, and therefore also to identify anomalies (Goffman, 1974). With reference to the contextual characteristics of a given system identity we may be able to describe deviants, displaying unexpected transactions with the environment. In a non-generalizable sense the analysis of system identity becomes a system classification built on a contextually confined typology. Each of the four windows constitutes a categorization of systemic characteristics in interaction with specified contingencies. On this basis, real-time studies may enable us to monitor the particular transactions between system and environment, and perhaps in the future also to discuss questions about system development and learning processes.

5.3.1. First Window

The first window opens up for a description of a systemic type characterized by loose coupling and low regulation. Complex adaptive systems falling under this category will be expected to show a high responsiveness towards changes in the immediate environment, and to demonstrate a low distinctiveness in the systemic responses. As such, the observed system will have a pronounced receptiveness to outside influences (often interpreted in steady-state analyses as indications of flexibility and adaptability); but it will be poor in internally transmitting the consequences of environmental impacts to other parts of the system (setting limitations to adaptability that will only be revealed in longitudinal, real-time studies).

5.3.2. Second Window

The second window opens up for a description of a systemic type characterized by tight coupling and low regulation. Complex adaptive systems falling under this category will be expected to show a low responsiveness towards changes in the immediate environment, and to demonstrate a low distinctiveness in the systemic responses. As such, the observed system will have a rigid and somewhat robust pattern of reactions to outside factors (often interpreted in steady-state analyses as indications of immunity and stability); but it will also be poor in internally transmitting any needs to compensate for external impacts (setting limitations to stability that will only be revealed in longitudinal, real-time studies).

5.3.3. Third Window

The third window opens up for a description of a systemic type characterized by loose coupling and high regulation. Complex adaptive systems falling into this category will be expected to show a high responsiveness towards changes in the immediate environment, and to demonstrate a high distinctiveness in the systemic responses. As such, the observed system will have a pronounced receptiveness to outside influences (often interpreted in steady-state analyses as indications of flexibility and adaptability); and it will also have a precise and effective way of internally transmitting the consequences caused by external impacts (stressing the flexibility of the system and the way this sustains the changing conditions of stability that will only be revealed in longitudinal, real time-studies).

5.3.4. Fourth Window

The fourth window opens up for a description of a systemic type characterized by tight coupling and high regulation. Complex adaptive systems falling into this category will be expected to show a low responsiveness towards changes in the immediate environment, and to demonstrate a high distinctiveness in the systemic responses. As such, the observed system will have a rigid and somewhat robust pattern of reactions to outside factors (often interpreted in steady-state analyses as indications of immunity and stability); but at the same time it will also have a precise and effective way of internally transmitting the consequences caused by external impacts that eventually are taken in (a fragile stability that builds on a relatively inflexible interchange with an environment that should not change too drastically; only to be revealed in longitudinal, real-time studies).

5.4. The Windows and Laboratory Biofilms

It is obvious from the presentation of the four windows, as emphasized, that the window-based identity descriptions in no way are to be considered an absolute taxonomy. The contextuality of the identity descriptions is the necessary basis for assigning the community systems to any of the four windows—without a direct connection to the external conditions the identity has no meaning.

In order to stress the contextuality we may consider simple laboratory biofilms consisting of mixed microbial communities established on surfaces under conditions that may be changed at will (flow chambers). In the early phases of such biofilms the individual organisms will live together in loose networks and respond to changing conditions as the systems belonging to the 1. window. In later phases, it is likely that tighter structural couplings develop or that non-structured well-controlled populations are formed (depending on the organisms and the prevailing conditions); these communities would be placed in windows 2 and 3, respectively. In late-phase biofilm communities it is possible that the organism profile is narrow, that rigid structures have formed, and that there are very distinct reaction patterns exisiting between the individual community members; such a community would seem to belong to the 4. window.

We can conclude from this example that system identity is relative, not absolute, and that the same community at different time points of its development may change identity. The description has been simplified deliberately; obviously, it will be common in studies of community identity to find that one part of the system may be identified as one type, and another part of the system as another. Once more contextuality appears as the basis for system characterization.

6. CASE Study—Research Designs

Throughout this article we have described the relationship between a paradigmatic platform on one hand, and a series of concepts, models, and methodological implications on the other. From an open systemic perspective, the principle arguments will be applicable to research from different disciplines, be it at macro-, mezo-, or microcommunity levels. Here we have taken our examples from microbial ecology and demonstrated how the proposed meta-theoretical framework may be implemented, employing the modern techniques available today to scientists in this field. Naturalistic inquiry has been conducted according to the four parameters of the SCIO model, and we have argued that a comprehensive understanding of complex adaptive systems ecology (CASE) may give us access to a classification of identity. Let us now close with a possible strategy for designing research that addresses the question of systemic identity. This scientific problem is obviously not the only one pertinent in microbial ecology, but it is placed centrally on our scientific agenda with its deep roots in fundamental biological questions relating to interactions, organization, and control. What we describe is a series of alternatives in the process of designing research studies that hold a greater chance of conducting a naturalistic inquiry.

6.1. Naturalistic Inquiry Paradigm

In this process we deliberately draw upon the fourteen operational characteristics of the emergent paradigm, which we described earlier (section 2), under the five headings: subjectivity, focus, methodology, analysis, and scientific control. The category, techniques, will now be added.

6.1.1. Subjectivity

The significance of scientific choice is tremendous. It should not be mistaken for a rational-choice perspective; we have argued for the researcher's obligation to construct the scientific agenda he wants to pursue. A multiple-reality perspective, as described, calls for outspoken choices and a conscious subjective role as scientist. The individual researcher becomes a human instrument with the responsibility to choose his or her focus and draw the necessary lines between system and environment. The first and perhaps most fundamental choice to be made by the individual researcher is the deliberate choice of being an actor in the process. The mutual interdependency between system and environment matches the mutual interdependency between researcher and study. The dialogue between system and environment, in fact, resembles the dialogue between researcher and study.

It is of great importance to reflect on the question of subjectivity. The choices made affect not only the actual research design that we may want to construct, but also the kinds of analyses that will be open to us and to the channels, language, and audience of the scientific community that we may be able to reach. If we choose a social constructivist view, and hereby accept the role of subjectivity, the consequences for designing research are radical, calling for radical choices concerning focus, methodology, techniques, analyses, and control. Table I is paradoxically ordered. The two columns of conventional and alternative designs are apparently antagonistic to each other. They highlight

Table I. Research Design

	Conventional	Alternative	Radical
Focus	known community	unknown community	tranfer and fittingness
	in vitro	*in vivo*	*in situ* (context)
Methodology	quantitative	qualitative	data integration
	tangential extrapolation	real-time study	longitudinal
	steady state/snap-shots	variation perturbation	system history
	sampling by representative or random	sampling by parallel or single case	pragmatic choices
Techniques	measurement equipment (gaschromatography)	monitoring devices (microscope)	multi-media technology
	graphical visualization	direct observation	image construction
Analyses	causality	relations	systemic identity: contextual knowledge
	effects	critical incidents	
	categorizing	patterns	
	characteristics	scripts	
Control	reproducability	histories of complex adaptive systems	?
	control experiments		criteria discussion

some of the fundamental issues to be addressed in the research design process, together with some of the possible radical choices to be made. By conventional we refer to research designs that fall well within the dominant scientific paradigm, by alternative we refer to designs that illustrate new and untested paradigmatic grounds, almost at the extreme, and by radical we refer to the contours of naturalistic designs that transcend the paradox of convention versus alternative.

6.1.2. Focus

The major issue here is the choice of target for scientific investigation. Microbiologists and ecologists often have distinct preferences for *in vitro* or *in vivo* designs, respectively. The radical but not necessarily controversial choice is *in situ* studies, in which the importance of contextuality is emphasized. The choice is not between the laboratory and nature, but between artificial and naturalistic settings.

6.1.3. Methodology

The major issue here is the relationship between methodology and research design. In conventional designs, the preferred methodology seems to have gained a life of its own. Whatever we study, and under whatever conditions it may be, our methodological choices are given by scientific tradition and demand no particular reflection or explicit argument. Under the conventional paradigm this implies the obvious path of quantitatively based tangential extrapolations from the steady-state observations, which will allow us to describe random (or representative) organisms or communities through snapshots of their activities and configuration. This has created a barrier separating quantitatively oriented scientists from qualitatively oriented scientists. In a larger perspective this has in fact separated most of the natural sciences from most of the social sciences. The radical choice in this paradoxical situation is pragmatic by nature, and integrative by virtue. The call for longitudinally based systemic histories may bridge between the quantitative/qualitative gap.

6.1.4. Techniques

The major issue here is documentation. We need empirical access to the target of our scientific study, and we need standards for the registration of our observations. This is normally achieved by the relevant equipment having both the refinements necessary to measure at the micro-level, and at the same time the capacity to transform, compile, and numerically sort the generated data. Until recently these techniques made direct observation and monitoring of living systems by microscopes obsolete. Now with modern three-dimensional microscopes in combination with advanced computer design technology, the radical choice is image construction. From a multiple reality point of departure we may profit by employing multi-media technology in attempts to construct systemic representations that serve as documentation in the future.

6.1.5. Analyses

The major issue here is the question of identification. Traditionally, identification means categorizing organisms/systems by predefined characteristics (often based on taxonomically ordered endogenous criteria). Such empirical studies lead to analyses and interpretations that focus on generalizable processes, functions, and features. With a contextual point of departure a more radical approach may be analyses of systemic relations and observations of patterns and scripts. In this case identification does not lead to descriptions according to fixed categories; instead, identification becomes the conscious scientific attempt to classify the system as a reflection of the researcher's scientific choices.

6.1.6. Control

The major issue here is "sense-making" (Weick, 1993). The conventional approach to control offers objective academic norms and procedures. Being taken seriously this becomes a question of upholding the conventional tradition for reproducibility and control experiments. Sensemaking is a radical and subjective approach, it offers no parallel framework for scientific communication and control. The choice may be to present ourselves as integral parts of our research.

Making sense then may be the possible interpretation of results that we and our peers have arrived at, knowing how the study was conducted, and by whom.

Future scientific debate will discuss how we may develop a format and language for communication about our naturalistic inquiry. Based on accounts of the histories of complex adaptive systems we need a dialogue with our peers that may provide us all with operational criteria for handling sense-making in the world of scientific study.

It becomes clear that there are no pregiven choices. No column in Table I is appropriate or correct in itself. As we have repeatedly argued in the text so far, much of the methodology and techniques that are conventionally applied will still be attractive as integral parts of naturalistic designs. The alternative measures and analyses are too fragile as platforms for discoveries to be meaningfully appreciated by other researchers. The radical choices give no solutions, but only offer the contours of a scientific challenge. The more we choose from the alternative and radical columns, the more we challenge the dominating paradigm that is conventionally applied by our colleagues. The more we want to integrate the conventional column, the greater are the efforts we must make to give it a meaningful role in the design, and the more we need to make sure that the requirements for transfer and fittingness are explicitly tested and present.

Not everything is allowed (Feuerabend, 1975), however, and not any mix across the conventional, alternative, and radical possibilities is feasible. The conventional tradition for representative or random sampling, for extrapolation and for analyses of causal effects, may in most cases be counterproductive to naturalistic inquiry, where image construction under norms of contextuality are central. Still, even these apparent antagonisms may in particular designs be complementary. Questions of incompatibility and exclusiveness are still to be debated on the grounds of concrete studies, exploring new controversial designs.

ACKNOWLEDGMENTS. We are grateful for the generous support of the Danish Research Councils over the years, without which our scientific work would not have been possible. This project was initiated during a 1-month stay in Provence, France and supported by Direktør Ib Henriksens Fond, for which we express our gratitude. Our many students, post-docs, and colleagues who have taken an interest in this work are thanked for their valuable contributions. In particular, we wish to thank Professor K. Marshall for encouragement and helpful suggestions concerning the manuscript.

References

Amann, R. I., Springer, N., Ludwig, W., Gortz, H. D., and Schleifer, K. H., 1991, Identification and phylogeny of uncultured bacterial endosymbionts, *Nature* **351**:161–165.

Amann, R. I., Ludwig, W., and Schleifer, K.-H., 1995, Phylogenetic identification and *in situ* detection of individual microbial cells without cultivation, *Microbiol. Rev.* **59**:143–169.

Atlas, R. M., and Bartha, R., 1993, *Microbial Ecology—Fundamentals and Applications*, Benjamin/Cummings Publishing, Redwood City, CA.

Bainton, N. J., Bycroft. B. W., Chhabra, S. R., Stead, P., Gledhill, L., Hill, P. J., Rees, C. E. D., Winson, M. K., Salmond, G. P. C.,Stewart, G. S. A. B., and Williams, P., 1992, A general role for the *lux* autoinducer in bacterial cell signalling: control of antibiotic synthesis in *Erwinia, Gene* **116**:87–91.

Bateson, G., 1972, *Steps to an Ecology of Mind*, Random House, New York.

Berger, P. L, and Luckmann, T., 1966, *The Social Construction of Reality,* Penguin Books, Baltimore.

Berkowitz, G. C., and Tschirgi, R. D., 1988, The biological foundation of space and the evolution of spatial dimension. *J. Soc. biolog. struct.,* **11**:323–335.

Bohm, D., 1983, Of matter and meaning: the super implicate order. A conversation with David Bohm, *Re-vision* **6**:34–44.

Boulding, K. E., 1953, *The Organizational Revolution - a Study in the Ethics of Economic Organization,* Harper & Brothers, New York.

Brakenhoff, G. J., van der Voort, H. T. M., Baarslag, M. W., Mans, B., Oud, J. L, Zwart, R., and van Driel, R., 1988, Visualization and analysis techniques for three dimensional information acquired by confocal microscopy, *Scanning Microsc.* **2**:1831–1838.

Brock, T. D., and Madigan, M., 1988, *The Biology of Microorganisms,* Prentice-Hall, Englewood Cliffs, N.J.

Bryant, M. P., Wolin E. A., Wolin, M. J., and Wolfe, R. S., 1967, *Methanobacillus omelianskii,* a symbiotic association of two species of bacteria, *Arch. Mikrobiol.* **59**:20–31.

Buckley, W, 1967, *Sociology and Modern Systems Theory,* Prentice Hall, Englewood Cliffs N.J.

Buckley, W., 1969, Society as an open complex system, in:*Modern Systems Research for the Behavioural Scientist* (W. Buckley, ed.), University of Chicago Press, Chicago

Bukau, B., 1993, Regulation of the *Escherichia coli* heat-shock response, *Mol. Microbiol.* **9**:671–680.

Caldwell, D. E., 1993, Steady-state microenvironments for subculture of steady-state consortia, communities, and microecosystems, in: *Trends in Microbial Ecology,* (R. Guerrero and C. Pedros-Alio, eds.), Spanish Society for Microbiology, Barcelona, pp. 123–128.

Caldwell, D. E., 1995, Cultivation and Study of Biofilm Communities, in: *An Introduction to Bacterial Biofilms* (H. M. Lappin-Scott, and J. W. Costerton, eds.), Cambridge University Press, Cambridge, pp. 1–15.

Caldwell, D. E., and Lawrence, J. R., 1988, Study of attached cells in continuous-flow slide culture, in: *CRC Handbook of Laboratory Model Systems for Microbial Ecology Research,* Volume 1 (J. W. T. Wimpenny, ed.), CRC press, Boca Raton, pp. 117–138.

Caldwell, D. E., Korber, D. R., and Lawrence, J. R., 1992, Confocal laser microscopy and computer image analysis, in: *Advances in Microbial Ecology,* Volume 12 (K. C. Marshall, ed.), Plenum Press, New York, pp. 1–67.

Caldwell, D. E., Korber, D. R., and Lawrence, J. R., 1993, Analysis of biofilm formation using 2-D versus 3-D digital imaging, in: *Microbial Cell Envelopes: Interactions and Biofilms* (L. B. Quesnel, P. Gilbert, and P. S. Handley, eds.), Blackwell Scientific Publications, Oxford, pp. 52S–66S.

Chalfie, M., Tu, Y., Euskirchen, G., Ward, W. W., and Prascher, D. C., 1994, Green fluorescent protein as a marker for gene expression, *Science* **263**:802–805.

Characklis, W. G., McFeters, G. A., and Marshall, K. C., 1990, Physiological ecology in biofilm systems, in: *Biofilms,* Chapter 10 (W. G. Characklis and K. C. Marshall, eds.), J. Wiley and Sons, New York, pp. 341–93.

Christensen S., and Molin, J., 1995, Origin and transformation of organizations: institutional analysis of the Danish Red Cross, in: *The Institutional Construction of Organizations,* (W. R. Scott and S. Christensen, eds.), Sage Publications, Thousand Oaks, California, pp. 67–91.
Costerton, J. W., Cheng, K.-J., Geesey, G. G., Ladd, T. I., Nickel, N. C., Dasgupta, M., and Marrie, T. J., 1987, Bacterial biofilms in nature and disease, *Annu. Rev. Microbiol.* **41**:435–464.
Costerton, J. W., Lewandowski, Z., DeBeer, D., Caldwell, D. E., Korber, D. R., and James, G. A., 1994, Biofilms, The customized microniche, *J. Bacteriol.* **176**:2137–2142
Delaquis, P. J., Caldwell, D. E., Lawrence, J. R., McCurdy, A. R., 1989, Detachment of *Pseudomonas fluorescens* from biofilms on glass surfaces in respons to nutrient stress, *Microb. Ecol.* **18**:199–210.
Dempsey, M. J., 1981, Marine bacterial fouling: A scanning electron microscope study, *Mar. Biol.* **61**:305–315.
DiMaggio, P., 1988, Interest and agency in institutional theory, in: *Institutional Patterns and Organizations: Culture and Environment,* (L. G. Zucker, ed.), Ballinger, Cambridge, pp. 3–21.
Dunlap, P. V., and Greenberg, E. P., 1991, Role of intercellular chemical communication in the *Virio fischeri*-monocentrid fish symbiosis, in: *Microbial Cell-Cell Interactions* (M. Dworkin, ed.), American Society for Microbiology, Washington, D. C., pp. 219–253.
Eberl, L., Givskov, M., Sternberg, C., Moller, S., Christiansen, G., and Molin, S., 1996, Physiological responses of *Pseudomonas putida* KT2442 to phosphate starvation. *Microbiol.* **142**:155–163.
Eberl, L., Winson, M. K., Sternberg, C., Stewart, G. S. A. B., Christiansen, G., Chabra, S. R., Bycroft, B., Williams, P., Molin, S., and Givskov, M., 1996, Involvement of N-acyl-L-homoserine lactone autoinducers in control of multicellular behavior of *Serratia liquefaciens, Molec. Microbiol.* **20**:127–136.
Feuerabend, P. K., 1975, *Against Method,* Verso, London.
Giddens, A., 1984, *The Constitution of Society,* University of California Press, Berkeley.
Giovannoni, S., 1991, The polymerase chain reaction, in: *Nucleic Acid Techniques in Bacterial Systematics* (E. Stackebrandt and M. Goodfellow, eds.), John Wiley & Sons, New York, pp. 177–203.
Givskov, M., Eberl, L., Moller, S., Poulsen, L. K., and Molin, S., 1994a, Responses to nutrient starvation in *Pseudomonas putida* strain KT2442: Analysis of general cross-protection, cell morphology, and macromolecular content, *J. Bacteriol.* **176**: 7–14
Givskov, M., Eberl, L., and Molin, S., 1994b, Responses to nutrient starvation in *Pseudomonas putida* strain KT2442: Two-dimensional electrophoretic analysis of starvation and stress inducible proteins, *J. Bacteriol.* **176**:4816–4824.
Glaser, B. G. and Strauss, A. L., 1967, *The Discovery of Grounded Theory,* De Gruyter, New York.
Glassman, R. B., 1973, Persistence and loose coupling in living systems, in: *Behavioural Science* vol. 18 pp. 83–98.
Goffman, E., 1974, *Frame Analysis,* Harper & Row, New York.
Greenberg, E. P., Ulitzur, S., and Hastings, J. W., 1979, Induction of luciferase synthesis in *Beneckea harveyi* by other marine bacteria, *Arch. Microbiol.* **120**:87–91.
Guba, E. G., 1985, The context of emergent paradigm research, in: *Organizational Theory and Enquiry* (Y. S. Lincoln, ed.), Beverly Hills.
Hall, A. D., and Fagan, R. E., 1969, Definition of system, in: *Modern Systems Research for the Behavioural Scientist* (W. Buckley, ed.), University of Chicago Press, Chicago, pp. 81–92.
Hegarty, C. P., and Weeks, O. B., 1940, Sensitivity of *Escherichia coli* to coldshock during the logarithmic growth phase, *J. Bacteriol.* **39**:4575–484.
Hejl, P. M., 1984, Towards a theory of social systems: self-organization and self-maintenance, self-reference and syn-reference, in: *Self-Organization and Management of Social Systems* (H. Ulrich and G. J. B. Probst, eds.), Springer Verlag, Berlin, pp. 60–78.

Hojberg, O., and Sorensen, J., 1993, Microgradients of microbial oxygen consumption in a barley rhizosphere model system, *Appl. Environ. Microbiol.* **59**:431–437.

Jaques, E., Gibson, R. O., and Isaac, D. J., 1978, *Levels of Abstraction in Logic and Human Action*, Heineman, London

Jepras, R. I., Carter, J., Pearson, S. C., Paul, F. E., and Wilkinson, M. J., 1995, Development of a robust flow cytometric assay for determining numbers of viable bacteria, *Appl. Environ. Microbiol.* **61**:2696–2701.

Jones, P. G., and Inouye, M., 1994, The cold shock response—a hot topic, *Mol. Microbiol.* **11**:811–818.

Kaiser, D., and Losick, R., 1993, How and why bacteria talk to each other, *Cell* **73**:873–885.

Kelly, G. A., 1955, *The Psychology of Personal Constructs*, W. W. Norton, New York.

Kjelleberg, S. (ed.), 1993, *Starvation in Bacteria*, Plenum Press, New York.

Korber, D. R., James, G. A., and J. W. Costerton, 1994, Evaluation of fleroxacin activity against established *Pseudomonas fluorescens* biofilms, *Appl. Environ. Microbiol.* **60**:1663–1669.

Kragelund, L, Christoffersen, B., Nybroe, O., and de Bruijn, F. J., 1995, Isolation of lux gene fusions in *Pseudomonas fluorescens* DF57 inducible by starvation for nitrogen or phosphorus, *FEMS Microbiol. Ecol.* **17**:95–105.

Kuhn, T. S., 1962, *The Structure of Scientific Revolution*, University of Chicago Press, Chicago.

Lawrence, J. R., Korber, D. R., Hoyle, B. D., Costerton, J. W., and Caldwell, D. E., 1991, Optical sectioning of microbial biofilms, *J. Bacteriol.* **173**:6558–6567.

Leigh, E. G., Jr., 1983, When does the good of the group override the advantage of the individual, *Proc. Nat. Acad. Sci.* **80**:2985–2989.

Lincoln, Y. S., and Guba, E. G., 1985, *Naturalistic Inquiry*, Sage Publications, Beverly Hills.

Lyotard, F., 1984, *The Postmodern Condition: A Report on Knowledge*, Manchester University Press, Manchester.

Maaloe, O., and Kjeldgaard, N. O., 1966, *Control of Macromolecular Synthesis*, W. A. Benjamin, New York.

Marshall, K. C., 1994, Microbial ecology: wither goest thou?, in: *Trends in Microbial Ecology* (R. Guerrero and C. Pedros-Alio, eds.), Spanish Society for Microbiology, Barcelona, pp. 5–8.

Marshall, P. A., Loeb, G. I., Cowan, M. M., and Fletcher, M., 1989, Response of microbial adhesives and biofilm matrix polymers to chemical treatments as determined by interference reflection microscopy and light section microscopy, *Appl. Environ. Microbiol.* **55**:2827–2831.

Maruyama, M., 1963, The Second Cybernetics: Deviation-Amplifying Mutual Causal Processes, in: *American Scientists*, **51**:164–179.

Massol-Dey, A. A., Whallon, J., Hickey, R. F., and Tiedje, J. M., 1995, Channel structure in aerobic biofilms of fixed-film reactors treating contaminated groundwater, *Appl. Environ. Microbiol.* **61**:769–777.

Mead, G. H., 1962, *Mind, Self and Society*, The University of Chicago Press, Chicago.

Molin, J., and Molin, S., 1988, *Den Iscenesatte Virkelighed*, Akademisk Forlag, Kobenhavn.

Molin, S., Boe, L., Jensen, L. B., Kristensen, C. S., Givskov, M., Ramos, J. L., and Bej, A. K., 1993, Suicidal genetic elements and their use in biological containment of bacteria, *Ann. Rev. Microbiol.* **47**:139–166.

Moller, S., Kristensen, C. S., Paulsen, L. K., Cartensen, J. M., and Molin, S., 1995, Bacterial growth on surfaces: Automated image analysis for quantification of growth-related parameters, *Appl. Environ. Microbiol.* **61**:741–748.

Moller, S., Pedersen, A. R., Arvin, E., and Molin, S., 1996, Activity and spatial distribution of a toluene degrading *Pseudomonas putida* in a multispecies biofilm assessed by quantitative in situ hybridization and scanning confocal laser microscopy, *Appl. Environ. Microbiol.* **62**:4632–4640.

Moller, S., Korber, D. R., Wolfaardt, G. M., Molin, S., and Caldwell, D. E., 1997, The impact of

protozoan grazing on the architecture of a degradative biofilm community, *Appl. Environ. Microbiol.* (submitted).
Morgan, G., 1986, *Images of Organizations,* Sage Publications, Beverly Hills.
Morgan, G., 1987, Knowledge, uncertainty, and choice; exploring choice, reframing the process of evaluation, in: *Beyond Method: Strategies for Social Research* (G. Morgan, ed.), Sage, Beverly Hills.
Neidhardt, F. C., Ingraham, J. L., and Schaechter, M., 1990, *Physiology of the Bacterial Cell,* Sinauer Associates, Sunderland, Mass.
Nybroe, O., 1995, Asssessment of bacterial metabolic activity—new developments in microcolony and dehydrogenase assays, review, *FEMS Microbiol. Ecol.* **17**:77–84.
Nystrom, T., Flardh, K., and Kjelleberg, S., 1990, Responses to multiple nutrient starvation in marine vibrio sp. strain CCUG 15956, *J. Bacteriol.* **172**:7085–7097.
Nystrom, T., Olsson, R. M., and Kjelleberg, S., 1992, Survival, stress resistance, and alterations in protein expression in the marine *Vibrio* sp. strain S14 during starvation for different individual nutrients. *Appl. Environ. Microbiol.* **58**:55–65.
Orton, J. D., and Weick, K. E., 1990, Loosely coupled systems: a reconceptualization, *Academy of Management Review,* **15**:203–223.
Pace, N. R., Stahl, D. A., Lane, D. J., and Olsen, G. J., 1986, The analysis of natural microbial populations by ribosomal RNA sequences, *Adv. Microb. Ecol.* **9**:1–55.
Passador, L., Cook, J. M., Cambello, J., Rust, L., and Iglewski, B. H., 1993, Expression of *Pseudomonas aeruginosa* virulence genes requires cell-to-cell communication. *Science* **260**:1127–1130.
Paster, B. J., Cooke, S., Dewhirst, F. E., and Breznak, J. A., 1994, Phylogeny of a novel *Treponema* species determined from PCR amplification of 16S rRNA gene, *Abst. Gen. Meet. Am. Soc. Microbiol.* **94**:313.
Patton, M. Q., 1980, *Qualitative Evaluation Methods,* Sage, Beverly Hills.
Pettigrew, A. M., 1995, Longitudinal research on change: Theory and practice, in: *Longitudinal Field Research Methods. Studying processes of organizational change* (G. P. Huber and A. H. Van de Ven, eds.), Sage, Thousand Oaks.
Pickup, R. W., 1991, Development of molecular methods for the detection of specific bacteria in the environment *J. Gen. Microbiol.* **137**:1009–1019.
Poulsen, L. K., Lan, F., Kristensen, C. S., Hobolth, P., Molin, S., and Krogfelt, K. A., 1994, Spatial distribution of *E. coli* in the mouse large intestine inferred from rRNA in situ hybridization, *Infect. Immun.* **62**:5191–5194.
Rapoport, A., 1986, *General Systems Theory,* Turnbridge Wells.
Revsbech, N. P., and Jorgensen, B. B., 1988. Microelectrodes: their use in microbial ecology, in: *Advances in Microbial Ecology,* Vol. 9 (K. C. Marshall, ed.), Plenum Press, New York, pp. 293–352.
Riedl, R., 1984, Self-organization: some theoretical cross-connections, in: *Self-Organization and Management of Social Systems,* (H. Ulrich and G. J. B. Probst, eds.), Springer Verlag, Berlin.
Robarts, R. D., and Zohar, T., 1993, Fact or fiction—bacterial growth rates and production as determined by [^{3}H-methyl]thymidine, in: *Advances in Microbial Ecology,* Vol. 13 (G.-F. Jones, ed.), Plenum Press, New York, pp. 371–418.
Robinson, R. W., Akin D. E., Nordstedt, R. A., Thomas, M. V., and Aldrich, H. C., 1984, Light and electron microscopic examninations of methane-producing biofilms from anaerobic fixed-bed reactors, *Appl. Environ. Microbiol.* **48**:127–136.
Rodrigues, G. G., Phipps, D., Ishiguro, K., and Ridgway, H. F., 1992, Use of a fluorescent redox probe for direct visualization of actively respiring bacteria, *Appl. Environ. Microbiol.* **58**:1801–1808.

Rorty, R., 1980, *Philosophy and the Mirror of Nature*, Princeton N.J.
Schwartz, P., and Ogilvy, J., 1979, *The Emergent Paradigm: Changing Patterns of Thought and Belief*, VALS report, no. 7, SRI International, Menlo Park, CA.
Scott, W. R., 1992, *Organizations: Rational, Natural, and Open Systems*, Prentice Hall, Englewood Cliffs, N.J.
Shapiro, J. A., 1988, Bacteria as multicellular organisms, *Sci. Am.* **256**:82–89.
Sjollema, J., Busscher, H. J., and Weerkamp, A. H., 1989, Real-time enumeration of adhering microorganisms in a parallel plate flow cell using automated image analysis, *J. Micobiol. Methods* **9**:73–78.
Slater, J. H., and Hartman, D. J., 1982, Microbial ecology in the laboratory:experimental systems, in: *Experimental Microbial Ecology* (R. G. Burns and J. H. Slater, eds.), Blackwell Scientific, Oxford, pp. 255–274.
Stahl, D. A., Flesher, B., Mansfield, H. R., and Montgomery, L., 1988, Use of phylogenetically based hybridization probes for studies of ruminal microbial ecology, *Appl. Environ. Microbiol.* **54**:1079–1084.
Stams, A. J. M., Grotenhuis, J. T. C., and Zehnder, A. J. B., 1989, Structur function relationship in granular sludge, in: *Recent advances in microbial ecology*, (T. Hattori, Y. Ishida, Y. Maruyama, R. Y. Morita, and A. Uchida, eds.), Japan Scientific Societies Press, Tokyo, pp. 440–445.
Torsvik, V., Goksryr, J., and Daae, F. L., 1990, High diversity of DNA of soil bacteria, *Appl. Environ. Microbiol.* **56**:782–787.
Toulmin, S., 1961, *Foresight and Understanding*, Cambridge University Press, Cambridge.
Tsoukas, H., 1989, The validity of idiographic research explanations, *Acad. Manage. Rev.* **14**:551–561.
Ulrich, H., and Probst, G. J. B., 1984, *Self-Organization and Management of Social Systems*, Springer Verlag, Berlin.
van Overbeek, L., Eberl, L., Givskov, M., Molin, S. and van Elsas, J. D., 1995, Survival of, and induced stress resistance in, *Pseudomonas fluorescens* cells residing in soil, *Appl. Envir. Microbiol.* **61**:4202–4208.
Van De Ven, A. H., 1992, Suggestions for studying strategy process: A research note, *Strateg. Manag. J.* **13**:169–188.
von Bertalanffy, L., 1968, *General Systems Theory*, George Braziller, New York.
von Foerster, H., 1984, Principles of self-organization in a socio-managerial context, in: *Self-Organization and Management of Social Systems* (H. Ulrich and G. J. B. Probst, eds.), Springer Verlag, Berlin.
Wagner, M., Amann, R., Lemmer, H., and Schleifer, K.-H., 1993, Probing activated sludge with oligonucleotides specific for protobacteria: Inadequacy of culture-dependent methods for describing microbial community structure, *Appl. Environ. Microbiol.* **59**:1520–1525.
Ward, D. M., Weller, R., and Bateson, M. M., 1990, 16S rRNA sequences reveal numerous uncultured microorganisms in a natural community, *Nature* **453**:63–65.
Ward, D. M., Bateson, M. M., Weller, R., and Ruff-Roberts, A. L., 1992, Ribosomal RNA analysis of microorganisms as they occur in nature, in: *Advances in Microbial Ecology*, Vol. 12 (K. C. Marshall, ed.), Plenum Press, New York, pp. 219–286.
Weick, K., 1979, *The Social Psychology of Organizing* (2nd edition) Addison Wesley, Reading, MA.
Weick, K. E., 1993, Sensemaking in organizations: Small structures with large consequences, in: *Social Psychology in Organizations*, Prentice Hall, London.
Whitehead, A. N., 1929, *Process and Reality*, Cambridge University Press, London.
Whitehead, A. N., 1967, *Science and the Modern World*, Free Press, New York.
Wilden, A., 1972, *System and Structure*, Tavistock Publications, London.
Woese, C. R., 1987, Bacterial evolution, *Microb. Rev.* **51**:221–271.

Wolfaardt, G. M., Lawrence, J. R., Robarts, R. D., and Caldwell, D. E., 1994a, Multicellular organization in a degradative biofilm community, *Appl. Environ. Microbiol.* **60:**434–446.

Wolfaardt, G. W., Lawrence, J. R., Robarts, R. D., and Caldwell, D. E., 1994b, The role of interactions, sessile growth and nutrient amendment on the degradative efficiency of a bacterial consortium, *Can. J. Microbiol.* **40:**331–340.

3

Roles of Submicron Particles and Colloids in Microbial Food Webs and Biogeochemical Cycles within Marine Environments

TOSHI NAGATA and DAVID L. KIRCHMAN

1. Introduction

The recent discovery of numerous detrital submicron particles in diverse marine environments (Koike *et al.*, 1990; Longhurst *et al.*, 1992; Wells and Goldberg, 1991, 1994) has stirred the interest of oceanographers and has spurred studies into the roles of these small particles in marine food webs and biogeochemical fluxes. The abundance of non-living submicron particles (10^7–10^{10} particles ml^{-1}) far exceeds the number of living particles of similar size dimensions, including phytoplankton, bacteria, and viruses (Koike *et al.*, 1990; Wells and Goldberg, 1991; see Table I). Bulk chemical measurements have confirmed that the "colloidal fraction" (size, 0.001–1 μm) represents a large fraction (10–50%) of total "dissolved" organic carbon (DOC) in seawater (Ogawa and Ogura, 1992; Benner *et al.*, 1992; Gau *et al.*, 1994). Several provocative hypotheses have been proposed to explain the roles of colloids and submicron particles in trophic dynamics (Sherr, 1988; Flood *et al.*, 1992), aggregate formation (Alldredge *et al.*, 1993; Kepkay, 1994), and condensation of organic matter (Nagata and Kirchman, 1992b, 1996; Keil and Kirchman, 1994).

Marine biogeochemists are keen to characterize colloids and to determine fluxes of their formation, transport, and degradation because DOC and colloids may impact regional (Carlson *et al.*, 1994; Peltzer and Hayward, 1996) and global (Siegenthaler and Sarmiento, 1993) carbon cycling, which contributes to

---

TOSHI NAGATA • Ocean Research Institute, University of Tokyo, Nakano, Tokyo 164, Japan.   **DAVID L. KIRCHMAN** • College of Marine Studies, University of Delaware, Lewes, Delaware 19958.
Advances in Microbial Ecology, Volume 15, edited by Jones. Plenum Press, New York, 1997.

Table I. Abundances of Planktonic Microbes and Submicron/Colloidal Particles in the Upper Ocean

Microbes and particles	Size (μm)	Abundance (# ml^{-1})	Study site	References
Cyanobacteria	0.5–1	10^3–10^5		Waterbury et al. (1979); Murphy and Haugen (1985)
Prochlorophytes	0.3–0.8	10^3–10^5		Chisholm et al. (1988)
Bacteria	0.3–0.8	10^5–10^6		Fuhrman et al. (1989)
Viruses	0.03–0.1	$<10^6$–10^8		Bergh et al. (1989); Hara et al. (1991)
Submicron particles[a]	0.38–0.7	1×10^7–8×10^7	Northwest Pacific	Koike et al. (1990)
		1×10^7	Northeast Pacific	Longhurst et al. (1992)
		1×10^7–3×10^7	Off Suruga Bay	T. Nagata (unpublished)
Colloids[b]	0.005–0.2	1×10^9–9×10^9	North Atlantic	Wells and Goldberg (1994)
		3×10^8–2×10^9	Southern Ocean	Wells and Goldberg (1994)
		$<10^7$–3×10^9	Santa Monica Basin	Wells and Goldberg (1991)

[a]Determined by particle counting.
[b]Determined by electron microscopy.

controlling atmospheric concentrations of carbon dioxide, the most important greenhouse gas on our planet (Siegenthaler and Sarmiento, 1993).

During the past 15 years, microbial ecologists have shown that heterotrophic bacteria consume most of the DOC in aquatic systems and that this carbon consumption represents approximately 50% of marine primary production when averaged globally (Cole et al., 1988; Ducklow and Carlson, 1992). Our knowledge about interactions between organic matter and microorganisms has increased substantially during the same period (Kirchman, 1993; Azam et al., 1994). However, there still remain several major unsolved questions. For instance, there is no general agreement about the mechanism of DOC production (Jumars et al., 1989; Baines and Pace, 1991; Nagata and Kirchman, 1992a). The turnover of large fractions of the DOC pool remains unknown.

This chapter presents recently emerging ideas on interactions of colloids and microorganisms and their implications for trophic dynamics and biogeochemical cycles in aquatic environments. We will emphasize interdisciplinary interactions between microbiology and geochemistry, where new perspectives are rapidly growing. Some challenging areas of future studies are also discussed. Although we mainly focus on marine environments, particularly open oceans, some basic processes should be similar in estuarine and freshwater environments (Massalski and Leppard, 1979; Leppard, 1984), except when effects of electrolyte concentrations are overwhelming.

2. Definition and Determination of Colloids and Submicron Particles

Particulate, dissolved, and colloidal organic matter are defined operationally in oceanography. Dissolved organic matter (DOM) is traditionally defined as organic matter that passes through glass fiber filters (typical nominal retention of 0.7 μm) and particulate organic matter (POM) as that which is retained on filters (Cauwet, 1981). With this definition, any small particles (even microorganisms) that pass through filters are included in the dissolved category. Ultrafiltration (molecular cut-off of 1000 Dalton) divides the DOM into high- (HMW) and low-molecular-weight (LMW) fractions (Ogawa and Ogura, 1992; Benner et al., 1992). Small particles are in the HMW-DOM pool, called colloidal organic matter (Santschi et al., 1995: Guo et al., 1994). Colloidal organic matter consists of small particles in the approximate size range of 0.001–1 μm, as depicted in Fig. 1.

There are some complications associated with size fractionation of organic matter by (ultra)filtration (Carlson et al., 1985). Flexible particles can squeeze through filter pores much smaller than their diameter. Fragile particles may be fragmented by shear associated with filtration, allowing the fragments to pass through pores smaller than their original size. On the other hand, aggregation and adsorption of DOM during filtration may result in artificial inclusion of truly dissolved organic matter in colloidal fractions.

Resistive-pulse counters are useful for determining the number and size distribution of submicron particles and large colloids (Fig. 1). By using a Elzone particle counter, Koike et al., (1990) examined submicron particles in the size range of 0.38–0.7 μm. Smaller submicron particles in the size range of 0.005–0.12 μm have been examined by electron microscopy (Wells and Goldberg, 1991). The limited "window" of each method has not allowed us to determine the entire size spectrum of colloidal particles (Fig. 1). Photon correlation spectroscopy is potentially useful to examine colloids in a wider size range (0.002–0.5 μm) (Rees, 1990; Newman et al., 1994), although this technique remains to be applied to oceanic environments where particle concentrations are very low. The abundance and volume of particles determined by resistive-pulse counters or electron microscopy have not been compared with bulk chemical measurements (carbon and nitrogen content).

A strict distinction in terminology between DOM and colloids has been

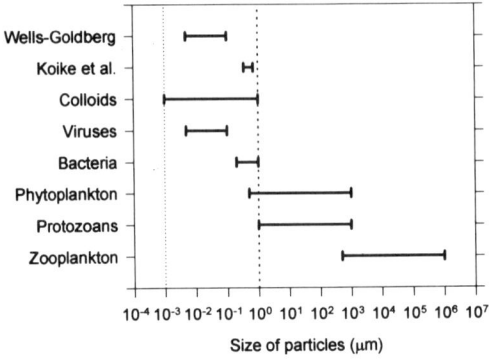

Figure 1. The size spectrum of colloidal particles and microorganisms in the ocean. Size ranges of submicron particles detected by particle counting (Koike et al., 1990) and those by electron microscopy (Wells and Goldberg, 1991) are also shown.

proposed on the basis of substantial differences in chemical behavior between truly dissolved and colloidal organic matter (e.g., Wells and Goldberg, 1994). However, the traditional definition of DOM, which includes colloids (or HMW-DOM), is still widely used in the oceanographic literature. In this chapter, we use DOM (DOC and DON for carbon and nitrogen) as traditionally defined. The term "submicron particles" is used where the particulate nature of the organic matter is important, e.g., collision models (see below).

3. Occurrences of Submicron Particles and Colloidal Organic Matter in Marine Environments

Table I compares the typical abundance of microorganisms with that of detrital submicron particles in upper oceans. The concentrations of large submicron particles (0.38–0.7 μm) determined by particle counting are on the order of 10^7 particles ml^{-1} (Koike et al., 1990; Longhurst et al., 1992). This abundance is roughly one to two order(s) of magnitude greater than the typical abundance of bacteria (5×10^5–10^6 cells ml^{-1}), picocyanobacteria (10^3–10^5 cells ml^{-1}), and prochlorophytes (10^3–10^5 cells ml^{-1}) (see Table I). Similarly, the concentration of small submicron particles (0.005–0.2 μm) by electron microscopy was on the order of 10^9 particles ml^{-1} (Wells and Goldberg, 1991, 1994), which is one to two order(s) of magnitude more abundant than the concentration of viruses (10^7–10^8 virions ml^{-1}; Bergh et al., 1989; Hara et al., 1991).

With transmission electron microscopy combined with energy disperse spectroscopy, Wells and Goldberg (1991, 1992) suggested that small submicron particles are mostly organic, although they also noted that there are some inorganic particles containing iron and alminosilicates. Koike et al. (1990) and Longhurst et al. (1992) suggested that larger submicron particles are highly flexible, fragile, and less dense than intact bacterial cells. These physical characteristics of submicron particles probably explain why many detrital particles cannot be seen by conventional epifluorescence microscopy. In fact, Koike et al. (1990) found that many submicron particles (0.35–0.5 μm) pass through 0.2 μm polycarbonate filters, which are commonly used for counting bacteria (Hobbie et al., 1977; Porter and Feig, 1983). Other data, however, suggest that a part of submicron particles can be collected on 0.2 μm filters. By combining epifluorescence microscopy with highly sensitive image analysis using a cool CCD (charge-coupled device) camera, Sieracki and Viles (1992) observed many dimly fluorescent particles in the Sargasso Sea; these particles were several times more abundant than bacteria counted by conventional epifluorescence microscopy after staining with DAPI.

Apparently, sources, sinks, and turnover rates for large and small submicron particles differ substantially, as suggested by strikingly different profiles of these particles in the oceans. Large submicron particles in the western north Pacific are more abundant in the upper layer than in the lower layer, covarying closely with biological variables such as chlorophyll a (Koike et al., 1990). In contrast, smaller particles distribute irregularly in several coastal and oceanic regions, with occasional maxima in the thermocline or in the deeper layer (Wells and Goldberg, 1991; 1994).

Recent studies with tangential-flow ultrafiltration techniques have shown that colloids or HMW-DOC constitute a large fraction (10–50%) of DOC in a wide variety of oceanic environments (Table II). An active involvement of HMW-DOC in biogeochemical cycles was suggested by Benner et al. (1992) who found that the HMW fraction contributed more to DOC in the upper layer than in the lower layer in the north Pacific. Benner et al. reported that about half

Table II. Contributions of Colloidal (or HMW-) Organic Carbon to DOC in the Upper Ocean

Study site	DOC (μM)	NMWCO[a] (KDa)	Approx. size range (μm)	% COC/DOC[b]	References
North Pacific	82	1	0.001–0.7	35	Benner et al. (1992)
Northwest Pacific	70–88	1	0.001–0.7	30–37	Ogawa and Ogura (1992)
		10	0.01–0.7	4–5	
North Atlantic coast	66–420	1	0.001–0.7	10–64	Carlson et al. (1985)
		30	0.03–0.7	0–34	
		100	0.1–0.7	0–22	
Gulf of Mexico	50–131	1	0.001–0.7	40–53	Guo et al. (1994)
		10	0.01–0.7	8–14	

[a]Nominal molecular weight cut-off.
[b]Percentage of HMW-DOC relative to total DOC concentration.

the HMW-DOC was polysaccharidelike in the upper layer. A large part of colloidal pool remains to be characterized chemically.

We know little about colloidal organic nitrogen in seawater. A large amount of organic nitrogen that passes through GF/F glass fiber filters (nominal retention, 0.7 μm), but is retained on 0.2 μm filters, has been found in the north Atlantic (Altabet, 1990) and the north Pacific (Libby and Wheeler, 1994), suggesting that submicron particles contain a great deal of nitrogen. One important component of DON is proteins (Keil and Kirchman, 1993). Recently, Tanoue (1995) and Tanoue et al. (1995) examined the composition of colloidal proteins in the Pacific Ocean by electrophoresis. Although there are thousands of protein species in living cells and POM, Tanoue et al. found only a limited number (<30) of protein molecules in the colloid fraction. The most dominant protein, with a molecular size of 48,000 Da, was widely found in both the subarctic and subtropical Pacific Ocean. Sequencing showed that this protein is a porin, a type of bacterial membrane protein (Tanoue et al., 1995).

4. Turnover Rates of DOC and Colloids

The turnover of the DOC pool and its importance in global carbon cycling have been subjects of considerable debate during the past decade (Williams and Druffel, 1988; Lee and Wakeham, 1992). Table III summarizes some of the major components of the DOC pool. Recent studies suggest that colloids and submicron particles may be important components of what is now known to be the "semi-labile" pool. Undoubtedly there is a great variety of compounds in the DOC pool, ranging in turnover times from minutes to millennium, but as a first approximation, it is convenient to divide DOC into three components: labile, semi-labile, and refractory. The refractory pool dominates deep-water DOC, which ^{14}C-dating indicates has an average age on the order of 1000–6000 years

Table III. The Classification of Oceanic DOC Based on its Turnover Rate[a]

Type	Turnover rate	Pool size (μM)	Chemically identified forms	Roles
Labile	<hours–days	<1	Dissolved free amino acids, proteins, free sugars	Fueling bacterial production
Semi-labile	months–100 years	10–30	Unknown	Transport of organic carbon
Refractory	>1000 years	40	Unknown	Storage of organic carbon

[a]Carlson et al. 1994; Carlson and Ducklow 1995.

(Williams and Druffel, 1987; Bauer et al., 1992)—time scale that far exceeds that of ecosystem processes and even the turnover of deep ocean waters. Williams and Druffel (1987) hypothesized that surface waters consisted of two components, one with turnover times like the deep water DOC and another, more labile component that turns over in less than 30 years, which is an upper estimate reflecting the limits of the ^{14}C method; more recent studies suggest turnover times on the week to month time scale (Kirchman et al., 1991; Amon and Benner, 1994; Zweifel et al., 1995; Carlson et al., 1994). Surface waters are well known to contain highly labile DOC with turnover times as fast as minutes (e.g., free amino acids), but the concentration of this labile component is very low (nanomolar or <1% of total DOC). As a first approximation, surface water DOC consists of roughly 50% refractory and 50% semi-labile DOC.

Still, we need more estimates of turnover times for HMW-DOM, colloids, and other submicron particles, but the estimates published to date suggest that these organic components can be classified with the semi-labile pool. ^{14}C dating of size-fractionated DOC have suggested that colloids (>10 kDa) are relatively young (in a geochemical sense), turning over on a time scale of <40 years (Santschi et al., 1995). The incubation experiments of Amon and Benner (1994) suggest even faster turnover times, on the order of days, as discussed in more detail below. Colloids and HMW-DOM, which can be produced in days, but not consumed immediately (i.e., semi-labile DOC), is available for export from surface waters.

Carlson et al. (1994) provided one of the clearest examples demonstrating the major role of DOC in the vertical transport of organic matter in open oceans. In the north Atlantic off Bermuda, they found that a DOC pool accumulates in the upper layer in spring and summer, which is transported down to deeper layers by winter mixing, and is then degraded there when the water column is stratified once again. The amount of organic carbon that is vertically transported by this mechanism was estimated to be equivalent to or exceeding the flux of sinking particles, suggesting a revision of the classical paradigm where the only significant mechanism of vertical transport of organic matter in the ocean is the sinking of large particles. In a review of older studies on DOC and DON concentrations, Williams (1995) also uncovered seasonal changes in DOM concentrations and further developed the hypothesis that DOM export could be important in controlling C export from the upper ocean.

The new concept emerging from the biogeochemical studies discussed should interest microbial ecologists. Historically, theory and experiments on the microbial loop have centered on a rapid and closely "coupled" system of carbon transfer via a labile DOC–bacteria–protozoan pathway (Azam et al., 1983). In fact, data have demonstrated that rapid consumption of labile monomers such as amino acids and sugars can account for a large part of bacterial production in diverse marine systems (Kirchman, 1994), indicating that the labile DOC–bacteria pathway dominates DOC fluxes. Here, the time scale of the transfer is less than days (often, less than hours; Fuhrman, 1987).

However, the existence of a less labile pool of DOM now changes our view of bacteria–DOM interactions. The eventual degradation of semi-labile DOM, especially high-molecular-weight DOM, may produce labile monomers (e.g., see Keil and Kirchman, 1994) and thus the turnover of labile and semi-labile DOC may be linked. But semi-labile DOC could fuel some bacterial production independent of the labile DOC pools and, thus, production of this DOC could be quite distant in time and space, i.e., uncoupled from degradation. Although complicating our view of microbial food webs, semi-labile DOC and its colloid and high-molecular-weight components may help explain unsolved problems in microbial ecology, such as the observation that bacterial production is often uncoupled from primary production, especially on short time and space scales (Ducklow, 1984; Ducklow et al., 1993; Kirchman et al., 1994).

Furthermore, the occurrence of semi-labile DOM suggests that a significant fraction of DOM can accumulate in seawater and is only later slowly consumed by microorganisms. Several questions arise after considering this DOM. By what mechanism(s) can organic matter escape from rapid microbial attack? What are the metabolic constraints that bacteria encounter during DOM consumption? Do colloids play important roles in the formation of semi-labile DOM or even refractory DOM? To discuss these questions and to understand the mechanistic basis of the turnover of DOC, we need to examine the sources and sinks of colloidal organic matter.

5. Sources of Submicron and Colloidal Particles

Phytoplankton and bacteria release a wide variety of dissolved organic matter, including colloids (Fogg, 1983; Decho, 1990). Microorganisms commonly produce extracellular polymeric materials called exopolymers, including polysaccharides, proteins, and lipopolysaccharides. These can be released in large amounts to surrounding waters depending on physiological conditions (Decho, 1990). Some exopolymeric materials are suggested to be highly labile. In support of this hypothesis, Amon and Benner (1994) found that more than half of the colloidal organic carbon produced during a diatom bloom in the Gulf of Mexico was rapidly (<6 days) degraded by bacteria. Consistent with this observation, Kirchman et al. (1991) found a high degradation rate of DOC during a diatom bloom in the North Atlantic, although they did not determine the size of DOC. Importantly, inorganic nitrogen was consumed rather than regenerated during the degradation (Amon and Benner, 1994; Kirchman et al., 1991), suggesting that colloids produced during blooms consist of mainly carbon-rich compounds, which are probably polysaccharides (Benner et al., 1992; Amon and Benner, 1994).

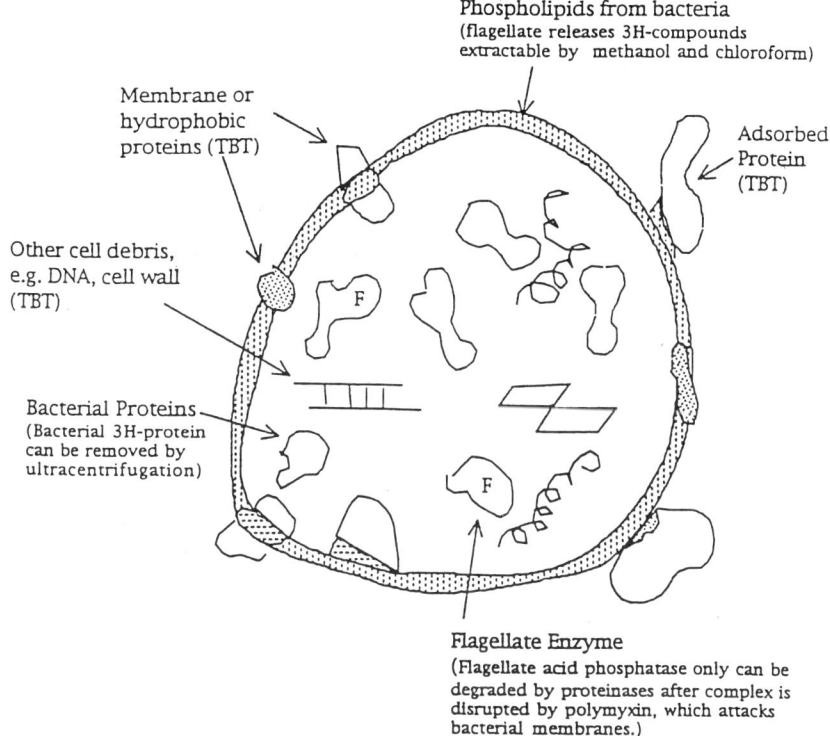

Figure 2. A hypothetical model of picofecal pellets released by bacterivorous flagellates (Nagata and Kirchman, 1992b). Phospholipids from incompletely digested bacterial membranes should spontaneously form liposomes as calculated by Nagata and Kirchman (1992b). Evidence for each structure is given in the parentheses. Structures remaining to be tested are indicated by TBT.

One important mechanism that produces colloids in surface seawater is egestion by protozoans of unassimilated materials or "picofecal pellets." In support of this, Koike *et al.* (1990) found that the number of submicron particles increased with the increase of flagellates grazing on bacteria in coastal seawaters. There is other support for the egestion hypothesis. Culture experiments demonstrated that flagellates release a wide variety of DOM, including macromolecules such as lipids, enzymes, and DNA (reviewed by Nagata and Kirchman, 1992a; also see Nagata and Kirchman, 1991, 1992b; Turk *et al.*, 1992; Tranvik, 1994). Experimental data (Nagata and Kirchman, 1991, 1992b) are consistent with model predictions from optimum digestion theory (Jumars *et al.*, 1989), suggesting that protozoans should produce more fecal pellets relative to ingestion rates when prey is more abundant. Further studies are required to parameterize rates and stoichiometry of the release of organic carbon and nitrogen by protozoans (Landry, 1993; Caron and Goldman, 1993).

Nagata and Kirchman (1992b) suggested that one type of flagellate fecal pellets, and potentially other colloids and submicron particles, are similar to liposomes. These structures, which consist of phospholipids, enzymes, and other compounds, would spontaneously form when high lipid concentrations are placed in an aqueous environment, e.g., during the excretion process (Fig. 2). In culture experiments with *Paraphysomonas impoerforata*, Nagata and Kirchman (1992b) found that acid phosphatase (a digestive enzyme of the flagellate) released along with egestion was resistant to hydrolysis by proteases because phospholipids from prey bacterial membranes protected the enzyme. Consistent with this hypothesis is the finding of bacterial membrane proteins surviving in the dissolved pool (Tanoue et al., 1995). An important implication of this hypothesis is that the liposome organic complexes produced by flagellates could provide a microenvironment for geochemical modifications of organic matter to proceed (see below).

We do not know much about the production of colloids by physical rupture, lysis, or natural death of planktonic organisms. Possible mechanisms include sloppy feeding (Lampert, 1978), viral infection (Bergh et al., 1989; Proctor and Fuhrman, 1990; Koike et al., 1993), and autolysis. Also, colloids could be released from macroaggregates and fecal pellets due to hydrolytic fragmentation of these particles (Smith et al., 1992). In Baltic and North Seas, Zweifel and Hagstrom (1995) suggested that a substantial fraction (70–98%) of total bacteria, counted by the conventional nucleic acid staining technique (DAPI method, Porter and Feig, 1980), was bacterial "ghosts," presumably bacterial cell residues containing no DNA. If proved to be true in general, this data suggest that a large fraction of bacterial production may be channelled into detrital submicron particles, perhaps due to lysis by viral infection or protozoan egestion. Regardless of their origin, these ghosts should also have a liposomelike structure, as Nagata and Kirchman (1992b) had suggested for flagellate fecal pellets.

6. Sinks of Submicron and Colloidal Particles

6.1. Bacterial Degradation and Its Constraints

We know very little about the mechanisms and regulation of bacterial degradation of colloids, but the available evidence indicates that heterotrophic bacteria are the most important sink of colloidal organic matter in the oceans (Amon and Benner, 1994). Generally, bacterial use of DOM can be limited by several factors, including availability of inorganic nutrients (Amon and Benner, 1994), low temperature (Pomeroy and Deibel, 1986), ultraviolet radiation (Herndl et al., 1993), viral infection (Bergh et al., 1989), and grazing pressure (Kirchman et al., 1991). Also, when easily utilizable monomeric compounds are available, the use of polymeric compounds can be inhibited because of the suppression of the synthesis of enzymes needed for the degradation of polymers (Chrost, 1991). These issues warrant a complete review (see Azam et al., 1994).

Here we focus on two critical processes specifically involved in bacterial–colloid interactions: transport and complexation. These physical/chemical constraints are important in discussions about potential mechanisms underlying the formation of semi-labile and refractory DOM pools in the oceans.

Johnson and Kepkay (1992) suggested that colloidal carbon could escape from bacterial degradation by virtue of the size characteristics of the colloids. Their hypothesis is based on model predictions for collision between bacteria and submicron particles. The rate of collision (dN/dt) is given by:

$$\frac{dN}{dt} = E\beta(r_1,r_2)N_1N_2$$

where E is the collision efficiency factor, β is a collision function for bacteria and colloids with diameters r_1 and r_2, and N_1 and N_2 are concentrations of bacteria and colloids. The collision function (β) accounts for Brownian motion (β_B):

$$\beta_B(r_1,r_2) = \frac{2kT(r_1 + r_2)^2}{3\mu(r_1r_2)}$$

and bacterial swimming (β_M):

$$\beta_M(r_1,r_2) = \Delta U\pi (r_1 + r_2)^2$$

where k is Bolzmann's constant, μ is absolute viscosity, T is absolute temperature, and ΔU is the difference in velocity of particle and bacterium. Turbulent shear affects collision of submicron particles (<1 μm) in the ocean to only a small degree. Fig. 3 shows an example of the curve of collision rate constants for the transport of particles (diameter: 0.002–10 μm) to bacterial cells (diameter: 1 μm).

A major finding from this model study is that the collision rate constant can reach a minimum. For non-swimming bacteria, the rate constant is lowest when the size of submicron particle is about the same as that of bacteria; in the case of Fig. 3, this size is 1 μm; however, in natural seawater, the rate is more likely to be around 0.5 μm because oceanic bacteria are smaller (Lee and Fuhrman, 1987) than suggested by Johnson and Kepkay (1992). Johnson and Kepkay suggest that bacteria have limited access to large submicron particles (0.4–0.5 μm) because of physical constraints.

Johnson and Kepkay's model also suggests that swimming can be a strategy used by bacteria to exploit colloidal carbon resources protected by physical restrictions. As shown in Fig. 2, if bacteria swim faster, bacteria collide with submicron particles more frequently. Marine bacterium can swim as fast as

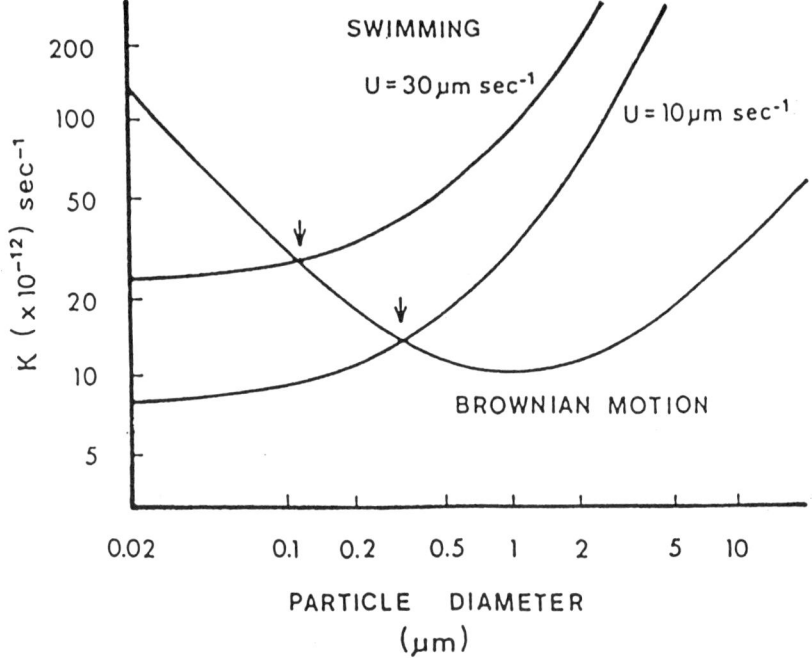

Figure 3. A theoretical curve of rate constants (k) for particle transport to bacterial cells (diameter, 1 μm) by Brownian motion and bacterial swimming (from Johnson and Kepkay, 1992). Rate constants are normalized to bacterial number. Note that rate constants increase with the increase of swimming speed (U). The transport minimum is indicated by an arrow.

100 μm sec^{-1}, suggesting that effects of swimming on collision could be even more dramatic than indicated in Fig. 3, although it is unknown what percentages of bacteria are actually motile in the sea.

Experimental results obtained by Nagata and Kirchman (1996) in part support Johnson and Kepkay's model. Nagata and Kirchman examined bacterial degradation of protein (^{3}H-labeled bovine serum albumin; MW:65 kD) adsorbed to model submicron particles (polystyrene latex beads) with different diameters (0.1–1.5 μm). They found that the rate constants for hydrolysis of adsorbed protein varied depending on the size of particles, with an apparent minimum at a size range of 0.8–1 μm, which was close to the average size of coastal bacterial assemblages used in the experiment (Fig. 4). This result is qualitatively consistent with Johnson and Kepkay's prediction.

However, Nagata and Kirchman's experiment also suggested that adsorption, rather than transport, can have large effects on protein degradation rates. They found that protein adsorbed to submicron particles is much less easily

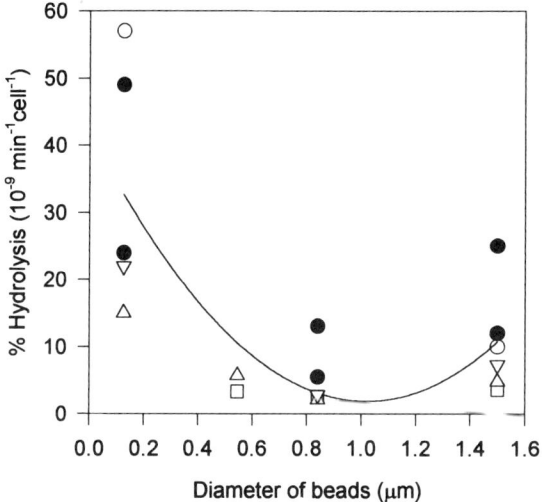

Figure 4. Percent hydrolysis of ^{3}H-protein adsorbed to polystyrene beads as a function of the diameter of beads. Different symbols represent different experiments with precultured marine bacterial assemblages from the Delaware Bay. Note that there is a minimum of hydrolysis rate constant, consistent with the prediction by Johnson and Kepkay's model (Fig. 3). From Nagata and Kirchman (1996).

degraded than free protein (Fig. 5). The degradation of adsorbed protein substantially differs among different bacterial strains, suggesting that bacterial enzymatic processes are an important factor in determining the degradation of adsorbed protein (Nagata and Kirchman, 1996). The critical factors may be the stickiness of bacterial surfaces and the affinity of membrane binding proteins that bacteria may use to bind to proteinous substrate.

These results may explain the experiments of Keil and Kirchman (1994). They found that fresh algal protein added to seawater became less easily degraded by bacteria in a short period. They suggested that the added protein adsorbed to colloids present in the estuarine water used in their experiments and that this protein is less easily degraded than free protein. Borch and Kirchman (unpublished) also found that protein adsorbed to or associated with phospholipid vesicles (liposomes) is protected from bacterial attack. We have already pointed out that there are potential mechanisms to produce some macromolecular organic complexes in seawaters (e.g., protozoan egestion; Nagata and Kirchman, 1992b). These studies support the general hypothesis that organic matter complexed with other organic matter is less easily degraded (Lee and Wakeham, 1992), and that these complexes are a potentially important source of the semi-labile and perhaps even the refractory DOM pool in the oceans (Keil and Kirch-

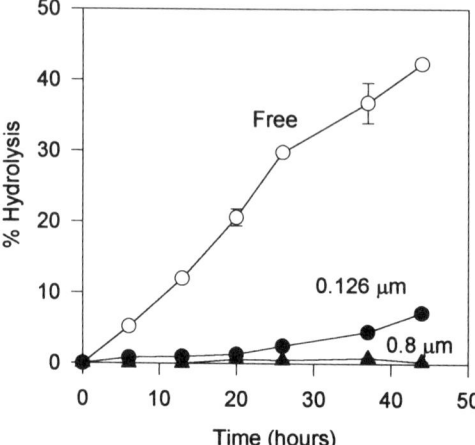

Figure 5. Hydrolysis of ^{3}H-protein freely dissolved and adsorbed to beads (diameters 0.126 and 0.8 μm) by a natural bacterial assemblage from the Delaware Bay. Y-axis indicates accumulation of LMW-^{3}H as percentages of total ^{3}H-protein added (38 ng ml^{-1}). The hydrolysis rate of adsorbed protein is much lower (<1/20) than that of free protein. Collision frequency (see Fig. 3) and biochemical interactions between bacteria and small particles are suggested to be important in determining the hydrolysis rate of adsorbed protein. From Nagata and Kirchman (1996).

man, 1994). Geochemical modifications of organic matter (e.g., Maillard reaction; Yamomoto and Ishiwatari, 1989; Hedges, 1988) may proceed rapidly on or inside the microenvironment of colloidal organic complexes. The modifications further inhibit bacterial degradation of organic matter (Keil and Kirchman, 1993) and may produce dissolved organic compounds that cannot be identified by standard analyses (Francois 1990).

6.2. Consumption of Colloidal Particles by Protozoan and Metazoan Grazers

It has been generally believed that heterotrophic protozoans are outcompeted by bacteria for labile DOM in natural marine environments where concentrations of DOM are low (Fenchel, 1987). The ecological niche of heterotrophic protozoans is thought to be only as a consumer of bacteria and small phytoplankton (Azam et al., 1983; Fenchel, 1987). This notion was challenged by Sherr (1988) who examined the uptake of HMW-DOM by heterotrophic nanoflagellates in estuarine and freshwater environments. By using epifluorescence microscopy, she observed that a significant fraction of flagellates (4–20%) took up fluorescently labeled dextran (a model for exopolymeric polysaccharides produced by microbes) of 500 or 2000 KDa into their food vacuoles during a 2-h incubation. Furthermore, culture experiments with mixed microbial assemblages showed that additions of 2000 KDa dextran at a concentration of 4 mg liter^{-1} significantly enhanced the growth of flagellates, but not that of bacteria. Sherr (1988) suggested that direct use of HMW-DOM by flagellates represents a "short-circuit" of the microbial loop (because it skips bacterial use of DOM), which increases the efficiency of trophic transfer from DOM to upper trophic levels, i.e., mesozooplankton and fish.

After the work of Sherr, a variety of fluorescently-labeled macromolecules have been used to examine uptake of colloidal organic matter by heterotrophic protozoans. These studies have shown that flagellates from freshwater, estuary, and coastal environments are apparently capable of taking up polysaccharides, lipids, and proteins of 4–2000 KDa (Tranvik et al., 1993; Marchant and Scott, 1993). No data are available about colloid consumption by flagellates in oligotrophic environments such as open oceans. Data are also scarce concerning consumption of colloidal organic matter by ciliates. Posch and Arndt (1996) observed that freshwater ciliates can take up detrital submicron particles (size, 0.2–1.3 μm).

One serious limitation of the fluorescence technique is that it provides only qualitative data about the capacity of flagellates to consume colloids. The presence or absence of fluorescent tracers in food vacuoles does not provide ingestion and clearance rates (volume of water swept clear per unit time per cell), which are key variables to examine rate and efficiency of protozoan consumption of each compound. This problem would be solved by determining the amounts of fluorescent substances in food vacuoles, perhaps with help of a highly sensitive image analysis system (Viles and Sieracki, 1992).

Models have been proposed to predict encounter rates (dimensionally identical to clearance rates) of submicron particles and protozoan grazers (reviewed by Shimeta and Jumars, 1991). Traditionally, suspension feeders, including protozoans, have been considered to graze food particles primarily by direct interception (Shimeta and Jumars, 1991). With this mechanism, smaller particles collide with flagellates less frequently than larger particles, consistent with experimental results showing that clearance rates for large bacteria are greater than those for smaller ones (Gonzalez et al., 1990). However, a new model proposed by Shimeta (1993) predicts that encounter rates increase significantly with the smaller diameters of submicron particles in a smaller size range (approx. <0.1 μm) because of collisions caused by diffusion (Brownian motion). Shimeta (1993) pointed out that these small particles are potentially a significant food resource for flagellates, although he also noted that model-based encounter rates are not necessarily equivalent to clearance rates because particles may be selected between encounter and ingestion (phagocytosis).

Limited data are available to test prediction from models with actual clearance rates of flagellates fed small colloidal particles. Gonzalez and Suttle (1993) used fluorescently labeled viral particles to measure clearance rates of flagellates fed viruses (head size, 50–70 nm). They found that clearance rates for viruses are only 4% of those for bacteria (0.5 μm). This experimental result is not consistent with Shimeta's model that predicts that flagellates encounter 50-nm particles 2.5 times more frequently than 0.5-μm particles. Thus, selectivity after encounter, rather than encounter rates, seems more important in determining clearance rates. Alternatively, there might be a threshold for the total volume of captured particles (particle number bound to phagocytotic site × volume of each particle) that triggers phagocytotic reactions. Selectivity associated with phagocytosis has not been well understood in feeding ecology of protozoans. Several studies have reported that flagellates ingest viruses and bacteria preferentially over inert polystyrene beads of similar sizes (e.g., Gonzales and Suttle, 1993), although the mechanism involved in this discrimination is unknown.

Some metazoan mesozooplankton can directly ingest bacteria-sized particles in seawater (Deibel and Lee, 1992; Flood et al., 1992). One outstanding example of zooplankton feeding on colloids has been reported for an apendicularian genus, *Oikopleura* (Flood et al., 1992). These animals pump water through mucous structures (called "houses") equipped with finely meshed filters (mesh, 0.2 µm) for food collection. These filters work as a tangential flow device to concentrate submicron particles in seawater up to 1000 × before ingestion. In fact, Flood et al. (1992) found that *O. vanhoeffeni* ingest colloidal melanin particles (size, 0.13 µm). The clearance rates for colloids were low; only 2–11% of the same rates for larger plastic particles (10–20 µm). Despite this low collection efficiency, Flood et al. suggested that colloidal organic matter could be an important food source for these animals because of high concentrations of colloidal organic carbon in the sea.

Given the lack of quantitative data on clearance rates by grazers of naturally occurring submicron particles, it is premature to suggest significant contributions of protozoan and metazoan grazers to colloid degradation. Digestion and assimilation of detrital colloids by protozoans is poorly understood. Enhancement of flagellate growth by addition of dextran (Sherr, 1988) suggests that flagellates can assimilate some form of colloidal exopolymers, but the generality of this finding and assimilation efficiencies remain to be confirmed. In those studies, the use of radiolabeled exopolymers (Decho and Lopez, 1993) is potentially useful.

The recent study of Barbeau et al. (1996) suggests a novel role for flagellates feeding on colloids. They observed that a protozoan grazer, which was able to ingest iron colloids, excreted a form of the iron that was more available for use by phytoplankton. Ingestion of the colloid was apparently required to transform the iron colloid to a more labile form because the transformation was not observed in bacteria-only cultures (the prey of the protozoa). The authors hypothesized that the acidic food vacuoles of protozoa carried out the refractory-labile iron transformation. It is intriguing to speculate if the organics, possibly in the form of liposomes (see above), that may have been released had any role in the subsequent fate of the transformed iron. In any event, this transformation associated with protozoan feeding could be quite important given the large contribution of protozoa to iron regeneration in oceanic waters (Hutchins and Bruland, 1994).

6.3. Aggregation

Colloids may coagulate to form larger particles (Wells and Goldberg, 1993). In coastal waters off California, Alldredge et al. (1993) found that polysaccharide-rich, transparent exopolymer particles (TEP) of size dimension 3–100 μm are abundant (28–5000 particles ml^{-1}). They suggest that TEP are formed by coagulation of submicron particles. In support of this, TEP were destroyed by ethylenediamine tetraacetic acid, a chelating agent that disrupts cation bridges. Apparently, carbohydrates are important components of TEP (Passow et al., 1994). Other studies on coagulation have indicated that bubbling may enhance coagulation of submicron particles (reviewed by Kepkay, 1994), suggesting that coagulation rates may be high in surface waters. These aggregates may stick to phytoplankton cells to form larger particles (marine snow) that enhances the sinking flux of organic carbon (Alldredge et al., 1993). Also, consumption of aggregated colloids by metazoan zooplankton may proride a direct pathway that channels colloidal carbon up to zooplankton. Alternatively, Alldredge et al. (1983) speculate that TEP may clog zooplankton feeding apparatuses, thus inhibiting feeding. Much remains to be studied about how colloid aggregation contributes to carbon transport and trophic dynamics.

7. Summary and Future Challenges

Recent research has suggested that submicron and colloidal particles are an integral part of marine food webs and biogeochemical cycles. Our view of microbial consumption of organic matter in the oceans should be revised to include explicitly colloid–microbes interactions. Assuming that the abundance of small (0.01–0.1 μm) and large (0.3–0.5 μm) submicron particles are 1×10^9 and 3×10^7 particles ml^{-1}, respectively, in oceanic waters (Koike et al., 1990; Wells and Goldberg, 1991, 1994) and that these particles collide with bacteria as predicted by Brownian collision model (Johnson and Kepkay, 1992), we estimate that one motile bacterium swimming at 30 μm sec^{-1} encounters small and large submicron particles every 20 seconds and 10 minutes, respectively. In order to consume these potentially significant resources, bacteria should trap (or bind) and hydrolyze small particles and then should transport hydrolyzate across membranes. Previous studies using fluorescent analogs (Chrost, 1991; Hoppe et al., 1993) have provided some insights on enzymatic hydrolysis, but much is still unknown about which process(es) can, in fact, regulate bacterial consumption of colloidal organic matter. Nagata and Kirchman (1996) have devised a model system using radiolabeled protein adsorbed to latex beads, which could be useful to examine in greater detail the relationship between degradation rate and physical/chemical characteristics of submicron particles.

In addition to uncertainty about how bacteria degrade colloids, we also know little about the role of other organisms in degrading colloids and small particles. Consumption of colloids and aggregates by protozoan and metazoan grazers would short-circuit the microbial loop and thus would increase the efficiency of trophic transfer (Sherr, 1988; Flood *et al.*, 1992). However, quantitative data in support of this hypothesis are still quite limited. We also need more information about the selectivity and assimilation efficiency of organisms grazing on colloidal and aggregated organic matter. Rates of coagulation reactions and fates of aggregates remain to be determined.

Although colloids and submicron particles are potentially important food sources for bacteria and other organisms, their most significant contribution to oceanic processes may be their role in organic matter cycling and the formation of refractory DOM. Submicron particles produced by bacterivorous flagellates (picofecal pellets), probably ghost bacteria, and potentially other submicron particles and colloids may have liposomelike structures. These structures and analogous ones that involve several types of organic compounds, potentially are important in geochemical modifications of organic matter, which would transform labile to semi-labile organic matter. It remains to be seen whether this pathway leads to the formation of truly refractory DOM (Nagata and Kirchman, 1992b). Adsorption of otherwise labile, small-molecular-weight DOM to colloids could also have a large impact on DOM degradation (Nagata and Kirchman, 1996; Keil *et al.*, 1994), although a different picture is suggested by Taylor (1995). However, very little is known about compositions and structure of submicron particles in oceanic waters. Obviously, chemical and physical characterization of submicron and colloidal particles in seawater is one of the major future challenges for marine biogeochemists. Also, further work should be aimed at determining rates and mechanisms of the turnover of colloidal organic matter if we are to better understand temporal accumulation and export of DOC and colloids, which have substantial impacts on the cycling of organic carbon in the ocean (Carlson *et al.*, 1994; Carlson and Ducklow, 1995; Peltzer and Hayward, 1996).

ACKNOWLEDGMENTS. This research was supported in part by the Monbusho International Scientific Research Program and by an NSF Grant. We thank I. Koike, R. Benner, R. Keil, B. Logan, K. Kogure, H. Ogawa, E. Tanoue, N. Borch, B. Sherr, and E. Sherr for discussions and information.

References

Alldredge, A. L., Passow, U., and Logan, B. E., 1993, The abundance and significance of a class of large, transparent organic particles in the ocean, *Deep-Sea Res.* **6:**1131–1140.

Altabet, M. A., 1990, Organic C, N, and stable isotopic composition of particulate matter collected on glass-fiber and aluminum oxide filters, *Limnol. Oceanogr.* **35**:902–909.

Amon, R. M. W., and Benner, R., 1994, Rapid cycling of high-molecular-weight dissolved organic matter in the ocean, *Nature* **369**:549–552.

Azam, F., Fenchel, T., Field, J. G., Gray, J. S., Meyer-Reil, L. A., and Thingstad, F., 1983, The ecological role of water column microbes in the sea, *Mar. Ecol. Prog. Ser.* **10**:257–263.

Azam, F., Smith, D. C., Steward, G. F., and Hagstrom, A., 1994, Bacteria-organic matter coupling and its significance for oceanic carbon cycling, *Microb. Ecol.* **28**:167–179.

Baines, S. B., and Pace, M. L., 1991, The production of dissolved organic matter by phytoplankton and its importance to bacteria: patterns across marine and freshwater systems, *Limnol. Oceanogr.* **36**:1078–1090.

Barbeau, K., Moffett, J. W., Caron, D. A., Croot, P. L., and Erdner, D. L., 1996, Role of protozoan grazing in relieving iron limitation of phytoplankton, *Nature* **380**:61–64.

Bauer, J. E., Williams, P. M., and Druffel, E. R. M., 1992, ^{14}C activity of dissolved organic carbon fractions in the north-central Pacific and Sargasso Sea, *Nature* **357**:667–670.

Benner, R., Pakulski, J. D., McCarthy, M., Hedges, J. T., and Hatcher, P. G., 1992, Bulk chemical characteristics of dissolved organic matter in the ocean, *Science* **255**:1561–1564.

Bergh, O., Borsheim, K. Y., Bratbak, G., and Heldal, M., 1989, High abundance of viruses found in aquatic environments, *Nature* **340**:467–468.

Carlson, C. A., and Ducklow, H. W., 1995, Dissolved organic carbon in the upper ocean of the central equatorial Pacific Ocean, 1992: daily and fine-scale vertical variations, *Deep-Sea Res.* **42**:639–656.

Carlson, C. A., Ducklow, H. W., and Michaels, A. F., 1994, Annual flux of dissolved organic carbon from the euphotic zone in the northwestern Sargasso Sea, *Nature* **371**:405–408.

Carlson, D. J., Brann, M. L., Mague, T. H., and Mayer, L. M., 1985, Molecular weight distribution of dissolved organic materials in seawater determined by ultrafiltration: A re-examination, *Mar. Chem.* **16**:155–171.

Caron, D. A., and Goldman, J. C., 1993, Predicting excretion rates of protozoa: reply to the comment by Landry, *Limnol. Oceanogr.* **38**:472–474.

Cauwet, G., 1981, Non-living particulate matter, in: *Marine Organic Chemistry* (E. K. Duursma and R. Dauson, eds.), Elsevier, New York, pp. 71–89.

Chisholm, S. W., Olsonm, R. J., Zettler, E. R., Goericke, R., Waterbury, J. B., and Welschmeyer, N. A., 1988, A novel free-living prochlorophyte abundant in the oceanic euphotic zone, *Nature* **334**:340–343.

Chrost, R. J., 1991, Environmental control of synthesis and activity of aquatic microbial ectoenzymes, in: *Microbial enzymes in aquatic environments* (R. J. Chrost, ed.), Springer-Verlag, New York, pp. 29–59.

Cole, J. J., Findlay, S., and Pace, M. R., 1988, Bacterial production in fresh and saltwater ecosystems: a cross-system overview, *Mar. Ecol. Prog. Ser.* **43**:1–10.

Decho, A. W., 1990, Microbial exopolymer secretions in ocean environments: their role(s) in food webs and marine processes, *Oceanogr. Mar. Biol. Annu. Rev.* **28**:73–153.

Decho, A. W., and Lopez, G. R., 1993, Exopolymer microenvironments of microbial flora: Multiple and interactive effects on trophic relationships, *Limnol. Oceanogr.* **38**:1633–1645.

Deibel, D., and Lee, S. H., 1992, Retention efficiency of sub-micrometer particles by the pharyngeal filter of the pelagic tunicate *Oikopleura vanhoeffeni*, *Mar. Ecol. Prog. Ser.* **81**:25–30.

Ducklow, H. W., 1984, Geographical ecology of marine bacteria: physical and biological variability at the mesoscale, in: *Current Perspectives in Microbial Ecology* (M. J. Klug and C. A. Reddy, eds.), American Society for Microbiology, Washington, DC, pp. 22–31.

Ducklow, H. W., and Carlson, C. A., 1992, Oceanic bacterial production, *Adv. Microb. Ecol.* **12**:113–181.

Ducklow, H. W., Kirchman, D. L., Quinby, H. L., Carlson, C. A., and Dam, H. G., 1993, Bacterioplankton carbon cycling during the spring bloom in the eastern North Atlantic Ocean, *Deep-Sea Res. II.* **40**:245–263.
Fenchel, T., 1987, *Ecology of Protozoa: The Biology of Free-Living Phagotrophic Protists*, Springer-Verlag, Berlin.
Flood, P. R., Deibel, D., and Morris, C. C., 1992, Filtration of colloidal melanin from sea water by planktonic tunicates, *Nature* **355**:630–632.
Fogg, G. E., 1983, The ecological significance of extracellular products of phytoplankton photosynthesis, *Bot. Marina* **26**:3–14.
Francois, R., 1990, Marine sedimentary humic substances: Structure, genesis and properties, *Rev. Aquat. Sci.* **3**:41–80.
Fuhrman, J., 1987, Close coupling between release and uptake of dissolved free amino acids in seawater studied by an isotope dilution approach, *Mar. Ecol. Prog. Ser.* **37**:45–52.
Fuhrman, J., Sleeter, T. D., Carlson, C. A., and Proctor, L. M., 1989, Dominance of bacterial biomass in the Sargasso sea and its ecological implications, *Mar. Ecol. Prog. Ser.* **57**:207–217.
Gonzales, J. M., and Suttle, C. A., 1993, Grazing by marine nanoflagellates on viruses and virus-sized particles: ingestion and digestion, *Mar. Ecol. Prog. Ser.* **94**:1–10.
Gonzales, J. M., Sherr, E. B., and Sherr, B. F., 1990, Size-selective grazing on bacteria by natural assemblages of estuarine flagellates and ciliates, *Appl. Environ. Microbiol.* **56**:583–589.
Guo, L., Coleman, C. H., Jr., and Santschi, P. H., 1994, The distribution of colloidal and dissolved organic carbon in the Gulf of Mexico, *Mar. Chem.* **45**:105–119.
Hara, S., Terauchi, K., and Koike, I., 1991, Abundance of viruses in marine waters: assessment by epifluorescence and transmission electron microscopy, *Appl. Environ. Microbiol.* **57**:2731–2734.
Hedges, J. I., 1988, Polymerization of humic substances in natural environments, in: *Humic Substances and Their Role in the Environment* (F. H. Frimmel and R. F. Christman, eds.), John Wiley & Sons, New York, pp. 45–58.
Herndl, G. J., Muller-Niklas, G., and Frick, J., 1993, Major role of ultraviolet-B in controlling bacterioplankton growth in the surface layer of the ocean, *Nature* **361**:717–719.
Hobbie, J. E., Daley, R. J., and Jasper, S., 1977, Use of Nuclepore filters for counting bacteria by fluorescence microscopy, *Appl. Environ. Microbiol.* **33**:1225–1228.
Hoppe, H. G., Ducklow, H., and Karrasch, B., 1993, Evidence for dependency of bacterial growth on enzymatic hydrolysis of particulate organic matter in the mesopelagic ocean, *Mar. Ecol. Prog. Ser.* **93**:277–283.
Hutchins, D. A., and Bruland, K. W., 1994, Grazer-mediated regeneration and assimilation of Fe, Zn and Mn from planktonic prey, *Mar. Ecol. Prog. Ser.* **110**:259–269.
Johnson, B. D., and Kepkay, P. E., 1992, Colloid transport and bacterial utilization of oceanic DOC, *Deep-Sea Res.* **39**:855–869.
Jumars, P. A., Penry, D. L., Baross, J. A., Perry, M. J., and Frost, B. W., 1989, Closing the microbial loop: dissolved carbon pathway to heterotrophic bacteria from incomplete ingestion, digestion and absorption in animals, *Deep-Sea Res.* **36**:483–495.
Keil, R. G., and Kirchman, D. L., 1993, Dissolved combined amino acids: chemical form and utilization by marine bacteria, *Limnol. Oceanogr.* **38**:1256–1270.
Keil, R. G., and Kirchman, D. L., 1994, Abiotic transformation of labile protein to refractory protein in seawater, *Mar. Chem.* **45**:187–196.
Keil, R. G., Montlucon, D. B., Prahl, F. G., and Hedges, J. I., 1994, Sorptive preservation of labile organic matter in marine sediments, *Nature* **370**:549–552.
Kepkay, P. E., 1994, Particle aggregation and the biological reactivity of colloids, *Mar. Ecol. Prog. Ser.* **109**:293–304.
Kirchman, D. L., 1993, Particulate detritus and bacteria in marine environments, in: *Microbiology—*

An ecological approach (T. E. Ford, ed.), Blackwell Scientific Publications, Boston, pp. 321–341.

Kirchman, D. L., 1994, The uptake of inorganic nutrients by heterotrophic bacteria, *Microb. Ecol.* **28**:255–271.

Kirchman, D. L., Suzuki, Y., Garside, C., and Ducklow, H. W., 1991, High turnover rates of dissolved organic carbon during a spring phytoplankton bloom, *Nature* **352**:612–614.

Kirchman, D. L., Ducklow, H. W., McCarthy, J. J. and Garside, C., 1994, Biomass and nitrogen uptake by heterotrophic bacteria during the spring phytoplankton bloom in the North Atlantic Ocean, *Deep-Sea Res.* **41**:879–895.

Koike, I., Hara, S., Terauchi, K. and Kogure, K., 1990, The role of submicrometer particles in the ocean, *Nature* **345**:242–244.

Koike, I., Hara, S., Terauchi, K., Shibata, A., and Kogure, K., 1993, Marine viruses—their role in upper ocean dissolved organic matter (DOM) dynamics, in: *Trends in Microbial Ecology* (R. Guerrero and C. Pedros-Alio, eds.), Spanish Society for Microbiology, Barcelona, pp. 311–314.

Lampert, W., 1978, Release of dissolved organic carbon by grazing zooplankton, *Limnol. Oceanogr.* **23**:831–834.

Landry, M. R., 1993, Predicting excretion rates of microzooplankton from carbon metabolism and elemental ratios, *Limnol. Oceanogr.* **38**:468–472.

Lee, S., and Fuhrman, J., 1987, Relationships between biovolume and biomass of naturally derived marine bacterioplankton, *Appl. Environ. Microbiol.* **53**:1298–1303.

Lee, C., and Wakeham, S. G., 1992, Organic matter in the water column: future research challenges, *Mar. Chem.* **39**:95–118.

Leppard, G. G., 1984, The ultrastructure of lacustrine sedimenting materials in the colloidal size range, *Arch. Hydrobiol.* **101**:521–530.

Libby, P. S., and Wheeler, P. A., 1994, A wet-oxidation method for determination of particulate organic nitrogen on glass fiber and 0.2 μm membrane filters, *Mar. Chem.* **48**:31–41.

Longhurst, A. R., Koike, I., Li, W. K. W., Rodriguez, J., Dickie, P., Kepkay, P., Partensky, F., Bautista, B., Ruiz, J., Wells, M., and Bird, D., 1992, Sub-micron particles in northernwest Atlantic shelf water, *Deep-Sea Res.* **39**:1–7.

Marchant, H. J., and Scott, F. J., 1993, Uptake of sub-micrometer particles and dissolved organic material by Antarctic choanoflagellates, *Mar. Ecol. Prog. Ser.* **92**:59–64.

Massalski, A., and Leppard, G. G., 1979, Morphological examination of fibrillar colloids associated with algae and bacteria in lakes., *Jour. Fish. Res. Board Canada* **36**: 922–938.

Murphy, L. S., and Haugen, E. M., 1985, The distribution and abundance of phototrophic ultraplankton in the North Atlantic, *Limnol. Oceanogr.* **30**:47–58.

Nagata, T., and Kirchman, D. L., 1991, Release of dissolved free and combined amino acids by bacterivorous marine flagellates, *Limnol. Oceanogr.* **36**:433–443.

Nagata, T., and Kirchman, D. L., 1992a, Release of dissolved organic matter by heterotrophic protozoa: implications for microbial food webs, *Arch. Hydrobiol.* **35**:99–109.

Nagata, T., and Kirchman, D. L., 1992b, Release of macromolecular organic complexes by heterotrophic marine flagellates, *Mar. Ecol. Prog. Ser.* **83**:233–240.

Nagata, T., and Kirchman, D. L., 1996, Bacterial degradation of protein adsorbed to model submicron particles in seawater, *Mar. Ecol. Prog. Ser.* **132**:241–248.

Newman, M. E., Filella, M., Chen, Y., Negre, J-C., Perret, D., and Buffle, J., 1994, Submicron particles in the Rhine River—II. Comparison of field observations and model predictions, *Wat. Res.* **28**:103–118.

Ogawa, H., and Ogura, N., 1992, Comparison of two methods for measuring dissolved organic carbon in the sea water, *Nature* **356**:696–698.

Passow, U., Alldredge, A. L., and Logan, B. E., 1994, The role of particulate carbohydrate exudates in flocculation of diatom blooms, *Deep-Sea Res.* **41**:335–357.

Peltzer, E. T., and Hayward, N. A., 1996, Spatial and temporal variability of total organic carbon along 140°W in the equatorial Pacific Ocean in 1992, *Deep-Sea Res.* **43**:1155–1180.

Pomeroy, L. R., and Deibel, D., 1986, Temperature regulation of bacterial activity during the spring bloom in Newfoundland coastal waters, *Science* **233**:359–361.

Porter, K., and Feig, Y. S., 1980, The use of DAPI for identifying and counting aquatic microflora, *Limnol. Oceanogr.* **25**:943–948.

Posch, T., and Arndt, H., 1996, Uptake of sub-micrometre and micrometre-sized detrital particles by bacterivorous and omnivorous ciliates, *Aquat. Microb. Ecol.* **10**:45–53.

Proctor, L. M., and Fuhrman, J. A., 1990, Viral mortality of marine bacteria and cyanobacteria, *Nature* **343**:60–62.

Rees, T. F., 1990, Comparison of photon correlation spectroscopy with photosedimentation analysis for the determination of aqueous colloid size distributions, *Water Resources Res.* **26**:2777–2781.

Santschi, P. H., Guo, L., Baskaran, M., Trumbore, S., Southon, J., Bianchi, T. S., Honeyman, B., and Cifuentes, L., 1995, Isotopic evidence for the contemporary origin 'of high-molecular weight organic matter in oceanic environments, *Geochim. Cosmochim. Act.* **59**:625–631.

Sherr, E. B., 1988, Direct use of high molecular weight polysaccharide by heterotrophic flagellates, *Nature* **335**:348–351.

Shimeta, J., 1993, Diffusional encounter of submicrometer particles and small cells by suspension feeders, *Limnol. Oceanogr.* **38**:456–465.

Shimeta, J., and Jumars, P. A., 1991, Physical mechanisms and rates of particle capture by suspension-feeders, *Oceanogr. Mar. Biol. Annu. Rev.* **29**:191–257.

Siegenthaler, U., and Sarmiento, J. L., 1993, Atmospheric carbon dioxide and the ocean, *Nature* **365**:119–125.

Sieracki, M. E., and Viles, C. L., 1992, Distributions and fluorochrome staining properties of submicrometer particles and bacteria in the North Atlantic, *Deep-Sea Res.* **39**:1919–1929.

Smith, D. C., Simon, M., Alldredge, A. L., and Azam, F., 1992, Intense hydrolytic enzyme activity on marine aggregates and implications for rapid particle dissolution, *Nature* **359**:139–142.

Tanoue, E., 1995, Detection of dissolved protein molecules in oceanic waters, *Mar. Chem.* **51**:239–252.

Tanoue, E., Nishiyama, S., Kamo, M., and Tsugita, A., 1995, Bacterial membranes: Possible source of a major dissolved protein in seawater, *Geochim. Cosmochim. Act.* **59**:2643–2648.

Taylor, G. T., 1995, Microbial degradation of sorbed and dissolved protein in seawater, *Limnol. Oceanogr.* **40**:875–885.

Tranvik, L. J., 1994, Colloidal and dissolved organic matter excreted by a mixotrophic flagellate during bacterivory and autotrophy, *Appl. Environ. Microbiol.* **60**:1884–1888.

Tranvik, L. J., Sherr, E. B., and Sherr, B. F., 1993, Uptake and utilization of "colloidal DOM" by heterotrophic flagellates in seawater, *Mar. Ecol. Prog. Ser.* **92**:301–309.

Turk, V., Rehnstam, A.-S., Lundberg, E., and Hagstrom, A., 1992, Release of bacterial DNA by marine nanoflagellates, an intermediate step in phosphorous regeneration, *Appl. Environ. Microbiol.* **58**:3744–3750.

Viles, C. L., and Sieracki, M. E., 1992, Measurement of marine picoplankton cell size by using a cooled, charge-coupled device camera with image-analyzed fluorescence microscopy, *Appl. Environ. Microbiol.* **58**:584–592.

Waterbury, J. B., Watson, S. W., Buillard, R. R. L., and Brand, L. E., 1979, Widespread occurrence of a unicellular, marine, planktonic, cyanobacterium, *Nature* **227**:293–294.

Wells, M. L., and Goldberg, E. D., 1991, Occurrence of small colloids in sea water, *Nature* **353**:342–344.

Wells, M. L., and Goldberg, E. D., 1992, Marine submicron particles, *Mar. Chem.* **40**:5–18.

Wells, M. L., and Goldberg, E. D., 1993, Colloid aggregation in seawater, *Mar. Chem.* **41**:353–358.
Wells, M. L., and Goldberg, E. D., 1994, The distribution of colloids in the North Atlantic and Southern Ocean, *Limnol. Oceanogr.* **39**:286–302.
Williams, P. J. leB., 1995, Evidence for the seasonal accumulation of carbon-rich dissolved organic material, its scale in comparison with changes in particulate material and consequential effect on net C/N assimilation ratios, *Mar. Chem.* **51**:17–29.
Williams, P. M., and Druffel, E. R. M., 1987, Radiocarbon in dissolved organic matter in the central North Pacific Ocean, *Nature* **330**:246–248.
Williams, P. M., and Druffel, E. R. M., 1988, Dissolved organic matter in the ocean: comments on a controversy, *Oceanography* **1**:14–17.
Yamamoto, S., and Ishiwatari, R., 1989, A study of the formation mechanisms of sedimentary humic substances II. protein-based melanoidin model, *Org. Geochem.* **14**:479–489.
Zweifel, U. L., and Hagstrom, A., 1995, Total counts of marine bacteria include a large fraction of non-nucleoid containing "ghosts," *Appl. Environ. Microbiol.* **61**:2180–2185.
Zweifel, U. L., Wikner, J., and Hagstrom, A., 1995, Dynamics of dissolved organic carbon in a coastal ecosystem, *Limnol. Oceanogr.* **40**:299–305.

4

Do Bacterial Communities Transcend Darwinism?

DOUGLAS E. CALDWELL, GIDEON M. WOLFAARDT, DARREN R. KORBER, and JOHN R. LAWRENCE

1. Introduction

Until the development of fluorescent molecular probes and confocal laser microscopy, there were few alternatives to isolating microorganisms from their communities prior to laboratory study. Isolation was necessary to obtain a sufficient amount of homogeneous cell material for chemical analyses, yet it constrained most laboratory work to the molecular, cellular, or organismal level. However, fluorescent probes and other molecular techniques now allow the analysis of individual microorganisms without isolation (Olsen *et al.*, 1986; Pace *et al.*, 1986; Caldwell *et al.*, 1992a). This affords the opportunity to perform community-level laboratory experiments that are not possible with plants and animals due to their large size. However, inconsistencies between evolutionary ecology (Mayr, 1993; Krassilov, 1994; Kauffman, 1993, 1995), ecosystem ecology (Maynard-Smith, 1991; Loehle and Pechman, 1988; Schulze and Mooney, 1993), microbial ecology (Margulis, 1990; Caldwell, 1993; Caldwell and Costerton, 1996), germ theory (Caldwell, 1995; Caldwell *et al.*, 1997a), and information theory (Rasmussen, 1988, 1991; Rasmussen *et al.*, 1990; Yockey, 1990, 1995; Kelly, 1994) make it difficult to formulate testable hypotheses that are relevant in understanding ecology at the community level. Consideration of communities as units of proliferation (and hence as units of evolution) requires a more generalized theory of life, amenable to the formulation of community-level hypotheses and tests.

DOUGLAS E. CALDWELL, GIDEON M. WOLFAARDT, and DARREN R. KORBER • Department of Applied Microbiology and Food Science, University of Saskatchewan, Saskatoon, Saskatchewan, S7N 5A8 Canada.   JOHN R. LAWRENCE • National Hydrology Research Institute, Saskatoon, Saskatchewan, S7N 3H5 Canada.
Advances in Microbial Ecology, Volume 15, edited by Jones. Plenum Press, New York, 1997.

Inconsistencies most frequently arise when attempting to apply ecological theory to groups, communities, and ecosystems (Loehle and Pechmann, 1988; Schwemmler, 1989; Goodnight, 1990a,b; Maynard-Smith, 1991; Mayr, 1993; Kauffman, 1993, 1995). In his monograph, The *Origin of Species by Means of Natural Selection or the Preservation of Favoured Races in the Struggle for Life*, Darwin applied the concept of selection to individual species, lineages, or races (Darwin, 1859). However, attempts to use selection alone to explain the full range of biological complexity at all levels of biological organization have not met with complete success (Rasmussen, 1988; Mayr, 1993; Kauffman, 1993, 1995; Kelly, 1994). As an alternative, we suggest a proliferation hypothesis that offers a simpler and more comprehensive explanation of ecology and evolution by recognizing the possibility of propagation and reproductive success at many different levels of biological organization simultaneously (genes, plasmids, cells, organisms, communities, ecosystems, etc.) rather than solely at the level of individual organisms (species populations). We also suggest that laboratory communities of bacteria may provide one of the few experimental systems readily amenable to the testing of this hypothesis.

Like the selection hypothesis, the proliferation hypothesis involves the assumption that organisms sometimes proliferate more effectively by adapting through genetic mutation and genetic recombination. However, it also assumes that self-replicating molecules sometimes proliferate more effectively if they associate and then propagate as macromolecules (through covalent and ionic bonding), that self-replicating macromolecules sometimes proliferate more effectively if they associate and then propagate as prokaryotic cells (through formation of cell membranes and walls), that prokaryotic cells sometimes proliferate more effectively if they associate and then propagate as eukaryotic organisms (through endosymbioses and attachment), that both prokaryotic and eukaryotic organisms sometimes proliferate more effectively if they associate and then propagate as communities (through behavioral adaptations), and that communities sometimes proliferate more effectively if they then associate to form ecosystems and the biosphere as a whole. The terms and definitions related to this proliferation hypothesis are discussed in Section 9.

The longer this process of self-organization continues, the more effective the associations become. Thus anaerobic bacterial systems (Dolfing and Beuroskens, 1995), which are among the oldest biological systems known, tend to show a higher degree of interaction, organization, and diversification than many aerobic bacterial systems. By comparison, plants and animals are still in the formative stages of associating with one another (Caldwell *et al.*, 1997a).

Proliferation through biological adaptation and association is thus postulated to be the primary source of biological diversification and complexity, as opposed to natural selection and competition between races or lineages. From the perspective of proliferation theory, the benefits of positive interactions directly affect reproductive success (proliferation) at each level of biological organization as well as at subordinate levels. Thus it is no longer necessary to explain traits and behavior that seem altruistic (detrimental to the individual possessing them but beneficial to other organisms or at some other level of organization) by envisioning a convolution of indirect competitive effects culminating in an effect upon reproductive success solely at the organismal level. Reproductive success occurs directly at any level of organization that is capable of propagating and proliferating, or of failing to propagate and proliferate. It is only through this less restrictive concept of ecology and evolution that the nature and existence of microbial communities can be adequately understood and investigated. Although it is possible to explain the same phenomena using selection theory, the explanations are unnecessarily complex and limited in scope as well as in application. Thus according to Occam's razor*, the ecology of community-level phenomena cannot be adequately understood using selection theory as summarized by Keller and Lloyd (1992).

The assumption that the environment can select individual organisms, but not groups of interacting organisms, is no longer plausible for prokaryotes. Genetic sequencing and other evidence shows that microbial associations and interactions provided the primary mechanism by which eukaryotic plant and animal cells evolved from prokaryotic lineages (Margulis, 1981, 1990, 1993; Schwemmler, 1989; Margulis and Fester, 1991; Maynard-Smith, 1991). This implies that evolution occurs not only through organismal evolution and speciation (with species optimizing within specific environments) but also through community-level evolution (with units of two or more organisms being necessary for optimal proliferation). Experimental studies of community-level evolution are difficult with multicellular plant and animal communities because of their large size and slow rates of growth. However, there are many examples of microbial communities that succeed or fail based on the network of interdependencies among their component populations. These include lichens, anaerobic digestor granules, dental plaque (Bradshaw *et al.*, 1989; Sissons *et al.*, 1995), degradative consortia, spoilage biofilms, kefir, corrosion biofilms, rumen communities, the *Chlorochromatium aggregatum* consortium, protoctista (protozoans), and other associations (Zeikus and Johnson, 1991; Margulis, 1990, 1995; Slater and Hartman, 1982), many of which remain poorly studied or uncultivated (Hirsch, 1980, 1984).

Lichens provide one of the best known and thoroughly studied examples of microbial associations that function as organisms. Each lichen has been given a separate genus and species designation and consists of algae, fungi, and in some

---

*Occam's razor, also known as the Law of Economy or the Law of Parsimony, is the logic used by Gallileo in defending the simplest hypothesis of the heavens. It assumes that the aim of logic is to represent the facts of nature using the simplest and most economic of all conceivable explanations.

cases, cyanobacteria (Ahmadjian and Hale, 1973; Farrar, 1976; Hawksworth, 1982, 1988; Petrini *et al.*, 1990; Kendrick, 1991). The habitat of free-living algae and cyanobacteria is normally aquatic, whereas the habitat of free-living fungi is soil. The habitat range for the communal association of these three unrelated lineages is primarily the lithosphere and phyllosphere. This is one of the clearest examples of extending habitat range through interactive association with other organisms. There is very little overlap between the habitat range of the free-living forms alone and their extended habitat range, which is obtained through the interactive associations they form. This lack of overlap between the habitat range of individuals and their extended range through interactive associations is one of several important criteria to be considered in establishing whether communities arise directly as discrete units of proliferation.

Within the lichen association, the alga photosynthesizes and supplies the fungus with organic substrates (Ahmadjian and Hale, 1973; Kendrick, 1991). The fungus protects the alga from desiccation and facilitates wetting by producing polyols. The cyanobacterium fixes nitrogen. Heterotrophic bacteria are often excluded from the lichen symbiosis through the production of fungal antibiotics. This may parallel the production of bacteriocins by bacteria as well as the production of lethal proteins by killer yeasts (Starmer *et al.*, 1987, 1992). The production of these antimicrobial agents may be an example of interactive mechanisms that control the composition of microbial communities and allow them to proliferate and evolve as a single unit.

Not all interactive microbial associations are tight enough to eventually become speciated (develop sexual reproductive mechanisms) during evolution. Neither are all communal associations loose enough so that the "survival of the fittest" organism or cell line need be the only consideration. A continuum exists in which the importance of communal interdependencies varies considerably but can never be ignored as insignificant. These communal interactions and associations are necessary to extend the habitat range of the community, make optimal use of environmental resources, and create favorable microenvironments within unfavorable macroenvironments. They arise through simultaneous evolution at all levels of organization. Through this process genes, individuals, communities, ecosystems, and the biosphere itself evolves. The characteristics of an individual that is seemingly unfit (from the perspective of competition and organismal selection) may be needed for the proliferation of the group in which it resides. Thus, groups containing this individual are better able to develop and expand, while those without it are unable to do so. Consequently, such groups would presumably evolve mechanisms to ensure the proliferation of this individual and its physical presence through the production of chemical signals, by making it an endosymbiont, by producing antibiotics effective against its predators, by selectively binding to it, or by feeding it (Zahavi and Ralt, 1984).

A mutation or recombination event within one member of a community affects the reproductive success and habitat range not only of the individual, but also of the association as a whole. Immigration and emigration are as important as mutation and recombination in organizing and distributing genetic information to optimize the probability of successful community development and expansion. Although communal associations adapt (Fulthorpe and Wyndham, 1992) and evolve much like a species population, they normally lack the complex sexual reproductive mechanisms necessary for the speciation of isolated gene pools. Nonetheless, they still succeed or fail based on their collective genetic resources, and thus propagate, adapt, and evolve as do individuals. Microorganisms are small and thus provide ideal models for developing and testing such community-level theories.

Although Maynard-Smith (1976) concludes that group selection is unlikely to have played a primary role in evolution, he has suggested it as the mechanism by which less virulent strains of myxomatosis were selected from virulent strains. However, to say that the environment "selects" may imply too much. The environment never makes a conscious decision to select the fit and kill or exclude the unfit. The environment is the habitat and resources necessary for life to develop, proliferate, and evolve. Organisms can proliferate within an environment as individuals, but in many cases they proliferate more effectively by interacting and forming communities. Some bacteria have proliferated as symbionts, consortia, and communities that eventually become recognizable as eukaryotic cells. However, in other cases it may be necessary for the members of a community or group to remain physically separate to interact synergistically.

One example is the highly specialized relationship between some insects and the specific plants that require them for pollination. Others include the relationships between plants and nitrogen-fixing bacteria, the associations between bacteria, protoctista (protozoa), and ruminants, the swarming behavior of bacterial communities (Mitchell *et al.*, 1995), as well as the relationships between fungi and leaf-cutting ants (Batra and Batra, 1967; Weber, 1966, 1972). The task is to determine how the conversion of physical resources into information resources has been optimized through the many associations between and among macromolecules, cells, organisms, communities, and ecosystems.

2. Conceptual Barriers to Community-Level Thought and Experimentation

Numerous conceptual barriers must be overcome for community-level thought to find greater utility in microbiology. These problematic concepts include the germ theory (Koch's postulates), Beijerinck's theory of selective enrichment culture, Darwin's theory of the origin of species, and a preoccupation with reductionism as the only conceivable approach to scientific investigation. Each of these views is a premise or assumption that works well only under a restricted set of conditions and that fails to explain a significant portion of the new information that has been obtained using molecular genetics, fluorescent molecular probes, electron microscopy, and laser microscopy. We must reevaluate the primary assumptions of microbiology and ecology with regard to their generality, economy, and utility in light of a century of new information.

2.1. Organismic Selection Theory

It is difficult for microbial ecology to move from the cell and molecular level to the community level without recognizing that evolution occurs both above (communities) and below (transposons) the species level. This is to say that the reproductive success of any gene, organism, or community depends upon its collective genetic resources. Darwin conceived of evolution and environmental selection as acting through the survival of "victorious" races (Darwin, 1859) rather than through the joint reproductive success and proliferation of genes, groups, communities, and species. He believed that "natural selection acts by life and death,—by the survival of the fittest and the destruction of the less well-fitted individuals." Most evolutionary ecologists still defend convolutions of this concept (Keller and Lloyd, 1992), which makes the individual organism the sole target of environmental selection (Lomnicki, 1978; Mayr, 1993) and which requires increasingly complex explanations of competitive interactions that result in cooperative and altruistic behavior (Zahavi, 1981). Competition may be the primary mechanism by which humanity harnesses and exploits the killer instinct to accomplish some higher objective. However, Darwin was mistaken when he extrapolated competition to all forms of life and invoked it as a primary mechanism of nature.

Darwin eventually realized that "Each living creature [plant or animal] must be looked at as a microcosm . . . a little universe, formed of a host of self-propagating organisms, inconceivably minute and as numerous as the stars in the heaven" (Darwin, 1868). However, his earlier generalizations are inconsistent with current understanding of prokaryotic organisms and smaller genetic elements, the nature and existence of which was unknown at that time. It is now apparent that microbial communities adapt through extensive interspecies transfer of homologous degradative genes (Fulthorpe and Wyndham, 1992; Fulthorpe et al., 1995). Analysis of 16s rRNA sequences and other evidence shows that all plants and animals were derived from symbiotic, parasitic, and other interactive associations among two or more prokaryotic lineages (Margulis, 1981, 1990; Schwemmler, 1989; Margulis and Fester, 1991; Margulis and Guerrero, 1991; Maynard-Smith, 1991). Thus there is no longer any debate as to whether prokaryotic groups became eukaryotic individuals at some point in evolution. The

Bacterial Communities 111

functional significance of this and other similar cooperative events in evolutionary history is lost if a life and death struggle for survival is considered to be the sole evolutionary force acting in nature. Limiting resources require synergistic associations if they are to be effectively exploited. If there was no possibility of synergistic interactions, then competition would be a more useful evolutionary concept, particularly in resource-limited situations. Under these circumstances the term "survival of the fittest" would become meaningful and the consequence would be numerous examples of natural environments containing a single lineage or race of organisms. However, there are no known examples of this.

During evolution, a lineage succeeds or fails through its success or failure in proliferating. This does not necessarily have anything to do with a life and death conflict between individuals that results in the survival of the fit and the death of the unfit. This has been shown in chemostat studies (Veldkamp and Jannasch, 1972; Veldkamp, 1977) in which two bacteria are grown together, one with a high K_s and high μ_{max} and the other with a low K_s and low μ_{max}. At high substrate concentrations the former proliferates; at low substrate concentrations the latter proliferates. One proliferates and the other is lost through attrition, but neither dies as a consequence of the other. There is a gradual displacement of one strategy by another as the system evolves. Neither death nor the direct inhibition of one organism by the other is necessary for this to occur.

The idea of evolution solely at the species level is most amenable to plants and animals, while the idea of community-level evolution may be most amenable to prokaryotes. Species-level evolution (speciation) is most important in organisms that are constrained both to sexual mechanisms of reproduction and genetic recombination. In this situation, genes must be randomly recombined with those of another individual of the same species to create each new generation of organisms. Genetic exchange with members of other species is prevented by specific sexual reproductive mechanisms. The concept of communities arising through community-level evolution (communitization) is more important in prokaryotic organisms than in eukaryotes because prokaryotes are not constrained either to sexual reproduction or to sexual genetic recombination. With generation times as short as six minutes and with several asexual mechanisms of genetic recombination, evolution can be very rapid in prokaryotic communities as compared to fungal, protoctistan, plant, and animal communities.

Bacteria diversify and evolve as do plants and animals, but they do not possess complex sexual reproductive mechanisms and thus cannot normally become speciated in the same sense that plants and animals are speciated (Sonea, 1991). This is evidence that speciation is not the sole evolutionary force at work in nature and that evolution must occur both above (communities) and below (genetic elements) the species level. If this is the case for bacteria it may also be probable for plants and animals. However, communal evolution (communitization) would be more difficult to demonstrate in plant and animal associations due to the enormous spatial and temporal requirements for community-level evolution to occur in these organisms as compared to prokaryotes.

Thus, communitization is more likely to find expression in bacteria because of their short generation time, their ability to exchange and reorganize genetic information without relying on sexual reproductive mechanisms (Shapiro, 1985b), the large populations that can be contained within an extremely small space (10^{12} cells per ml), the time period over which bacterial communities have evolved as compared to plants and animals (3.5 billion years for unicellular organisms as compared to 0.6 billion for marine animals and 0.4 billion for vertebrates), and their limited physical capacity for storing large quantities of genetic information. Their small size also prevents the formation of diffusion gradients around individual cells (Koch, 1991). This necessitates the formation of multicellular aggregates if they are to effectively control their microenvironment and optimize the probability of reproductive success.

The limited genetic resources of bacterial cells may be the single most important reason that communitization is more likely to have occurred in bacteria than in plants and animals. Bacteria are limited not only by environmental resources, but also by the amount of genetic programming they can contain. They possess only 0.1% of the DNA found within the cells of vertebrate animals (Brock and Madigan, 1988) and this occupies as much as 50% of their volume. Consequently, vertebrate animals can afford to retain larger quantities of somewhat obsolete and redundant genetic coding that may be of limited value at one point in time, but might be of periodic value later. In contrast, bacteria delete the coding for non-essential traits more quickly. Traits essential for bacterial reproductive success *in situ*, such as gas vacuoles in planktonic organisms (Fig. 1), heterocysts in nitrogen-fixing cyanobacteria (Evenboom et al., 1981), or filamentous growth and flocking behavior in predation-resistant bacteria (Guede, 1979) or other traits (Costerton et al., 1978, 1994) are lost when *in situ* environmental pressures are relieved. Consequently, those prokaryotic communities that share genetic resources among collaborators (even those that may be only distantly related), have the advantage over communities that do not. This sharing can occur either through the formation of associations, consortia, and communities (emigration and immigration), or through genetic exchange via transformation, transduction, and conjugation.

Bacteria frequently form associations during the biodegradation of chlorinated hydrocarbons (Rajogopal et al., 1984; Neilson et al., 1988; Dietrich and Winter, 1990; Distefano et al., 1991; Ney et al., 1991; Allard et al., 1992; Madsen and Aamand, 1992; Bagley and Gossett, 1995). Degradative plasmids move among genetically unrelated bacterial cell lines (beta and gamma subgroups of the proteobacteria) when a strong environmental constraint, such as the use of 3-chlorobenzoate as a carbon source, is applied (Fulthorpe and Wyndham, 1991, Fulthorpe et al., 1995). Although the original strain supplying the plasmid

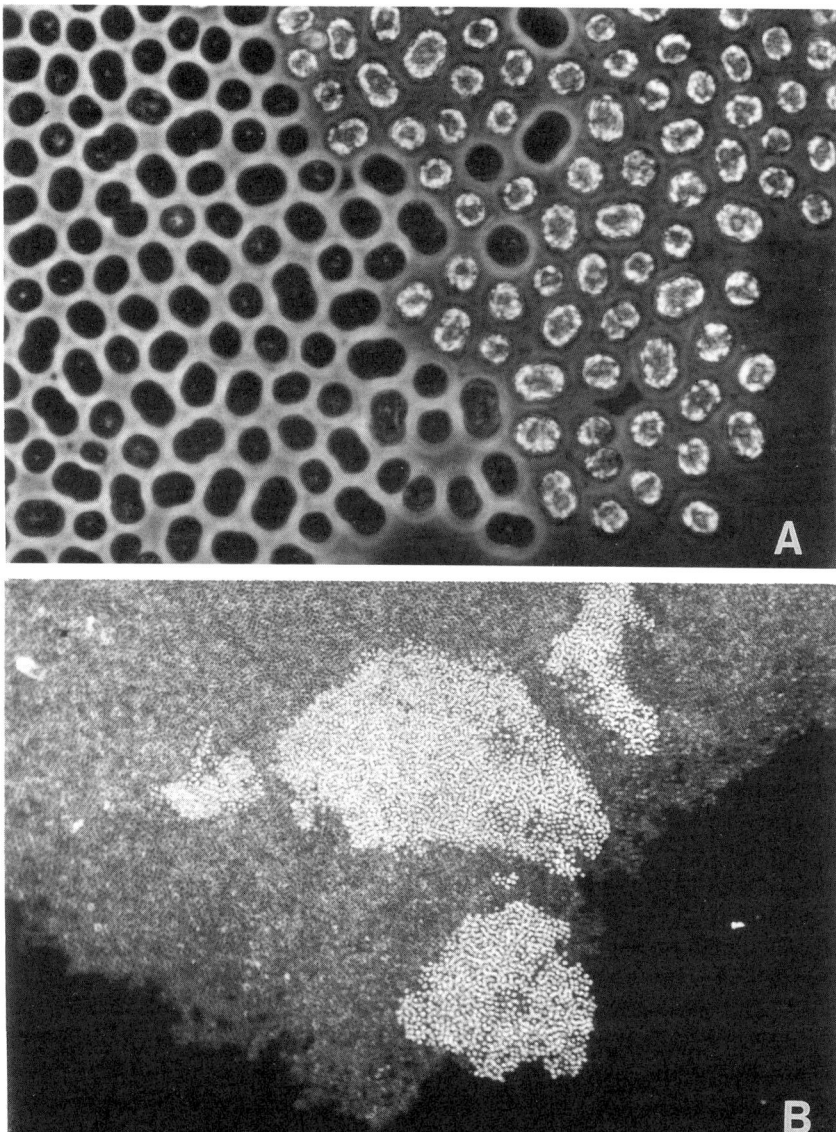

Figure 1. Loss of gas vesicles (the highly refractile cell inclusions) in sectors of *Microcystis aeruginosa* colonies during laboratory cultivation. The loss of this and other essential traits frequently occurs when bacteria are isolated from their native community and removed from the environmental stresses associated with their original habitat. A, low-power darkfield image (width of field = 1 mm) and B, high-power image (width of field = 88 μm).

may die out, its highly mobile catabolic plasmid proliferates by moving to other community members (Fulthorpe and Wyndham, 1989, 1992). This movement has resulted in extensive interspecies transfer of a variety of homologous degradative genes during the evolution of degradative bacteria (Fulthorpe et al., 1995). Other studies of biodegradation have shown that bacterial communities respond to chlorinated hydrocarbons by forming degradative consortia consisting of populations that are unable to degrade the hydrocarbon as isolated populations (Dawson, 1980; Aftring et al., 1981; Lappin et al., 1985), which degrade only at a reduced rate (Wolfaardt et al., 1994b), or which degrade only at reduced toxicant concentrations (Bagley and Gossett 1995). In the case of dechlorination, these consortia have been shown to mediate only the most thermodynamically favorable (yielding the highest energy release) of many possible alternative dechlorination reactions (Beurskens et al., 1994). This suggests that bacterial communitization optimizes the use of limiting environmental resources rather than being a random process.

These and other observations confirm that evolution occurs both above and below the species level despite the dominant view among evolutionary ecologists that species-level evolution is the sole mechanism of evolution (Keller and Lloyd, 1992 and that: ". . .advances in population biology show that no selection on the level higher than selection among individuals is possible. Therefore, the super organisms, i.e., higher ecological units, should be ruled out not only from ecology but also from evolutionary theory" (Lomnicki, 1978). More recently, a growing body of evidence suggests that group evolution might be demonstrated not only within microbial communities but also in plant and animal communities (Maynard-Smith, 1976; Aoki, 1982, 1986; Leigh, 1983; Fix, 1984; Damuth, 1985; Nunney, 1985; Aviles, 1986; Wilson, 1987a, 1992; Goodnight, 1990a,b; Goodnight et al., 1992; Peck, 1992; Kauffman, 1993, 1995). These concepts can be tested as mathematical models (Leigh, 1983; Fix, 1984; Aoki, 1986; Szathmary and Demeter, 1987; Wilson, 1987b; Kauffman, 1993, 1995). However, bacterial communities offer the possibility of providing ecological theory with a substantially larger body of experimental evidence at the community level (Caldwell, 1993, 1995).

Chloroplasts, mitochondria, bacteroids, sulfur-oxidizing bacteria, methanogens, methane-oxidizing bacteria, and other intracellular units within eukaryotic cells are endosymbiotic prokaryotes or their descendants (Margulis, 1981, 1990, 1993; Margulis and Fester, 1991). Consequently, at some point in evolution, groups of two or more organisms became functional units of evolution despite their independent lineages. Similar events have been observed in laboratory studies of bacteria and amoebae (Jeon, 1972; Jeon and Jeon, 1976; Jeon and Ahn, 1978). A neo-Darwinian response to this discovery is that although groups may function as units of selection, they normally do so only if they are enclosed within a compartment. Consequently, group selection could not occur in the case of most communities and ecosystems (Maynard-Smith, 1991) although it could occur in the case of endosymbionts. However, the collective reproductive success of a group of interdependent individuals is not determined exclusively by whether they are within one another, attached to one another, or only periodically in contact. In facing an environmental challenge, the success of community development and expansion depends solely on the collective effect of community-level interactions. Physical positioning would be only one of many factors determining the success of an interactive association.

It has often been useful to conceive of selection at various levels within communities (Wade, 1978; Doolittle and Sapienza, 1980; Wilson, 1980; Hickey, 1982; Sober, 1984). Selection at two different levels may occur in the same direction or different directions. A trait that is beneficial to an individual may be detrimental to its community and vice versa (Aviles, 1986; Peck, 1992). These various levels of biological organization have been referred to by several terms including: groups (Maynard-Smith, 1991; Goodnight *et al.*, 1992), species (Mayr, 1993), hamlets (Fix, 1984), settlements (Fix, 1984), demes (Nunney, 1985; Aoki, 1986), avatars (Damuth, 1985), clades (Damuth, 1985), neighborhoods (Wilson, 1987b), teams (Sonea and Panisset, 1983; Sonea, 1991), and symbioses (Margulis, 1981, 1990, 1993; Margulis and Fester, 1991; Maynard-Smith, 1991). However, it is the reproductive success of the community (Wilson, 1992; Kauffman 1993, 1995) or ecosystem (Dunbar, 1971; Loehle and Pechmann, 1988) itself that represents the final culmination of all the genetic programming and other information that resides within.

Evolution at the level of community has thus directly resulted in specific ecological interactions between community members. Presumably, these interactions optimize resource utilization (Dunbar, 1971; Loehle and Pechmann, 1988) and extend the habitat range of the biosphere. These same interactions have also arisen, in part, through simultaneous evolution at numerous subordinate levels of organization. It is important to emphasize that if communities and other levels of organization are units of evolution, they are also units of extinction. This underscores the relevance of community-level thought in conserving biodiversity. The extinction of a single species may also result in a reduced ability to produce a self-sustaining community network.

2.2. Germ Theory

The germ theory and Koch's postulates will always be useful in determining the causative agents of disease (Koch, 1881, 1884). However, they fail to provide an adequate theoretical or experimental basis for understanding ecological processes in which microbial communities or associations are causative agents (rather than isolated cell lines). Basing microbiology on the germ theory requires that microorganisms be isolated in "pure" culture to establish that they, and they alone, are the causative agents responsible. This approach is effective in medicine when an isolated cell line slips past a host's defenses and causes a disease. It is not necessarily appropriate in other situations related to disease, agriculture, industry, and the environment (Caldwell, 1993, 1995; Caldwell *et al.*, 1995, 1996). If a community of microorganisms is responsible for a process, then an isolated cell line alone may not be capable of serving as the causative agent. In this situation, it may be an interactive association that is the active agent to be cultivated and studied under defined laboratory conditions (if the process mechanism is to be adequately understood and controlled).

The isolation and cultivation of a single bacterial cell produces a cell line that evolves toward an interactive association of individuals to utilize environmental resources more efficiently than the parental cell. Consequently, the isolation of a bacterium fails to produce the well-defined "perfect pure culture" envisioned by Koch (1881). Instead, it produces a mixture of differentiated and mutated cells (Shapiro, 1984, 1988, 1992; Shapiro and Higgins, 1988; Shapiro and Trubatch, 1991; Maenhaut-Michel and Shapiro, 1994). In some cases these isolated cell lines function as though they were components of a multicellular organism (Rosenberg, 1984; Dworkin, 1985; Dworkin and Kaiser, 1985). To conceive of isolating and purifying a homogeneous preparation of bacteria, in the same sense that an enzyme might be purified, is a misconception derived from the simplifying assumption that bacteria do not significantly evolve or differentiate during cultivation. The term "pure culture" is thus a misnomer. Terms such as isolate, isolated cell line, strain, and cell clone are more explicit and accurate. The production of genetic variants within isolated cell lines has been extensively documented and quantified in continuous culture (Novick and Silard, 1950; Drake, 1970; Harder *et al.*, 1977; Evenboom *et al.*, 1981; Brown and Oliver, 1982; Claassen *et al.*, 1986; Noack, 1986; Upton *et al.*, 1990). Assuming that there is a spontaneous mutation rate of one in 10^6 cell divisions and that there are 10^{12} cells in a large bacterial colony on agar means that even before the first subculture, a bacterial cell line has already become a heterogeneous group including nearly 10^6 closely related, but genetically unique, individuals. The progeny of mutants can sometimes be seen as radial sectors within bacterial colonies if the mutation results in a visible change in the color, morphology, or texture of the colony (Shapiro, 1985a).

Community culture (Fig. 2) helps to avoid artifacts that arise through the degeneration of bacterial strains grown in "pure" culture. When a cell line is isolated from its *in situ* community, it immediately begins to diversify. It forms a new group of recombinants and mutants, which evolve in response to environmental constraints associated with laboratory cultivation. Consequently, the environment is always involved in defining a culture, and it can not be assumed that a culture is adequately defined based solely on the use of an isolated cell as the inoculum. The *in vitro* evolution of laboratory strains is based primarily on rapid growth through the deletion of traits necessary for *in situ* proliferation, but unnecessary for proliferation in laboratory isolation, as shown by Evenboom *et al.* (1981) for heterocystis cyanobacteria and as illustrated in Fig. 1 for gas

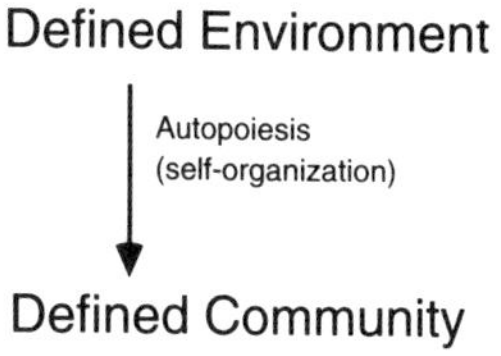

Defined Environment

| Autopoiesis
(self-organization)
▼

Defined Community

Figure 2. Use of defined environments to produce defined communities. Adequately defining the environment requires that all of the relevant physicochemical conditions have been controlled and quantified (concentrations, fluxes, spatial variations, and temporal variations). In this way a community can be defined by defining the environment (rather than by isolating cells and using aseptic technique, as in the case of isolated cell lines). The criteria for determining whether a community has been obtained in laboratory culture include autopoiesis, synergy, communality, and homeostasis as explained in section 7 and by Caldwell et al. (1997b).

vacuolate cyanobacteria. Traits that were essential *in situ* become obsolete or even detrimental *in vitro*. This can be due to the lack of predators, lack of chemical cues and chemoreceptors needed to interact with missing community associates, and excessively high nutrient levels in "pure" laboratory cultures. Cells that disperse (rather than adhering or aggregating) proliferate due to the strong environmental pressure in favor of cells that are easily diluted in plating procedures and dilution tubes. *In situ*, it is more likely that cells will aggregate to create favorable microenvironments within less favorable macroenvironments. Consequently, laboratory cultivation has a tendency to redefine the characteristics of bacteria such that they no longer accurately represent the capabilities of their ancestors.

Examples of communities as causative agents include dental plaque, mixed bacterial infections, spoilage biofilms, silage, fermented foods (cheese, yogart, beer, wine, sourdough bread, sausage, etc.), nitrification, nitrosofication, microbial corrosion, degradative consortia, biofouling, industrial and municipal wastewater treatment (activated sludge, septic tanks, anaerobic digestor granules), and numerous other situations in which a process occurs at the level of a consortium, community, or microecosystem. In these situations both the community and its component cell lines must be cultivated and understood in the laboratory (Caldwell, 1993, 1995). Although isolated cell lines are still necessary in these circumstances, they are insufficient to elucidate the mechanisms responsible for the phenomenon or process of interest. Understanding communities as causative agents requires that the mechanisms of interaction within the community be understood, and that the performance of the community be compared to the performance of individual organisms, as well as to other communities and sets of organisms. This generally involves the steps used in the study of diclofop-methyl degrading consortia by Wolfaardt et al. (1993, 1994 a–d, 1995). A defined environment is used to establish a defined community (Fig. 2) through autopoiesis. Environmental conditions determine the specific combination of genetic elements, organisms, consortia, predators, parasites, and/or communities that proliferate. The community members are isolated and the community is then reconstructed as a mixed culture. This is followed by comparisons of interactive mechanisms, internal microenvironments, process performance, habitat range, and resource utilization for each of the isolates and for the individual isolates as members of mixed cultures, as well as for the individuals as members of the native cultured community. Better process performance, extended habitat range, creation of favorable microenvironments, and effective resource utilization for the cultured community as compared to the isolates and mixed culture are all consistent with the concept of community-level evolution (communitization). Inferior performance by the community would be consistent with the idea of individual selection. If selection occurs only at the individual level, then the performance of individuals should be optimized at the expense of community performance whenever the interests of the individual and community are in conflict.

The following quotes are typical of the strong bias in favor of pure cultures as opposed to communities used when conducting laboratory experiments:

> Work with impure cultures yields nothing but nonsense. (Brefeld, 1881)

> Progress in microbiology as a result of the adoption of pure-culture methods, has been so striking that in the opinion of some biologists a [algal] culture not free of bacteria is not a real culture at all. (Pringsheim, 1946)

> Using an undefined microbial culture . . . is too vague . . . reject. (anonymous journal referee, 1992)

> The study of pure cultures remains the most reliable source of basic information for understanding the properties and evolution of the vast majority of bacteria. (Gest, 1993)

2.3. Selective Enrichment Theory

Beijerinck used selective enrichment as a means of isolating and culturing the bacteria responsible for nutrient cycling. Reasoning logically from both the germ theory and from selection theory, he assumed that the result of selection in an enrichment culture would be an isolated species responsible for the specific biogeochemical transformation upon which the enrichment was based. If urea was the sole substrate, then the ideal urea-degrading species would be selected. His theory of selective enrichment is summarized briefly in the following quote:

> Because of our very imperfect understanding of the environmental requirements of the majority of microbes, it is impossible in most enrichment culture experiments to go further than to bring about a relative increase in the numbers of a desired form without leading to a complete disappearance of the other species present. Often this partial enrichment only occurs at a particular stage of the experiment, whereas earlier and later other forms predominate. Because of this, enrichment culture experiments can be called "perfect" or "imperfect". In a perfect experiment a single species is isolated in all its varieties (Beijerinck, 1901).

Beijerinck reasoned that the product of selective enrichment should logically be a single species. When he observed that this was not the case, he explained the discrepancy by assuming that the various organisms selected were "varieties" of the same species or that the enrichment was "imperfect." In fact, he had confirmed that the product of environmental constraints is a group rather than an isolated species. It may appear that an early colonizer is an isolated species because it is numerically dominant during a phase of succession (particularly in cases of r-selection during the log phase of microbial growth in batch enrichments). However, given sufficient time, an interactive community containing a diversity of organisms (including predators and parasites) and optimizing the utilization of environmental resources, is always the final product of biological adaptation to environmental pressures (Novick and Silard, 1950; Drake, 1970; Harder et al., 1977; Evenboom et al., 1981; Brown and Oliver, 1982; Claasen et al., 1986; Noack, 1986; Upton et al., 1990).

The consequence of environmental pressures in enrichment culture is the establishment of a mature community through a process of colonization and succession. Studies of evolution in controlled chemostat environments reveal that communities self-organize through a successional process in which periodic environmental variations generally increase biodiversity (Upton et al., 1990). Of all bacterial enrichment cultures, degradative enrichments are the most thoroughly studied. Chlorinated hydrocarbons are used as substrates and degradative consortia generally proliferate under these conditions rather than isolated species. Mobile genetic elements (encoding degradative function) spread among the species populations of degradative communities (Fulthorpe and Wyndham, 1989, 1992; Fulthorpe et al., 1995) and the utilization of environmental resources is optimized at the community level in terms of the most thermodynamically favorable dechlorination reactions (Beurskens et al., 1994).

Some studies of evolution within isolated cell lines ignore the possibility of community-level evolution and recognize only the significance of fit individuals (Lenski and Travisano, 1994). However, as an isolated cell line proliferates, it begins to form a community of interacting individuals over as few as ten generations. This occurs primarily through the production of mutants and through cellular differentiation. During the growth of bacterial colonies on agar, a pattern of concentric rings and radial sectors develops and normally corresponds to the differentiation (rings) and genetic mutations (sectors) arising within the developing population (Shapiro, 1984, 1985a, 1988, 1992, Shapiro and Trubatch, 1991). Studies of colony formation by genetically engineered organisms, in which gene expression is visualized through beta-galactosidase insertions, have confirmed this via genetic analyses of the sectors and rings. Similar phenomena occur in liquid culture as well as on solid media. Continuous culture of the fungus *Aspergillus oryzae* for 39 d at a dilution rate of 0.18 per hour (doubling time of 4 h) results in the formation of a stable group of organisms consisting of several morphological mutants (Harder *et al.*, 1977). Continuous culture of an antibiotic-producing strain of *Streptomyces noursei* resulted in the formation of a set of differentiation mutants altered in the pattern of sporulation, antibiotic biosynthesis, and antibiotic resistance (Noack, 1986) and there are numerous other similar examples (Novick and Silard, 1950; Drake, 1970; Evenboom *et al.*, 1981; Brown and Oliver, 1982; Claassen et al., 1986; Noack, 1986; Upton *et al.*, 1990).

The significance of adaptive mutations in the reproductive success of an individual population or microbial community should not be underestimated (Foster, 1993). A point mutation may turn on a set of genes coding for traits that were previously repressed but that may be of strategic ecological importance under conditions of environmental stress. However, much longer periods of evolution would be required to generate sufficient biodiversity within an isolated cell line to produce meaningful community interactions (providing communal as well as individual benefit) than to produce new individuals. From these observations it is apparent that when evolution occurs within an isolated population the result is seldom a homogeneous cell line. It is a mixture of individuals that function collectively, often interactively, to make optimal use of environmental resources and habitat (Sonea and Panisset, 1983; Rosenberg, 1984; Shapiro, 1988; Sonea, 1991). There are no laboratory experiments in which the endpoint of evolution has been a pure culture of homogeneous cells, as envisioned by Beijerinck. All known natural environments are occupied by communities rather than a single lineage or race of organisms. However, enrichment cultures sampled during the early phases of succession can be easily misinterpreted as pure cultures or lineages. During this period of r-selection, a single transient opportunist may greatly outnumber other organisms and seem to be a pure culture (based on plate counts and direct microscopy). Given sufficient time for k-selection to occur, and as resources become limiting, a diverse community eventually begins to develop.

2.4. Reductionism

Scientific reductionism, as used here, refers to the process by which a phenomenon is understood through the study of its components. However, knowledge can also be obtained by considering a phenomenon as a component of some larger mechanism. For example, Darwin formulated the theory of evolution and the origin of species by thinking in non-reductive terms. He understood the biological diversity of individual organisms as the product of a larger evolutionary process. In contrast, the vast majority of understanding in microbiology has been obtained through reductive experimentation and thought. Interactive associations of microorganisms are reduced to isolated cell lines and stored in culture collections. Their enzymes and genes are then isolated, sequenced, and put in computer databases and gene libraries. However, microorganisms are not static, independent objects that can be adequately understood through isolation. Each is part of a dynamic ecological network that optimizes the utilization of environmental resources (conversion of abiotic resources to biotic resources). If the mechanisms of these interactions are to be understood, then the community must be sustained, cultivated, and studied as a causative agent under controlled laboratory conditions—complete with predators and parasites.

Just as each biological adaptation in a species population is necessary to optimize its reproductive success (or was necessary for the reproductive success of its ancestors), each community member is necessary to increase the habitat range, effectiveness, and proliferation of its community. Unnecessary organisms, like unnecessary biological adaptations, are lost through the gradual displacement of less effective networks by more effective networks. Thus, when thinking non-reductively, it is not sufficient to itemize the traits of a bacterium and to elucidate its molecular mechanisms. Nor is it enough to determine the relationship between each trait and the reproductive success of the species. If the design theory of life is to be understood, the relationship between environmental stress, reproductive success, and the expression of each trait must be determined experimentally. In the same sense, it is not enough to know the species composition of a community; it is essential to also know the relationship of each organism (and each of its traits) to the formation, reproductive success, efficiency, internal interactions, and habitat range of its community. It is also essential to elucidate these relationships in experimental laboratory studies under defined environmental conditions.

Before the advent of molecular genetics, fluorescent molecular probes, and laser microscopy there were few alternatives to reductive studies in bacteriology. Cells had to be isolated from their *in situ* associations and cultivated in pure culture to obtain enough homogeneous material for biochemical analysis and identification. The assumption of purity and homogeneity was a necessity for the development of the discipline. However, microbiology is no longer methodologically limited by Petri-dish technology. It can be done non-destructively, *in situ*, and without the need to isolate organisms from their native communities (Caldwell *et al.*, 1992a). This includes the description of new strains without the need to cultivate them (Amann *et al.*, 1991, 1994).

Much of the work done thus far in microbial ecology has focused on molecules as mechanisms of cells. Work done in evolutionary ecology has focused on individuals as mechanisms of evolution. Studies in ecosystem ecology have focused on nutrient cycling, energy transfer, and mass transfer as mechanisms of ecosystems. However, confocal laser microscopy and community culture methods open a new frontier that integrates all of these approaches. It potentially allows the cultivation of microbial ecosystems within well-defined environments and environmental gradients, while also permitting simultaneous analyses of the spatial and temporal distribution of genes, gene expression, metabolic rates, chemistry, growth rates of populations, habitat range, and all of the mechanisms that have normally been studied independently or only at a single level of biological organization. This is one of the few approaches that can lead to a comprehensive experimental understanding of the mechanisms by which molecules and individuals have come to function as integral components of communities and ecosystems.

3. Community Theory

If the environment selects only individual organisms, then any beneficial interactions between individual organisms are coincidental or indirect. A simpler and more comprehensive hypothesis is that the reproductive success of the community and individual are interdependent and equally important. Consequently, the loss of a species through extinction may endanger the community, and, conversely, the loss of community integrity may endanger the species within. If the search for adaptive reproductive strategies is confined to the organismal level, then crucial strategies based on interactions between organisms (Hirsch, 1980, 1984) may be overlooked at the level of consortia, communities, and ecosystems. Consequently, it is useful to consider both community-level adaptations (Fulthorpe and Wyndham, 1992; Fulthorpe *et al* 1995) as well as the adaptation of individuals in developing evolutionary theory. A theory that fails to recognize the direct evolution of interactive associations is unlikely to provide a broad enough theoretical basis to understand the network of interactions which maintains the habitability of the biosphere.

3.1. Are There Community-Level Reproductive Strategies?

The possibility of reproductive strategies extending beyond and within species boundaries to other levels of biological organization (including plasmids, transposons, symbioses, communities, and microecosystems) implies that evolution not only gives rise to species populations, as envisioned by Darwin, but that it also gives rise to self-organized communities and ecosystems (Wade, 1978; Doolittle and Sapienza, 1980; Wilson, 1980; Hickey, 1982; Sober, 1984) that maximize habitat range by making optimal use of environmental resources and by creating favorable microenvironments within hostile macroenvironments (Brown *et al.*, 1988; Gilbert *et al.*, 1990; Eng *et al.*, 1991). This occurs through community-level processes that are of reproductive value to the community as a whole, as well as to individual organisms and genetic elements. As mentioned previously, degradative microbial consortia serve as useful models of community-level processes. In some cases, for example, the aerobic degradation of DDT, pure cultures are sufficient for complete mineralization (Skryabin *et al.*, 1978). However, under anaerobic conditions, substrates such as phthalic acids and oxalates are degraded only by associations of interacting organisms and not by pure cultures of individual organisms (Aftring *et al.*, 1981).

If evolution acts only on individuals and not on associations, then the product of evolution might be a single organism for the degradation of each substrate in each environment. Instead, a diverse community, rather than a single individual, results from evolution. Some members of the group may sorb the substrate, others may cleave specific bonds within the substrate, still others may protect the group from predation. Through the successful proliferation of the most effective cooperative network (as opposed to less effective networks) the interactive association not only acquires the ability to degrade toxicants, it becomes more durable and able to function successfully as a component of the biosphere. Isolated cell clones, even if they can degrade a toxicant independently in the laboratory, are seldom able to function successfully when released in the environment unless they are adapted to function as part of a community network. However, the mobile genetic elements they contribute to the community may be able to proliferate very well within the native members of the community (Fulthorpe and Wyndham, 1989, 1991, 1992) although the original parental cells themselves never become established.

Initially, evolution may be based solely on individual bacterial lineages. However, individual bacteria proliferate more effectively through diversification and the subsequent development of cooperative interactions, allowing more effective utilization of habitat and resources. If these interactions are highly effective and require continuous physical contact, a set of individuals coalesces and becomes physically recognizable as a distinct individual, the components of which proliferate as a group. However, the formation of an interaction that requires continuous physical contact can preclude beneficial interactions. Some associations require that associates periodically or continually disassociate, as in the case of the association between flowering plants and pollinating insects. Another example is nitrogen fixation in plants. Although it is commonly said that plants are incapable of nitrogen fixation, leguminous plants do fix nitrogen. However, the organelle involved (the bacteroid) is not continuously associated with the plant and evolves simultaneously as an independent organism and as an intracellular component of legumes. If a prokaryote remains permanently within a plant it is considered to be part of the plant organism (chloroplasts and mitochondria). If it does not remain permanently inside the plant (*Rhizobium, Bradyrhizobium,* and *Frankia* spp.), it is regarded as an independent organism. However, each of these bacteria, whether associated with the plant by permanent endosymbiosis or periodic infection, evolves simultaneously as part of the plant organism and as an independent organism. It also evolves simultaneously as an ecosystem component and as a biosphere component. The latter are not as apparent or as easy to conceptualize or study, but they are vital to understanding global habitability and life as a whole.

3.2. A Mini Gaia Hypothesis

Degradative microbial consortia on a grain of soil, lichens on rock surfaces, spoilage biofilms on food products, plaque communities on teeth, and many fermented foods could easily be used as experimental systems to test the concept that community-level reproductive strategies exist as a consequence of biological evolution (Margulis, 1995). This community-level strategy hypothesis might also be referred to as a "mini Gaia hypothesis." The Gaia hypothesis suggests that there is a global biological strategy that ensures that the chemistry of the Earth remains relatively constant and habitable, as opposed to that of neighboring planets whose atmospheric chemistry has varied dramatically over geological time periods (Lovelock, 1979; Margulis, 1990, 1993; Lovelock and Margulis, 1974). Lovelock has referred to the Earth's biosphere as "the largest living organism" and as "a total planetary being" (Lovelock, 1988). The development of the Gaia concept has been briefly summarized both by Fairbairn (1994) as well as Margulis and West (1993).

It is apparent that the evolution of microbial associations over time has resulted in an amalgamation of interactive reproductive strategies that now evolve together as integrated units. The consequences of initially localized strategies may thus become global (Caldwell *et al.*, 1985). This may have resulted in the formation of a planetary prokaryotic genome as envisioned by Sonea (Sonea and Panisset, 1983; Sonea, 1991). Periodic global environmental stresses over periods of millions of years may also play a role in the genesis of genetically encoded strategies for biological proliferation on a planetary scale.

This idea is in conflict with that of Darwin, who's view of the natural world focused on the survival of the fittest individual organisms and races (Darwin, 1859). Efforts to explain planetary and ecosystem ecology by restricting evolution to the level of individuals (Lomnicki, 1978; Mayr, 1993) may eventually fail because of the unnecessary complexity that is required to explain relatively simple evolutionary relationships at the community level. It is only through a more general and less restrictive evolutionary theory that microbial ecology, information theory, germ theory, evolutionary ecology, and ecosystem ecology can become a unified discipline (Loehle and Pechmann, 1988; Caldwell, 1993; Mayr, 1993; Yockey, 1990, 1995; Kelly, 1994; Kauffman, 1995).

3.3. Individual Traits vs. Community Traits

The experiments of Goodnight, who used flour beetles, offer evidence for community-level evolution among animals. The following quotations provide a good example of the group evolution concept and explain the rationale of experiments often used to test it.

> Coevolution generally refers to the process of two or more organisms adapting to each other as a result of individual selection. Another possibility, however, is that coevolu-

Bacterial Communities 125

tion may result from selection acting directly at the community level. Certain types of multispecies associations, such as lichens, which are a symbiotic association between an alga and a fungus, are examples of simple two species communities that may be units of selection. . . . I demonstrate that community selection, defined as the differential survival and/or reproduction of communities, can result in significant changes in the phenotype of a community. The observed changes in the phenotype of a community as a result of community selection included changes in the trait under selection (direct effects of selection), as well as changes in traits that are not under selection (correlated responses to selection). Furthermore, two types of correlated responses to selection were observed. The first, within-species correlated responses to selection, are changes in a trait measured in one species as a result of community selection acting on another trait measured in the same species. The second, between-species correlated responses to selection, are changes in a trait measured in one species as a result of community selection acting on a trait measured in another species. Between-species correlated responses to selection are of particular interest because they cannot be mediated by pathways of gene action that are internal to an individual, rather they can be mediated only through ecological pathways. In other words, between-species correlated responses to selection suggest that genetically based interactions among individuals are contributing to the response to community selection. These among-species ecological pathways of gene action cannot contribute to response to selection at a lower level; thus community selection may be able to bring about a response to selection that is qualitatively different from the response selection that would occur as a result of selection acting at a lower level. (Goodnight, 1990a).

Community selection, defined as the differential proliferation and/or extinction of communities, can bring about a response that may be qualitatively different from the response to selection acting at lower levels. This is because community selection can result in genetic changes in all of the species within the community by acting on the interaction among species. (Goodnight, 1990b)

Thus if community-level evolution is being considered, it is simplest and most accurate to consider community-level interactions as community traits without regard for whether the necessary information resides within an organism, within mobile genetic elements, in genetic fragments distributed among several organisms, in the spatial distribution of organisms (genomes) within a community, or the distribution of communities within geographical boundaries.

3.4. Testing the Mini Gaia Hypothesis and Community-Level Strategy Hypothesis

The challenge for microbial ecologists is to more fully evaluate the community-level strategy hypothesis. One relatively simple approach is based on the relationship between dilution and plating efficiency. If the isolation of bacteria from their native communities uncouples community interactions, then at lower dilutions the mean distance between colonies would be reduced and community interactions would be less uncoupled. This density-dependent growth effect has been observed by Stevens and Holbert (1990) when plating bacterial isolates from terrestrial subsurface environments. By quantifying this density-dependent plating effect using specific media, it is possible to screen for the specific community-level mechanisms of *in situ* interactions. This approach has been used to identify the protector guilds first postulated by Brannan (1995). In density-dependent plating experiments (Caldwell *et al.*, 1997a), Karthikeyan has detected these guilds as central colonies surrounded by satellites that were protected from the toxic effects of benzoate.

There are also other questions, each of which will require the formulation of appropriate tests. Have microbial communities evolved to create favorable microenvironments within unfavorable macroenvironments? Have they evolved to optimize the efficiency of resource utilization? Is the habitat range of the biosphere extended through the formation of interactive associations within communities? Are there communal strategies encoded within the biosphere that optimize the reproductive success not only of individual species but of communities? Is communitization a part of evolution as is speciation? Have environmental stresses culminated in a planetary ecological strategy that is widely distributed across the genetic resources of the biosphere? If so, what is this planetary strategy and how does it relate to sustainable human development and the habitability of the planet? What portion of this strategy resides within the microbial world as opposed to plants and animals?

4. Methods of Community-Level Cultivation

Marshall (1994) has suggested that it is important to focus on community processes and interactions, including: 1) defining the population dynamics in living communities, 2) defining the physicochemical characteristics of the microbial microenvironment, and 3) understanding the metabolic processes carried out by the individual bacteria. This approach allows examination of the growth and behavior of a group of organisms rather than fragments of a group. The inherent complexity of *in situ* microbial ecosystems, complex microcosms, and mixed consortia have tended to limit the success of the first two criteria, due primarily to the lack of methodology by which the details of microbial interrelationships could effectively be examined (Robarts and Zohary, 1993). However, new technologies are gradually removing these barriers (as discussed in section 5). Nonreductive study of microbial communities also requires the experimental analysis of evolution at the community level. This makes it necessary to envision communities as evolving networks, and includes the determination of: 1) relative stability of various community networks under defined laboratory conditions, 2) whether there is only one community network that can "lock into" and occupy a specific environment (or several environments), 3) whether the presence of one community precludes the development or encroachment of another, 4) the habitat range of organisms alone and in association with their community, 5) the ecotones between communities and whether they represent sharp boundaries or gradual transitions, 6) whether the communities occupying a specific environment have been optimized through evolution in terms of their ability to proliferate and to make most efficient utilization of environmental resources and habitat (optimize the conversion of the abiotic to the biotic), 7) whether spatial or temporal pathways are necessary for the development of specific microbial communities and associations, 8) whether communities create favorable microenvironments within unfavorable microenvironments (mini Gaia hypothesis and homeostasis), 9) whether the composition of the community depends upon the timing of immigration by specific species, and 10) whether some traits of organisms specifically affect the proliferation of the community directly, while affecting the proliferation of the individual only indirectly (habitat range of parental strains and adaptation-negative mutants alone, versus their respective range in combination with the other members of their community). These experiments require microbiological systems in which the environment is well defined and in which controlled environmental gradients permit the study of habitat range (Caldwell and Hirsch, 1973; Caldwell et al., 1973, 1975; Lovitt and Wimpenny, 1981; Wimpenny and Waters, 1984; Wimpenny et al., 1986, 1988, 1992; Emerson et al., 1994) as well as providing the spatial (Caldwell et al., 1975; Caldwell, 1992a, 1993, 1995; Wolfaardt et al., 1993 and temporal (Rutgers et al., 1993) pathways necessary for communities to develop.

One objective of microbial ecology is to determine relationships between microbial adaptation, the living and abiotic environment, and reproductive success. Examination of microbial adaptation at the organismal level often involves the enrichment and/or isolation of an organism bearing a specific trait or adaptation, construction of an adaptation-minus mutant from the wild-type strain, and demonstration that, *in vitro*, the mutant organism is no longer selected under the conditions of the enrichment (Haefele and Lindow, 1987; Korber et al., 1989; Lawrence et al., 1992). This approach should be extended to the community level by studying the effect of adaptation-negative community members on the proliferation of communities as well as species populations. However, removal of the organism from it's community has disadvantages, including the loss of cryptic traits following repeated lab culture (Evenboom et al., 1981; Guede, 1979; Costerton et al., 1978, 1994), inhibition of interspecies interactions, and breakdown of the complex network of chemical sources and sinks normally active in nature.

Prokaryotic communities offer an exceptional opportunity to expand the experimental database needed to test community theory. Millions of individuals and dozens of communities can be accommodated within an experimental laboratory system occupying only a few milliliters. The rapid growth rate of microbial communities, their ability to exchange and reorganize genetic information without relying on sexual reproductive mechanisms, the extended time period over which bacterial communitization has occurred, and their limited physical capacity for storing genetic information make them particularly useful systems for the formulation and testing of community-level theory as discussed above. Numerous experimental systems have applicability in studying communities and some of these are described below. In cultivating and studying a microbial community, the first step is to define the environment in terms of the concentrations, fluxes, spatial variations, and temporal variations. Given a defined environment or habitat, a reproducible community of organisms will develop as a consequence of environment pressures (Fig. 2). It is thus not appropriate to use aseptic technique in maintaining and adequately defining community cultures (although this may sometimes be desirable). Community culture makes it possible to elucidate the role of communities as causative agents in ecological and other processes (Caldwell, 1993, 1995; Caldwell et al., 1997b).

Most microbial communities arise through dynamic interactions. They have inputs and outputs that link them with their macroenvironment and that are necessary to sustain them. Consequently the discussion below is restricted to methods of cultivation which provide a well-defined, quasi-steady-state environment. Batch enrichment cultures do not adequately control the environment, and thus do not adequately define the communities which arise in response to environmental constraints.* They are useful in the enrichment of organisms to facilitate isolation, but the depletion of substrate and accumulation of metabolic products often makes them unsuitable for the cultivation of microbial communities under controlled environmental conditions. The chemostat provides control of substrate flux. The nutristat provides control of the flux, concentration, and spatial variation of substrate. Continuous-flow slide culture provides control of the flux, concentration, and temporal variation of substrate. The microstat provides control of substrate flux, concentration, temporal variation, and spatial variation of substrate.

4.1. The Chemostat, Nutristat, and Gradostat

Continuous culture is one of the more widely used systems for the study and cultivation of microbial communities (Senior et al., 1976; Harder et al., 1977; Slater and Hartman, 1982; Gottschal and Dijkhuizen, 1988; Rutgers et al., 1993; Wolfaardt et al., 1994b). It provides an input and output sufficient to maintain a dynamic network of organisms. It also provides a quasi-steady-state condition over prolonged time periods, sufficient to study the time course of evolutionary adaptation (Senior et al., 1976). By using the chemostat, a constant flux of a recalcitrant toxicant can be maintained at a low level for a long period of time. This is important due to the time often required for the development of stable

---

*Refer to the definitions of community culture and enrichment culture in Section 9.

community associations, particularly those that degrade recalcitrant compounds (Senior *et al.*, 1976; Lappin *et al.*, 1985; Prosser, 1989; Wolfaardt *et al.*, 1994b). As stated by Senior *et al.* (1976) "It is reasonable to postulate that interacting microbial communities are common in nature; however, very few microbial associations of this type have been defined, principally because inappropriate enrichment and selection techniques have been used to isolate them from natural environments". As stated by Bhatnagar and Fathepure (1991), "Suitable techniques are now available to isolate and characterize stable microbial communities. It still remains an open question whether such isolated microbial communities perform differently in their natural habitats."

Within the chemostat, cell growth is restricted by the availability of one growth-limiting factor. Thus growth within an idealized chemostat follows the two relationships established by Monod (1942, 1949) and shown below in Eqs. 1 and 2. These assume that the uptake of substrate is the single factor limiting the rate of growth, that substrate uptake is mediated by a permease obeying the same kinetics used for enzyme-catalyzed reactions, and that cells are acquired solely by growth and lost only by dilution:

$$\mu = \mu_{max} \left(\frac{S}{K_s + S} \right) \quad (1)$$

where μ = specific growth rate, μ_{max} = maximum specific growth rate, S = concentration of the limiting substrate, and K_s = saturation constant equal to the substrate concentration when $\mu = \frac{1}{2} \mu_{max}$. The growth rate is normally described by the relationship above and the change in the density of organisms is described by the relationship below. Assuming that the cells are uniformly distributed (no wall growth or bioaggregation), then

$$dN/dt = \mu N - DN \quad (2)$$

where N = cell density, t = time, and D = dilution rate, the rate of flow divided by the volume of the culture. dN/dt is the instantaneous rate of change in cell density, μN is the instantaneous rate of increase in cell density due to growth, and DN is the instantaneous rate of decrease in cell density due to dilution. If dN/dt is greater than 0, then the cell density increases, the concentration of substrate thus decreases (Eq. 1), and dN/dt decreases to 0 (Eq. 2). If dN/dt is less than 0, then the cell density decreases, the concentration of substrate increases, and dN/dt increases to 0. Thus a chemostat always approaches a steady state in which $dN/dt = 0$, at which the growth rate, μ, equals the dilution rate, D. Setting the dilution rate thus sets and controls the growth rate. Although it is not possible to directly control the chemistry within a chemostat, the concentration of substrate and other environmental conditions remain relatively constant once a quasi-steady-state condition has been obtained. The concentration of substrate must be determined empirically and is a consequence of the steady-state dynamics between substrate consumption, growth, and dilution. The same general principles apply to all cultures grown in a substrate-limited continuous culture, regardless of whether they are pure, mixed, or community cultures.

Those organisms or associations unable to grow at or above the chemostat dilution rate are gradually lost from the system as a consequence of dilution. Those that grow most rapidly at the prevailing substrate concentration gradually displace others. This commonly decreases the species diversity of the system (Bouwer, 1989) and produces what appears to be a "pure culture," unless the incubation period is extended to allow sufficient time for bioaggregates and biofilm communities to form. However, if the growth rate of one organism is dependent upon the growth of others through ecological interactions, then the chemostat can support a stable mixture of planktonic cells (Bungay, 1995) as well as individual cell lines.

Chemostat studies of mixed cultures are normally based on substrate limitation. In this case, the measure of an organism's performance is its half-saturation constant (K_s) and maximum growth rate (μ_{max}). This assumes that substrate concentration and specific growth rate are sufficient to predict the outcome at any given dilution rate. Jannasch (1967) demonstrated this experimentally using a *Pseudomonas* and *Spirillum* spp. isolated at high and low dilution rates (high and low substrate levels), respectively. When isolated cell lines of each organism were mixed together and cultured at low dilution rates, the *Spirillum* sp. (low K_s, low μ_{max}) displaced the *Pseudomonas* sp. (high K_s, high μ_{max}), whereas at high dilution rates, the *Pseudomonas* sp. predominated. In similar chemostat enrichment studies using marine communities, low dilution rates yield organisms from the genera *Spirillum, Achromobacter, Vibrio,* or *Micrococcus*, while high dilution rates yield organisms from the genera *Pseudomonas* or *Aerobacter*.

The dynamics of metabolic interactions between individuals can be studied in chemostat cultures of mixed cell lines. In one study of a mixed culture containing an *Anabaena* spp. and *Zoogloea* spp. isolated from the epilimnion, cell yields were increased through the complementary metabolism of the phototroph (*Anabaena* spp.) and heterotroph (*Zoogloea* spp.), which spontaneously formed phototroph/heterotroph bioaggregates (Schiefer and Caldwell, 1982). In the chemostat, the displacement of one organism by another occurs by attrition through dilution. Consequently, cells must have an equal probability of being diluted from the culture as individuals or as components of bioaggregates (cell aggregates) if evolution and adaptation are to be based solely on substrate concentration and growth rate. Suspended cell aggregates do not move as freely through the outlet of a chemostat as do unicells. Consequently, the design of the outlet mechanism must be modified to ensure that both individual cells and cell aggregates are diluted at the same rate. When grown alone, the phototroph supersaturates the microenvironment with oxygen, raises the pH, and depletes the concentration of free carbon dioxide. The presence of the heterotroph alleviates these stresses through the consumption of oxygen, production of carbon dioxide, and reduction of pH. This reduces the rate of photorespiration, thus making more effective use of environmental resources through recycling. It also creates a more favorable microenvironment within which metabolism is balanced. A similar metabolic interrelationship of this kind may have served as the basis for the evolution of algal and plant cells from heterotrophic bacteria and cyanelles.*

In another chemostat study, a stable four member bacterial association was maintained using methane as the sole substrate. This included a *Pseudomonas* sp. (a methane-utilizing strain), a *Hyphomicrobium* sp. (a methanol-utilizing strain), a *Flavobacterium* sp., and an *Acinetobacter* sp. (Wilkinson et al., 1974). The *Hyphomicrobium* sp. scavenged methanol produced through the metabolism of methane by the pseudomonad, whereas the *Flavobacterium* and *Acinetobacter* spp. scavenged other metabolites. The success of each organism was linked to the success of its cohabitants and dilution resulted in proliferation of a set of individuals rather than leading to the predominance of a single homogeneous lineage.

The large gene pool associated with microbial groups facilitates adaptive change to periodic fluxes of various substrates within the environment, suggesting that a group of generalist organisms may out-compete less-flexible specialist individuals having higher rates of growth on a particular substrate (Gottschal and Dijkhuizen, 1988). Metabolic versatility and diversity also enables a range of organisms to flourish on metabolic end products resulting from the growth of key community members when only a single refractory carbon source is available, as is frequently the case in degradation studies.

A number of chemostat studies have been performed involving mixed microbial populations, however few of these have characterized the consortium members and the significance of the interactive mechanisms involved. Detailed examinations include those performed on organisms that degrade complex haloorganic herbicides or pesticides. For example, Lappin et al. (1985) isolated and characterized a five member consortium (two *Pseudomonas* spp., an *Alcaligenes* sp., an *Acinetobacter* sp., and a *Flavobacterium* sp.) using the herbicide, mecoprop, as sole carbon source. The authors determined that no pure culture was capable of growth using mecoprop (even in the presence of a labile substrate), however certain combinations of two community members were capable of growth. Although two of the community members could grow in the absence of the other three members, the community still sustained all five members during cultivation over relatively long time intervals (4 months). Thus some of the interactive mechanisms involved in this association must not yet be understood.

*Cyanelles are cyanobacteria as chloroplasts within eukaryotic cells. The origin and significance of cyanelles in eukaryotes is discussed by Herdman (1977) and Scott (1985).

Wolfaardt et al., (1994b) isolated a nine member bacterial community in continuous culture using diclofop methyl as the sole energy source. This community appeared to consist of several degradative consortia and was stable over a period of 18 months. The community members included eight gram negative and one gram positive bacterium, as well as a *Chlorococcum* sp. (alga) (Wolfaardt et al., 1994a). Presumably, the recalcitrance of the halogenated carbon source provided the impetus for sustained community involvement within this system. No single consortium member could mineralize diclofop methyl as sole carbon source (determined using ^{14}C-diclofop). However two strains could degrade the herbicide following amendment with an alternate carbon source (tryptic soy broth). During the degradation of either mecoprop or diclofop methyl, the success of the degradative association as a whole appeared dependent on the success of key consortium members.

Continuous cultures have also been used to study the effects of temperature, pH, light, predation, and substrate composition (e.g., gradual addition of recalcitrant xenobiotics) (Wirsen and Jannasch, 1970; Harder and Veldkamp, 1971; Ratnam et al., 1982; Rothmel et al., 1989; Wolfaardt et al., 1994b). For example, Ghosal et al. (1985) used chemostat enrichment to study the evolution of degradative genes in a *Pseudomonas cepacia* strain exposed to increasing concentrations of 2,4,5-trichlorophenoxyacetic acid. Rutgers et al. (1993) devised a new culture device, the nutristat, to maintain controlled concentrations of pentachlorophenol (PCP) in continuous culture. When the PCP concentration was gradually increased, this resulted in higher growth rates and shorter acclimation times than in batch or traditional chemostat systems. This demonstrates the potential value of temporal substrate gradients (temporal pathways) in the cultivation of degradative communities.

Chemostats may also be linked in combination, either unidirectionally or bidirectionally, more closely mimicking the interconnected chemical and cellular sources and sinks found in nature (Veldkamp and Jannasch, 1972; Veldkamp, 1977; Herbert, 1988; Parkes and Senior, 1988; Wimpenny, 1988, 1992) and increasing the probability that a greater diversity of organisms will be sustained in continuous culture. Bidirectionally-linked chemostats such as the gradostat (Lovitt and Wimpenny, 1981; Wimpenny, 1988), have been used to model naturally occurring gradient systems (e.g., salinity, oxygen). For example, when *Escherichia coli* was cultured in a stepwise, linear gradient of oxygen (from aerobic to anaerobic), the organism rapidly adapted to a wide range of environmental conditions, with cells isolated from discrete gradostat vessels being metabolically distinct (i.e., in the presence of oxygen, cells produced nitrate reductase, whereas in the absence of oxygen, hydrogenase was detected). Subsequent coculture experiments were performed using *E. coli* in combination with *Pseudomonas aeruginosa* and *Clostridium acetobutylicum*. These demonstrated the preferential enrichment of each strain under favorable conditions. This also demonstrated the potential of gradients as spatial pathways of community development just as the studies of Rutgers *et al.* (1993) demonstrated the importance of temporal pathways in the development of degradative communities.

4.2. The Dual-Dilution Continuous Culture

Although planktonic growth is common in low-nutrient natural systems, a larger and less transient community of organisms colonizes solid–liquid interfaces (Costerton *et al.*, 1978; Geesey *et al.*, 1978; Kjelleberg, 1984). Bacterial adherence, along with the formation of a stable biofilm community, are factors that are notably absent (or not accounted for) in most chemostat systems (the surface to volume ratio of the chemostat is generally low and sampling is performed solely on planktonic cells) (Ratnam *et al.*, 1982; Characklis, 1988). To overcome this, continuous culture systems have been modified to incorporate surfaces on which stable populations of bacteria may develop (Bryers, 1993). Artificial surfaces (e.g., glass beads, glass slides, mineral crystals, etc.) have been added to continuous culture systems to compare the growth rate of attached and suspended organisms (Kieft and Caldwell, 1984), evaluate attachment effects on the success of mixtures of bacteria, enrich for adherent organisms, or prevent the loss of strains due to low growth rate (Maigetter and Pfister, 1974; Trulear and Characklis, 1982; Caldwell and Lawrence, 1986; Characklis, 1988; Wolfaardt *et al.*, 1994a).

Korber *et al.* (1994c) used dual-dilution continuous culture (DDCC), in which planktonic and surface-associated organisms (on glass beads) were diluted independently by continually replacing the surfaces in the culture as well as the liquid phase of the culture. A mixture of motile and nonmotile *P. fluorescens* were used in colonization experiments to evaluate the significance of polar flagellation during the colonization process. In the DDCC study, the persistence of bacteria was based on the ability of organisms to continually recolonize glass beads rather than their ability to persist in the planktonic phase. This uncoupled reproductive success from rate of growth. In process studies such developments have significant impact. For example, Wolfaardt *et al.* (1994a) determined that addition of a glass bead perfusion column to a continuous culture system resulted in 36% greater rates of diclofop methyl loss by a microbial consortium than where surfaces were absent, suggesting that spatial interactions between consortium members play an important role during *in situ* degradation.

4.3. Rototorque Annular Bioreactor

One of the best characterized continuous culture systems for studying surface-associated organisms is the Rototorque annular bioreactor (Characklis *et al.*, 1990). The Rototorque consists of two (one fixed and one rotatable) concentrically mounted cylinders, incorporates a high surface area to volume ratio, and operates at steady state in a gradient-free condition due to efficient mixing. Integrated with torque transducers, the resistance to cylinder rotation resulting from biofilm growth provides for detailed information on hydrodynamic parameters (e.g., friction, shear), and is thus of industrial significance. If only biofilm accumulation is of interest, then rates of D set for Rototorque experiments exceed the growth rate of the organisms in question, and only sloughed or sheared cells may be detected in the planktonic phase. Alternatively, if both biofilm and planktonic growth are of interest, D is set below the maximum specific growth rate.

The Rototorque has been applied to process-oriented studies (substrate removal or flux, biocide efficacy, etc.), incorporation of removable plates also permit quantitative measurements of biofilm biomass, biofilm thickness, as well as chemical determinations. Heterogeneities within both pure culture biofilms, as well as mixed-species biofilms, have been studied by removing coupons (Trulear and Characklis, 1982; Stewart *et al.*, 1993).

4.4. Continuous-Flow Slide Cultures

Continuous-flow slide culture (CFSC) and on-line digital image analysis has been used for direct analysis of growth and development of biofilm communities on surfaces (Caldwell and Lawrence, 1986, 1988; Korber *et al.*, 1989; Lawrence *et al.*, 1992; Wolfaardt *et al.*, 1994a) as well as for the study of isolated cell lines (Caldwell and Lawrence, 1986; Sjollema *et al.*, 1989a,b; Busscher *et al.*, 1990). CFSC provides well-defined chemical conditions that can be changed almost instantaneously (Caldwell and Lawrence, 1986; Korber *et al.*, 1994a). The hydrodynamic conditions within CFSC systems may also be carefully controlled, with known laminar flow velocities (flow rate divided by cross sectional area) and nutrient fluxes (the product of flow rate and concentration) at the solid–liquid interface (Caldwell and Lawrence, 1986; Lawrence *et al.*, 1987; Busscher *et al.*, 1990; Korber *et al.*, 1990). Using this approach, Caldwell and Lawrence (1986) demonstrated flow-dependent growth kinetics of a *Pseudomonas fluorescens* strain in continuous-flow slide culture. At high nutrient concentrations (1 g L^{-1} glucose), the growth rate of microcolonies was independent of flow rate. At lower nutrient concentrations (100 mg L^{-1} glucose), microcolony growth ceased when flow was stopped. This suggested that microcolonies become glucose limited as opposed to oxygen limited during the early stages of biofilm development, and that there is a mechanism by which substrates can be concentrated from flowing solutions (thus increasing their effective concentration).

Surface colonization involves the transport of cells to the surface microenvironment, reversible and irreversible attachment of cells, cell growth to form microcolonies, redistribution of cells from microcolonies to new attachment sites, and formation of a confluent biofilm (Lawrence *et al.*, 1987; Korber *et al.*, 1989). Maintenance of quasi-steady-state environmental conditions can then result in biofilms that reach a quasi-steady state, where cell loss (due to sloughing, emigration, or predation) is balanced with continual regrowth of biofilm material (Characklis *et al.*, 1990; Korber *et al.*, 1994b; Wolfaardt *et al.*, 1994a). One limitation of CFSC is that substrate depletion and the accumulation of metabolic wastes may occur downstream within flow lamina at the surface of the biofilm. This may result in sloughing or detachment. To minimize downstream effects, a sterile flow cell can be inoculated using a syringe from the downstream side to provide a biofilm "front" or "leading edge" where the irrigation medium is unaffected due to the absence of an upstream biofilm. Using computer-enhanced microscopy to quantify bacterial growth and behavior at the front also permits direct evaluation of flow effects on motile organisms, as used previously to study the development of this leading edge against the direction of flow (backgrowth) for motile versus nonmotile *P. fluorescens* strains (Korber *et al.*, 1989) at laminar flow velocities above and below the maximum rate of motility for this strain.

Lawrence and Caldwell (1987) have also used CFSC and digital image analysis to document the diversity of microbial colonization strategies in a natural stream community. CFSC has further been applied to a range of ecological studies, including evaluation of the role of bacterial motility during surface colonization events (Korber *et al.*, 1989), examination of cell sedimentation effects for flagellar mutants (Korber *et al.*, 1990), testing of physicochemical attachment theory (Christersson *et al.*, 1988; Sjollema *et al.*, 1989a,b, 1990), determination of the behavior of swarming bacteria in high and low viscosity environments (Lawrence *et al.*, 1992), effects of antimicrobial agents on fully-developed biofilms (Korber *et al.*, 1994a), and the study of growth, behavior, and architecture of more complex, mixed-species biofilms (Lawrence and Caldwell, 1987; Korber et al., 1994b, 1995; Wolfaardt et al., 1994c). More recently, Wolfaardt has designed multi-channel flow cells constructed from a single block of polycarbonate which have been used rather than single-channel flow cells. This has advantages in terms of experimental replication, experimental controls, use of several different treatments (flow velocity, substrate concentration), and the sacrifice of multiple channels during the destructive analysis of biofilm communities (Korber *et al.*, 1994a; Wolfaardt *et al.*, 1994c).

4.5. The Microstat

The microstat is a flow cell used to study the response of bacterial communities to environmental gradients. The response is visualized using fluorescent molecular probes and confocal laser microscopy. It maintains constant physicochemical conditions within the microenvironment, and provides control of the concentration and flux of substrate, as well as spatial and temporal variations in concentration and flux.

Microstat gradients are normally produced by diffusion in gels. The pores permeating agar-type gels are large enough to permit molecular diffusion through the matrix at rates approximating those achieved in pure water (Takana et al., 1984; Koch, 1991). The development and application of these diffusion-based methods of cultivation began a century ago with the auxanographic methods of Beijerinck (1889). A more quantitative version of the traditional agar plate was devised by Szybalski (1952), who poured a wedge of agar containing antibiotics and then overlaid this wedge with agar containing no antibiotic. This produces a linear concentration gradient that is relatively stable over time. This method has since been used to demonstrate synergism between pH and streptomycin efficacy against an *E. coli* strain (Szybalski and Bryson, 1953), for general screening of antimicrobial agents (Weinberg, 1957), and for the isolation of spontaneous antibiotic-resistant organisms. It has also been used by Wimpenny and co-workers (1986, 1988) for examining the response of bacteria to gradients of salinity, pH, and antibiotic concentrations.

Multiple agar wedges (each containing a different test compound poured at right angles) have been used to produce two-dimensional gradients and response surfaces representing the spatial growth domain of organisms to as many as four different compounds (Wimpenny, 1992). However, gel wedges do not maintain constant conditions over time (e.g., the chemical gradient continually shifts). Thus this system is somewhat analogous to a semi-solid batch system where the chemistry is in constant flux (Wimpenny et al., 1988).

Caldwell and Hirsch (1973; Caldwell et al., 1973) developed the first steady-state gradient system with quantitative gradients that did not shift during the time course of experiments. Numerous modifications of these two-dimensional steady-state diffusion gradients have been developed (Caldwell 1993). Most recently, Wolfaardt et al. (1993) developed a one-dimensional microstat for the study of degradative consortia, and used a finite-element transport model (FRACTRAN) to predict the flux and concentration of test compounds at different rates of flow, gel thicknesses, flow velocities, and number of reservoirs. These simulation experiments resulted in modification of the single reservoir system to a multiple reservoir system to linearize the diffusion gradients. The simulations were also used to predict the time required to reach steady state using gels of varying thicknesses (Fig. 3). The multiple reservoir system provided an extended, linear gradient (40 mm), and was used to compare community growth responses to a range of diclofop methyl concentrations. It also provided a spatial pathway for development of functional consortia at higher concentrations than those found *in situ* or in batch culture. Other microstats have incorporated modifications (Wolfaardt et al., 1993) including porous metals, porous hydrophilic plastics (Porex™), or sintered glass (Fig. 4 and Fig. 5).

Chemical gradients along the surface of the gel can be measured using conservative fluorescent tracers (e.g., fluorescein) and either a fluorometer, a SCLM, or by measuring the concentrations of test compounds (e.g., chloride or radioisotopic tracers) in gel plugs removed from defined xy gel locations (Cald

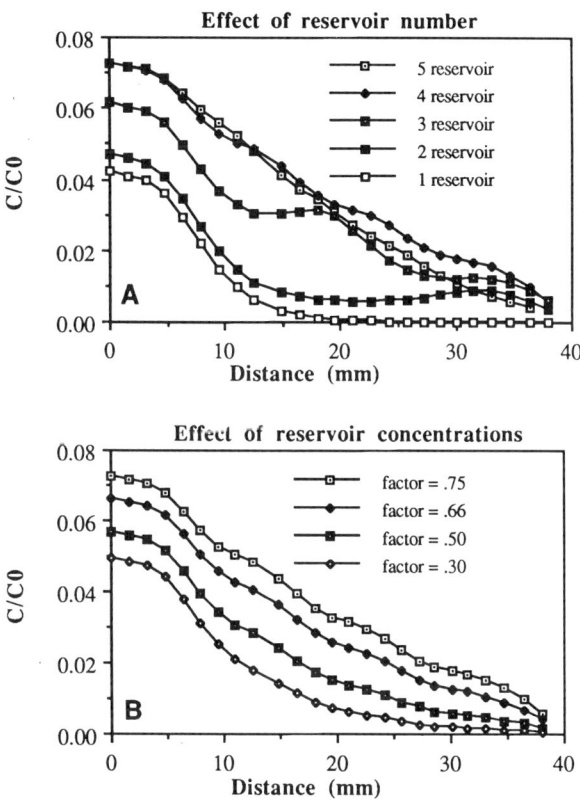

Figure 3. The effect of changes in diffusion chamber design on concentration gradients at the gel/water interface as predicted in computer simulations of diffusion processes. The model consists of a gel (dimensions: 38 mm × 16 mm × 5 mm) with five reservoirs, each containing a 0.75-fold increase in concentration over that in the adjacent reservoir. Flow of the irrigation solution over the surface of the gel was maintained at a velocity of 368 mm h^{-1}. Changes in the number of reservoirs (A) show that a minimum of 4 reservoirs were required to create a smooth gradient extending over the length of the gel. The effect of concentration ratios in the 4 equally spaced reservoirs is illustrated in (B). A profile resembling the one reservoir physical model resulted when the factor was smaller than 0.3 (each reservoir contained solute at 0.3× the concentration of the preceding reservoir). The range of the detectable concentration gradient was limited when the factor was greater than 0.8. (C) Effect of gel thickness on the concentration gradient; a minimum thickness of 5 mm was required to ensure an even concentration gradient at the gel/water interface. (D) Effect of water velocity over the gel; a low velocity (37 mm h^{-1}) resulted in an extended detectable concentration range. However, solute flux from the gel at these low flow rates did not differ significantly from those at higher flow rates. (Reprinted with permission from Wolfaardt et al., 1993.)

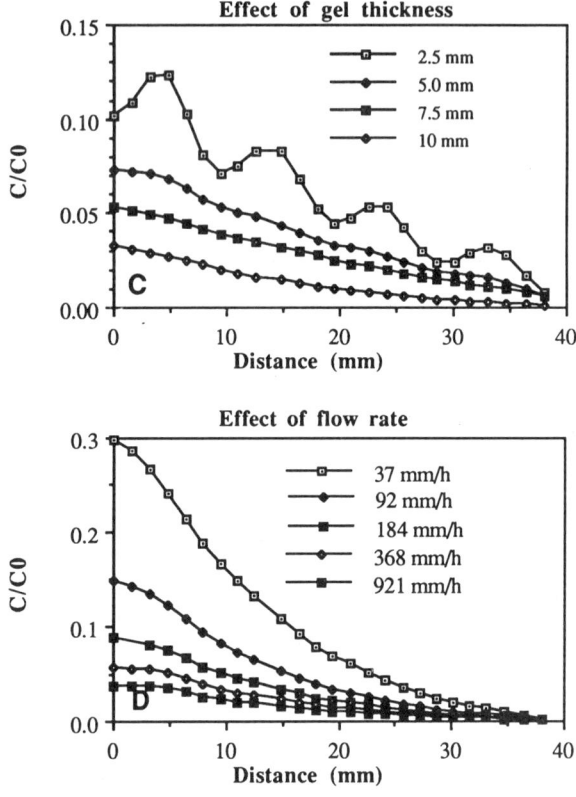

Figure 3. *(Continued)*.

well *et al.*, 1973; Wolfaardt *et al.*, 1993; Emerson *et al.*, 1994). The flux of substrate at the surface of the gel can also be measured using fluorescence recovery after photobleaching (FRAP) as described in more detail below (Korber *et al.*, 1994b; Lawrence *et al.*, 1994). These data are then used to produce a contour map or 3-D relief map of concentration or flux throughout the surface of the gel (Peters, 1990). By overlying the chemical map on the growth responses of a particular bacterial strain or community, an environmental response surface is obtained, showing any interaction effects between the biological response to each of the two environmental gradients (Caldwell, 1995). Depending on the fluorescent molecular probe used, several different response surfaces can be obtained. These include quantitation of microbial biomass, microbial activity, metabolic condition, viability, biofilm architecture, and biodiversity.

Reproducible cultivation of microbial communities requires that the environment be precisely defined and controlled. The microstat is one of several effective methods of achieving this. It also provides well-defined environmental

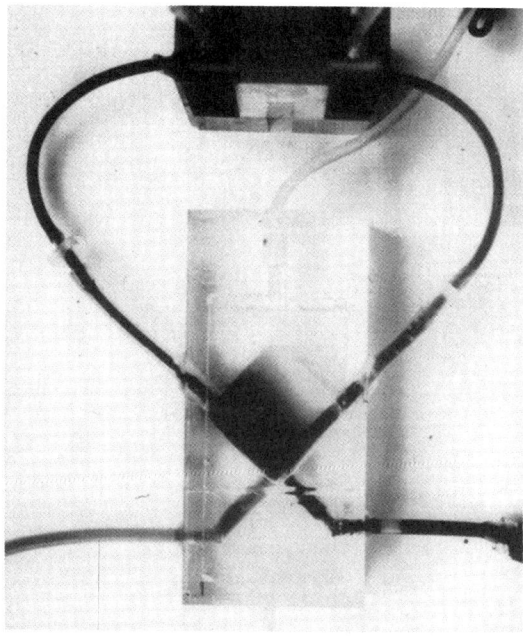

Figure 4. Photograph of the Wolfaardt microstat. Porous materials (e.g., porous plastic, glass, or metal) may be used to deliver test compounds into the gel matrix upon which bacterial communities have colonized. The constant flux of the test compounds through the gel results in a steady-state 2-D concentration gradient at the gel–liquid interface. Adaptation of the microbial community to the test compounds may then be determined nondestructively over time using SCLM monitoring and molecular probes. For a color representation of this figure, see the color insert facing page 140.

gradients necessary to determine the habitat range of individuals vs. the habitat range of the interactive associations among individuals, whether the ecotones between communities represent sharp boundaries or gradual transitions, whether the presence of one community precludes the development of other communities, and whether spatial/temporal gradients or pathways are necessary for the development of specific types of microbial associations and communities.

5. Methods of Community-Level Analysis

5.1. Scanning Confocal Laser Microscopy (SCLM) and Fluorescent Molecular Probes

Analysis of microbial interactions within communities has been hindered by a lack of direct methods that are nondestructive and quantitative. However,

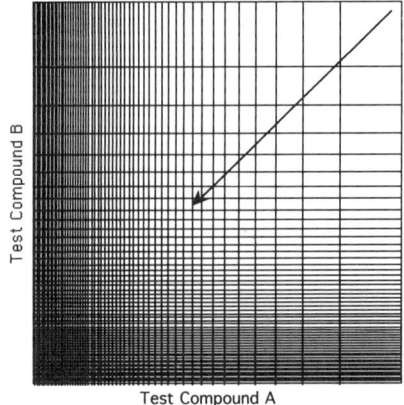

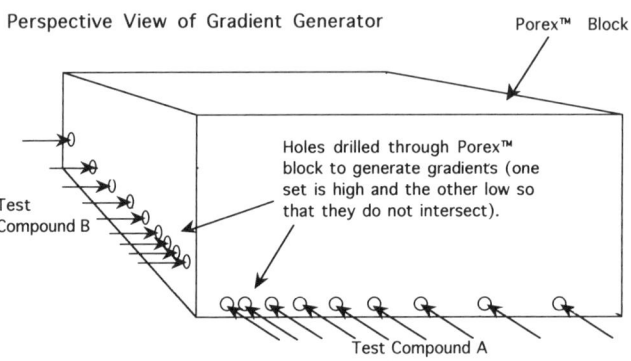

Figure 5. A diagram of the Porex™ microstat showing the direction of flow within the irrigation solution and the configuration of generator holes supplying the test compounds used to generate the quasi-steady-state diffusion gradients.

advances in microscopy have resulted in the development of confocal imaging methods based on laser microscopy and applicable to biological specimens (White et al., 1987; Brackenhoff et al., 1988; Shotton, 1989). The features that make SCLM applicable for the analysis of biofilm communities include: 1) nondestructive *in situ* analysis of fully hydrated biofilms, 2) precise evaluation of complex *xy* and *xz* spatial relationships among biofilm bacteria (through the elimination of unfocused stray light), 3) elimination of the need for harsh

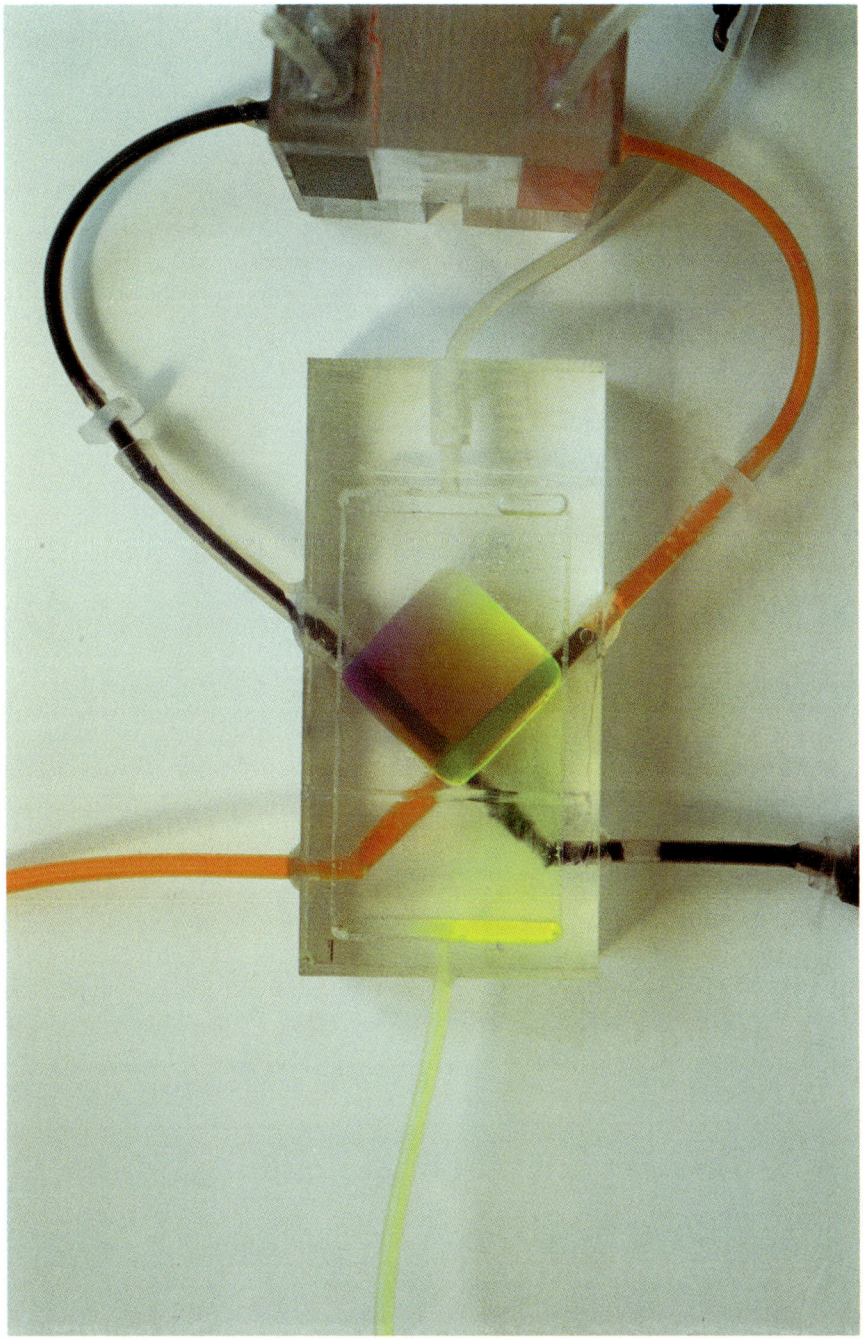

Figure 4. Photograph of the Wolfaardt microstat. Porous materials (e.g., porous plastic, glass, or metal) may be used to deliver test compounds into the gel matrix upon which bacterial communities have colonized. The constant flux of the test compounds through the gel results in a steady-state 2-D concentration gradient at the gel–liquid interface. Adaptation of the microbial community to the test compounds may then be determined nondestructively over time using SCLM monitoring and molecular probes.

transparent surfaces (e.g., minerals, metals, synthetics), and 5) visualization of physicochemical, behavioral, metabolic, and genetic information as well as morphological information (Lawrence et al., 1991; Caldwell et al., 1992a,b, 1993). Environmental and biochemical probes (Stahl et al., 1984; Delong et al., 1989; Amann et al., 1990; Tsien and Waggoner, 1990; Haugland, 1992; Rodriguez et al., 1992), multi-dimensional image analysis and processing, and SCLM must be used together to most effectively exploit the potential of laser microscopy in community studies.

5.1.1. Fluorescence Exclusion to Determine Cell Distribution

Fluorescent probes that allow evaluation of the architecture of microbial biofilms (the three dimensional arrangement of cells, polymer, and void space) reveal biofilm heterogeneity (Lawrence et al., 1991; Caldwell et al., 1992a,b; Korber et al., 1993; Lawrence and Korber, 1994). Fluorescence exclusion (negative staining) is one technique that may be used to image bacteria embedded within thick biofilms (Caldwell et al., 1992b). Using fluorescein as a negative stain, bacteria are seen as dark objects against a bright background. This is a technique successfully used to identify and quantify bacteria in both pure and mixed-species biofilms (Lawrence et al., 1991; Korber et al., 1992, 1993; Wolfaardt et al., 1994a). Fig. 6 shows a negatively stained, horizontal, optical thin section from the base of a 10 μm thick anaerobic sludge-degrading biofilm (Korber et al., 1994b). It illustrates the use of SCLM to determine the arrangement and morphology of the consortium members present at the glass–liquid interface. Fluorescein, resazurin, or fluor-dextran conjugates may be used as non-toxic negative stains. When applied as a pulse or used in conjunction with probes that fluoresce at different wavelengths, they allow temporal studies of growth and development that may later be merged or compared with secondary indicators of cell physiology or microenvironment chemistry (Korber et al., 1994a; Wolfaardt et al., 1994c).

The fluorescence of fluorescein is highly pH-dependent (Haugland, 1992) being brightest at high pH and lowest at low pH. At a pH of 7.2, intracellular fluorescence is less than that of the ambient environment (due to the exclusion of fluorescein from the cells). However when the pH of the bulk phase is less than that of the cell cytoplasm (<6.0), cells appear as bright objects on a black background (positively stained). This results from the relatively bright fluorescence of traces of fluorescein that penetrate the cell, as compared to the weak fluorescence of the higher concentrations of fluorescein in the cell's relatively acidic extracellular environment. This effect can be used to eliminate background (non-specific) fluorescence, as non-living particles that normally appear negatively stained do not fluoresce at low pH. Manipulation of the pH of the macroenvironment can also be used to help identify the presence of homeostatic pH microenvironments present *in situ* within biofilm communities.

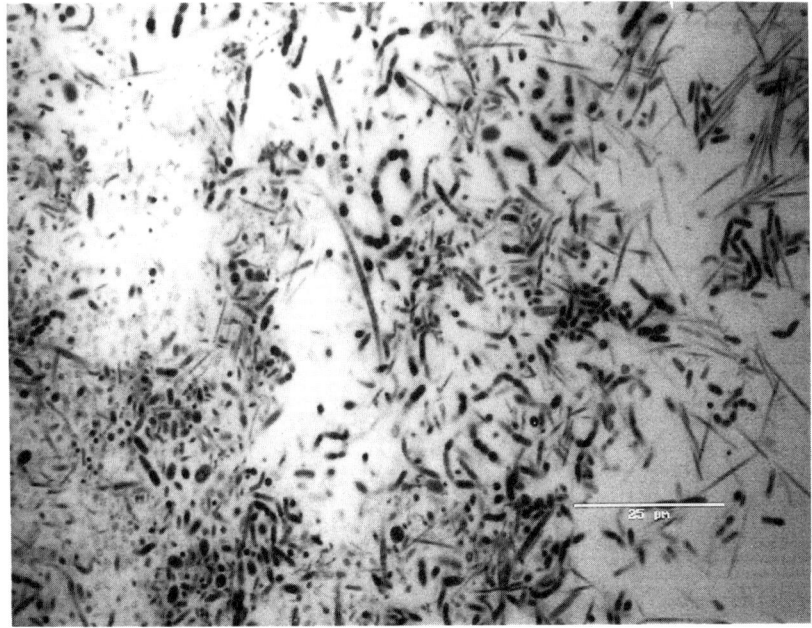

Figure 6. SCLM optical thin section obtained from the basal region of an anaerobic sludge biofilm 10 μm thick, showing the distribution of cells and diversity of cell morphology near the attachment surface.

5.1.2. FITC-Conjugated Dextrans to Determine Diffusion Coefficients

Barriers to diffusion imposed by a cell's exopolymer matrix may also be probed using fluorescent compounds and SCLM. Fluorescein, as well as fluorescent derivatives of dextrans varying in molecular weight, have been used to determine effective diffusion coefficients (D_e, $\times$ 10^6 cm^2 sec^{-1}) for both pure and mixed-species biofilms using either fluorescence monitoring or fluorescence recovery after photobleaching (FRAP) (Lawrence *et al.*, 1994; Korber *et al.*, 1994b). Fig. 7 illustrates the effect of time on the migration of a fluorescent-labeled dextran (2000 KD) into a degradative biofilm. Using a range of molecular weights (from 300 to 2000 KD), D_e values were determined to vary with the hydrated radius of the test compound (from 2% to 13% of D_{aq} values for water), as well as with specific locations within the mixed biofilm community (e.g., regions of high and low cell densities) (Fig. 8) (Lawrence *et al.*, 1994). These studies have contributed to an increased understanding of the heterogeneous nature of both pure and mixed-species microbial communities, features that

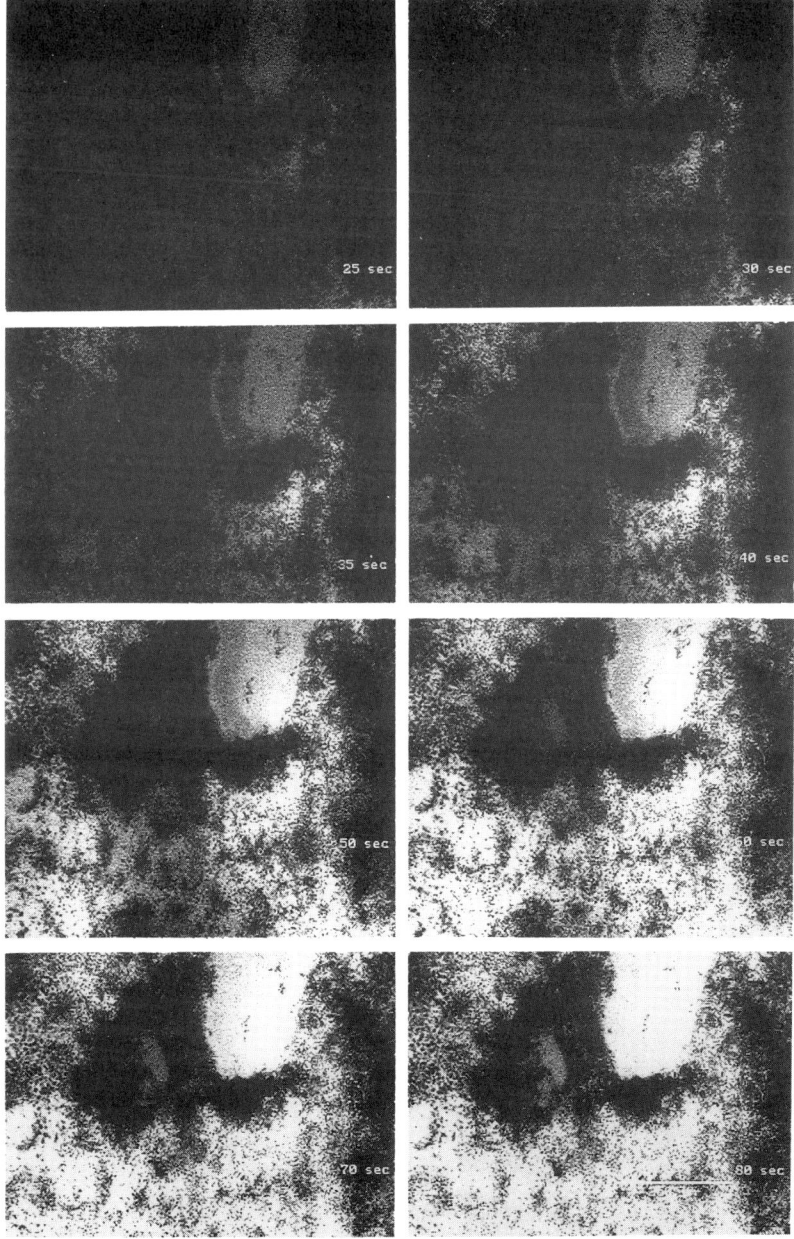

Figure 7. FITC-conjugated dextrans as probes of exopolymer permeation. Time series from 25–80 s, showing the variation in penetration of a 2000 K dextran within a mixed-species biofilm. Images are xy optical sections at a depth of 5 μm from the glass surface. Gray scale code: white indicates maximum intensity; black indicates lowest intensity. (Reprinted with permission, Lawrence *et al.*, 1994.)

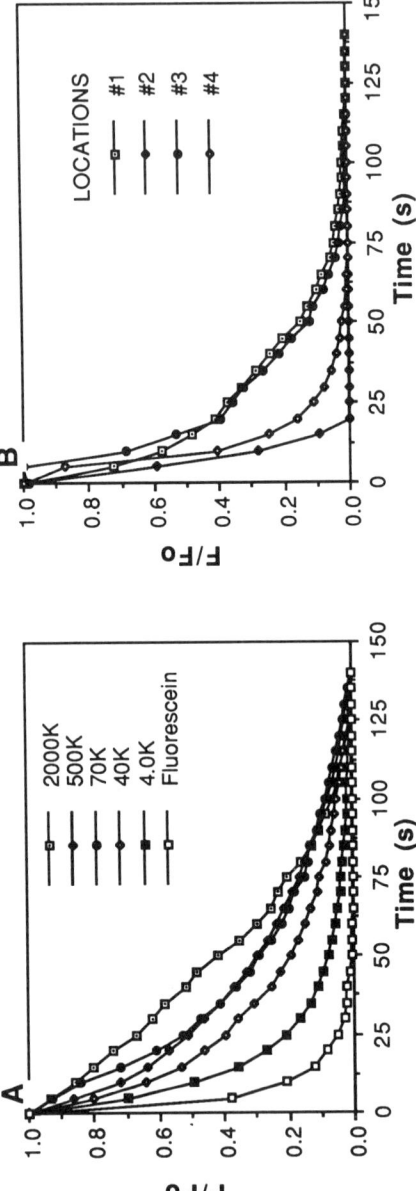

Figure 8. FITC-conjugated dextrans as probes of exopolymer permeation by elution. Curves show the decrease in fluorescence after elution of fluorescein from the bulk-phase irrigation solution of biofilms similar to those shown in Fig. 7. Note the effect of molecular weight on the profiles for the FITC-conjugated dextrans and fluorescein in a mixed-species biofilm (A). (B) illustrates the effect of location on the rate of decrease in fluorescence at four different locations within a series of monitoring images using the 2000K dextran. (Reprinted with permission, Lawrence *et al.*, 1994.)

undoubtedly contribute to the durability and versatility of biofilm communities (Costerton et al., 1994).

5.1.3. Polyionic Dextrans to Determine Charge Distribution

Most bacteria produce exopolysaccharides (EPS), either as capsules attached to the cell wall or as mucoid secretions in the extracellular environment (Costerton et al., 1978; Troy, 1979; Shen et al., 1993). They have a functional role during the formation of microcolonies (Allison and Sutherland, 1987), in attachment to surfaces (Marshall et al., 1989), in the accumulation of metals (Brown and Lester, 1982), in increased resistance to antimicrobial agents (Uhlinger and White, 1983), and other mechanisms that improve microbial performance. EPS may also play a role in the bioaccumulation of substrates from flowing solutions (Caldwell and Lawrence, 1986; Wolfaardt et al., 1994c). In order to elucidate the mechanisms responsible for these processes, as well as their importance in the development and expansion of communities, it is important to define the physical and chemical characteristics of EPS *in situ*. Commercially available fluorescent dextrans bearing defined charge (e.g., polyanionic, cationic, neutral) have been used to document some of the chemical heterogeneity within complex biofilm systems. During studies of sewage- and herbicide-degrading mixed-species biofilms grown in continuous-flow slide culture, considerable spatial variability in the binding patterns of polyanionically- and cationically-charged dextrans have been identified (Korber et al., 1994b; Lawrence et al., 1995) despite reports that biofilm cells and polymer carry a net negative charge. In a more detailed study, Wolfaardt et al. (1995) demonstrated that within a diclofop-degrading consortium, specific binding sites for the fluorescent herbicide were predominantly cationic and hydrophobic (determined using Nile Red) in nature, even though adjacent sites bound cationically charged probes.

5.1.4. FITC-Conjugated Lectins to Determine Oligopolysaccharide Distribution

Fluorescent-conjugated lectins can also reveal differences between the prevalent saccharide moieties expressed within the EPS of various community members (Neu and Marshall, 1991; Caldwell et al., 1992a). Wolfaardt et al. (1995) applied a panel of fluorescent lectins (specific for either a-D-man, b-D-gal(1-4)-D-glcNAc, b-D-gal(1-3)-D-galNAc, a-D-man; a-D-glc, (D-glcNAc)$_3$, a-L-fucose, sialic acid, and D-galNAc) to their diclofop-degrading biofilm consortium, demonstrating that considerable vertical and horizontal variability existed within their system (the relative abundance of these components varied between 0 and 67 percent of biofilm area at any depth). Regions in the EPS where diclofop fluorescence was highest had a high fucose content, particularly in the lower 6 μm where the diclofop:fucose ratio was nearly one. Notably, a more uniform distribution of lectin binding sites was observed in biofilms formed by the same consortium, but grown on a labile growth medium.

5.1.5. FITC-Conjugated Polyclonal Antibodies for Identification

Immunological approaches (Schmidt, 1972; Brayton et al., 1987; Hoff, 1988; Mackie et al., 1989), have broad potential for locating and differentiating between morphologically-similar consortium members present in complex interactive associations. For example, James et al. (1993) used a fluor-labeled polyclonal antibody and SCLM to quantify the colonization success of an *Acinetobacter* sp. from within a triculture biofilm. Genetic probes have similar specificity and potential in applied and ecological research. However, they can not as yet be used non-destructively as in the case of fluorescent antibodies.

5.1.6. Carboxyfluorescein to Determine Hydrogen Ion Distribution

It is apparent from the direct consequences of biofilm growth (e.g., formation of anaerobic zones, corrosion, food spoilage) that there is a need for the nondestructive definition and measurement of the chemical microenvironments associated with active bacteria. For example, *P. fluorescens* is normally rod shaped, but can be induced to form filaments when oxygen conditions become limiting (Jensen and Woolfolk, 1985; Wright et al., 1988). It is possible that filaments forming deep within *P. fluorescens* biofilms are the result of oxygen depletion (Korber et al., 1993). Microelectrode studies indicate that chemical gradients result from active microbial metabolism in biofilms and bioflocs (Lens et al., 1993; Costerton et al., 1994). However these physical devices are not yet small enough to evaluate chemical gradients associated with individual cells. Many fluorescent probes are sensitive to solution redox potential, pH, and concentration of free ions, and thus may be utilized with SCLM to define these gradients *in situ*. For example, 5 and 6 carboxy-fluorescein has previously been used to visualize biogenic pH gradients associated with 24h *Vibrio parahaemolyticus* biofilms, as well as zones of low pH associated with the corrosion of mild steel (Caldwell et al., 1992a). Fig. 9 illustrates the decrease in fluorescence associated with the activity of *Vibrio parahaemolyticus* cells as compared with the bulk phase fluorescence within a water channel.

Fluorescein-based ratiometric pH indicators, which emit fluorescence at a pH-dependent wavelength and also at a separate pH-independent wavelength, may also be used to quantify these zones (Luby-Phelps et al., 1988; Tsien, 1989; Tsien and Waggoner, 1990; Haugland, 1992). Fluorescent probes for ions other than hydrogen (calcium, magnesium, etc.) may also be used to evaluate other aspects of biofilm chemistry (Tsien, 1989; Tsien and Waggoner, 1990; Haugland, 1992). Significantly, the majority of work using environmentally sensitive

Bacterial Communities

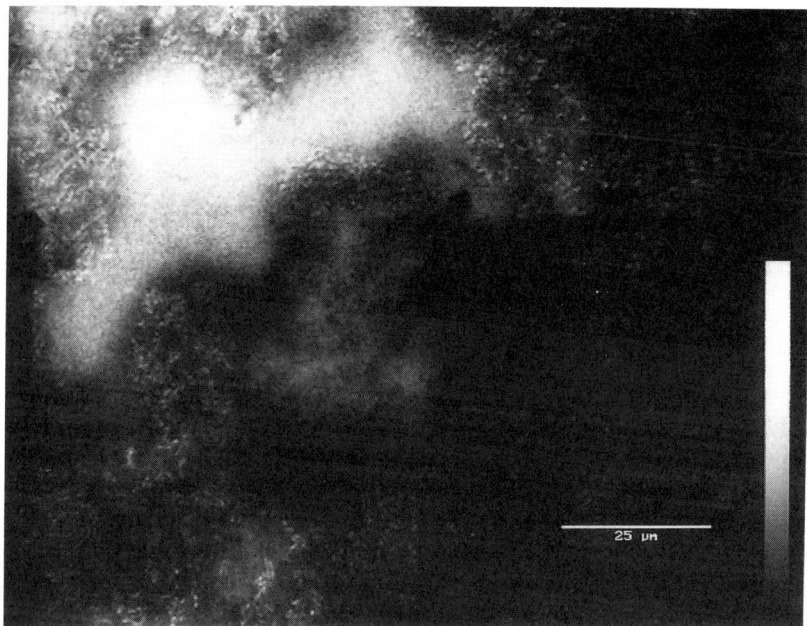

Figure 9. Carboxyfluorescein as a pH probe. SCLM optical thin section (10 μm section depth) of a *Vibrio parahaemolyticus* biofilm stained with a pH-sensitive molecular probe (5, 6-carboxyfluorescein). Note the decreased fluorescence associated with the activity of *V. parahaemolyticus* cells as compared with the bulk phase fluorescence of the fluor within an adjacent water channel. Dark regions correlate with regions of decreased pH, whereas bright regions correlate with relatively higher pH values.

probes has been conducted using eukaryotic systems, where the pH remains constant and the environment is relatively stable (Hernandez-Cruz *et al.*, 1990; Lattanzio, 1990; Saavedra-Molina *et al.*, 1990; Lattanzio and Bartschat, 1991). There is a need to further define and calibrate these probes for use in microbiological systems where metabolic activity alters the cell's ambient chemistry.

5.1.7. CTC and Resorufin to Determine Viability

The ability to identify viable cells is of major concern to industrial and medical researchers involved in the management and control of biofilm communities. Thus numerous fluorescent probes have been specifically designed (or derivatized) to provide information about the metabolic condition of individual bacteria. Fluorescent indictors of bacterial viability have been based on cytoplasmic redox potential, electron transport chain activity, enzymatic activity, cell membrane potential, membrane integrity, and various combinations of the above

(Tabor and Neihof, 1982; Marxsen, 1988; Betts *et al.*, 1989; Caldwell and Lawrence, 1989; Bottomly and Maggard, 1990; Back and Kroll, 1991; Nix and Daykin, 1992; Rodriguez *et al.*, 1992). In most cases, the primary objective has been to link one physiological facet of the cellular condition (or a combination of elements) to overall cell viability. Such probes have similar potential in evaluating the natural variability of cells from natural communities.

A number of viability probes have been based on production of internal reducing equivalents by actively metabolizing cells. For example, Rodriguez *et al.* (1992) used cyanoditolyl–tetrazolium chloride (CTC) to visualize actively respiring cells in mixed bacterial populations. Oxidized CTC is soluble and nonfluorescent, however when reduced by an active electron transport chain, it becomes an insoluble CTC-formazan crystal that accumulates intracellularly and fluoresces in the red wavelengths. Rodriguez *et al.* (1992) compared the total number of bacteria obtained from environmental samples (which fluoresced blue following application of DAPI) with those that were metabolically active (which reduced CTC and fluoresced red), and found that CTC counts were less than DAPI control counts, but equal to or greater than plate count results. Yu and McFeters (1994) subsequently used CTC for estimating cell injury resulting from chlorination of biofilms in a model pipeline system, and found that the results obtained using CTC were in general agreement with those of Rodriguez *et al.* (e.g., plate counting methodology generally overestimated the efficacy of disinfectant treatment).

Resorufin has also been used as an indicator of cytoplasmic reducing potential (Caldwell and Lawrence, 1989; Caldwell *et al.*, 1992a). Following application of resorufin to a mixture of living and killed yeast cells, actively metabolizing cells appeared dark using SCLM (cells with reducing intracellular conditions) whereas dead cells were fluorescent (indicating non-reducing intracellular conditions) (Fig. 10). However, natural assemblages of organisms demonstrate a variable permeability to resorufin, making it difficult to reliably define cell viability using this approach. Thus a second probe used in combination with CTC or other redox-dependent fluors may be required as an internal permeation control.

5.1.8. Membrane and Nucleic Acid Probes to Determine Viability

Another approach to analyzing bacterial viability is based on the need for a functional cell membrane in all viable cells (with the exception of spores and cysts). A cell membrane is required to maintain the integrity of the cell's cytoplasm and genome. Once it has been compromised, it can no longer act as a selective barrier to exclude extrinsic compounds or retain intracellular cytoplasmic molecules. Consequently, cell components (e.g., proteins, enzymes, nucleic acids) leak out of the cell, hydrolytic enzymes enter and degrade cell

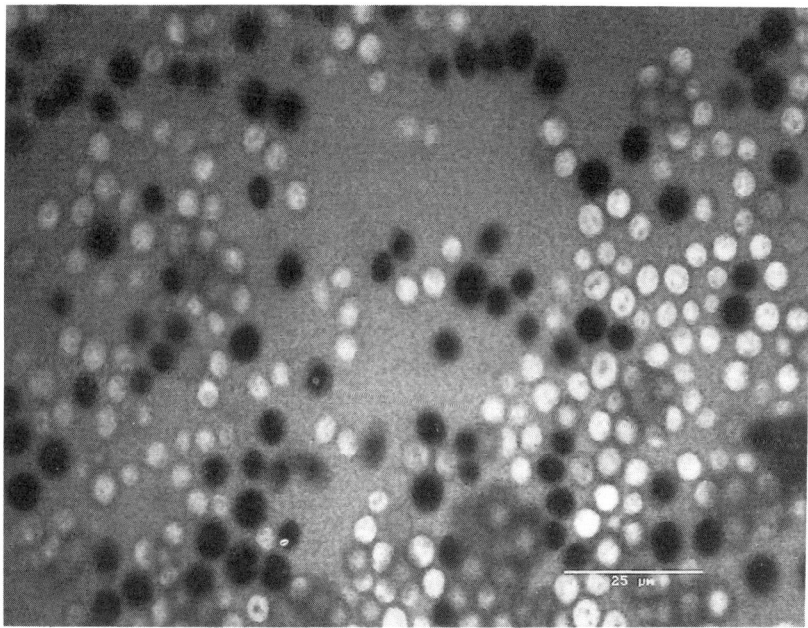

Figure 10. Resorufin as a viability probe. SCLM optical thin section of a mixture of living and killed yeast cells treated with 0.1% resorufin. Actively metabolizing cells appeared negatively stained as a result of the reduction of resazurin to a colorless, non-fluorescent form within the cytoplasm of cells deprived of oxygen. Dead cells continue to fluoresce due to the lack of sufficient biochemical oxygen demand for reduction.

components, and the transmembrane potential (proton gradient) responsible for ATP synthesis becomes nonfunctional.

Probes dependent on the integrity of the bacterial membrane include the nucleic acid stains propidium iodide (PI) and ethidium bromide (EB). PI and EB are relatively membrane impermeable, and do not cross fully functional cell membranes as readily as membranes that have been ruptured (Sauch et al., 1991; Haugland, 1992). Cells with an active or functional cell membrane thus exclude these probes and cells fluoresce weakly. Once inside the cell cytoplasm and bound to nucleic acids, these probes undergo a large Stoke's shift (to the red wavelengths) and an increase in fluorescence emission intensity. However, this method will not indicate death if the bacterium has already lost the DNA from within its rigid cell wall (peptidoglycan sacculus) through the action of degradative enzymes.

In some studies, membrane integrity probes have been used in conjunction with green-emitting fluorogenic substrates (such as fluorescein diacetate) to determine cell death in mammalian cells (Labatiuk et al., 1991; Sauch et al., 1991). In this situation, cell viability is indicated by an intact cell membrane and a complement of intracellular enzymes. Similarly, multiple-fluor preparations include the Baclite™ viability probe (offered by Molecular Probes, Inc). The composition of this mixture is proprietary; however, it results in living cells appearing green and dead cells appearing red. The green fluorescence results from a membrane permeable probe that penetrates both the membrane and cytoplasm. Red fluorescence is due to a membrane impermeable probe that does not penetrate the cell unless there is a loss of membrane integrity. However, the red probe also accumulates on the surface of living cells and produces a faint red "ghost." When used in conjunction with argon ion lasers and commercial filter sets, subtraction of green fluorescence bleed-through from the red channel (and vice versa) can be used to improve image quality. A dual-channel (bleed-through corrected) SCLM image of a biofilm community from soil stained using the Baclite™ viability probe is shown in Fig. 11. It shows green-fluorescing living cells (left) and red-fluorescing dead cells (right). It also shows that microcolonies of living cells that were embedded in exopolymer were not penetrated adequately by the green probe (which is presumably hydrophobic). However, they were penetrated by the red probe (which is presumably hydrophilic). This could result in the underestimation of direct viable counts. Cells may also lose their cytoplasmic contents following death and fail to fluoresce red, although they are thoroughly penetrated by the viability probe. Consequently, these methods should be verified via physical methods such as the physical viability assays described below or difference imaging as described by Caldwell and Germida (1985).

5.1.9. Physical Methods of Determining Viability

There are also several physical methods for determining the functionality of the bacterial membrane. One of these is based on the physical response (plasmolysis) of the cell when pulsed with an osmotic shock (Caldwell and Lawrence, 1989, Korber et al., 1996). Cells with a semi-permeable membrane plasmolyze, while those with a damaged membrane do not. This method has the advantage of being inexpensive, simple, and amenable to a range of microscopic techniques (darkfield, phase, SCLM). It can also be used in combination with a variety of fluorescent probes. Fig. 12 illustrates the difference between living and dead *Salmonella enteritidis* cells subjected to 1.5 M NaCl.

As mentioned above in the discussion of fluorescence exclusion (section 5.1.1), fluorescein provides another physical method for determining the functionality of the cell membrane and protoplast. Cells with a functional membrane

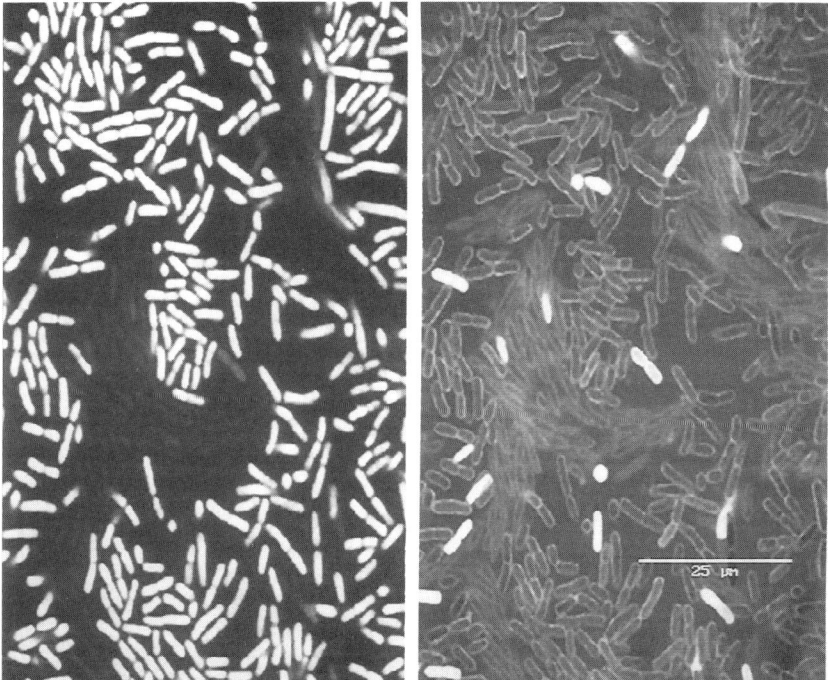

Figure 11. The Baclite™ viability probe. Dual-channel (red-green) SCLM optical thin section of a 24 h biofilm containing natural soil flora cultivated on trypticase soy broth and stained with a commercial viability probe (Baclite™ viability probe; Molecular Probes, Inc.). The left channel shows living cells, which emit green fluorescence, whereas the right channel (corrected for green fluorescence bleed-through) shows dead cells, which emit a strong red signal. Note that in the red channel the fluorescent indicator can be seen to accumulate at the surface of the living cells (ghosts), resulting in a ring of low-intensity fluorescence. Also note that some cells, which appear to be actively growing in microcolonies, emit much less green fluorescence than surrounding bacteria and consequently appear to be neither living nor dead. These cells, which have "fuzzy" cell boundaries when viewed in the red channel, grew embedded in exopolymers that prevented adequate penetration of the green fluor, possibly resulting in underestimates of direct viable counts.

maintain their intracellular pH. Thus, by dropping the pH in the ambient environment, viable cells can be distinguished from non-viable cells based on their internal pH, which remains at approximately 7.0. Consequently, flooding a sample with 0.01% fluorescein at pH 6.0 (or less) normally results in a dark background (the fluorescence of fluorescein is inhibited at low pH) and fluorescent cells (a small amount of fluorescein diffuses through the membrane and fluoresces).

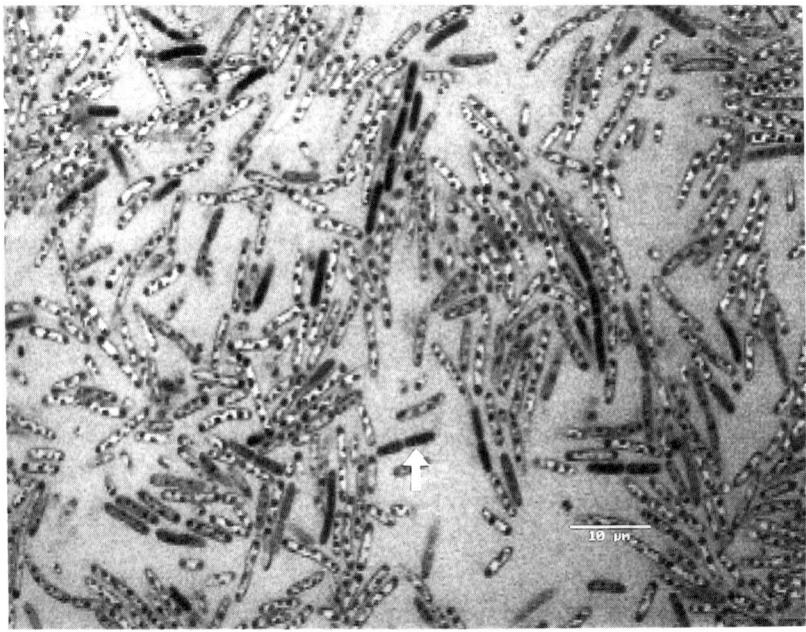

Figure 12. Plasmolysis as a measure of viability. SCLM optical thin section of a negatively-stained *Salmonella enteritidis* biofilm exposed to mild heat treatment. Following heating, a pulse of 1.5 M NaCl was applied. Living cells plasmolyzed following salt stress, indicating that these cells possessed an intact cell membrane and protoplast. Note that plasmolyzed cells concentrated fluorescein, resulting in increased fluorescence between the cytoplasmic membrane and the rigid but highly permeable cell wall. Dead cells were not responsive to salt stress, and appeared as non-plasmolyzed, negatively-stained cells (see arrow). (Photograph provided by Anita Choi.)

5.1.10. Other Fluorescent Molecular Probes

In addition to the probes listed above, other compounds also have potential for microbiological studies using SCLM, including fluorogenic enzymatic substrates (fluorescein- and carboxy fluorescein-diacetate, calcein AM), probes sensitive to DNA damage (Hoechst 33342), fluorescent gram stains (Baclite™ gram stain), and membrane-potential-sensitive probes (rhodamine 123, RH-795, and carbocyanine derivatives). In specific cases involving antimicrobial agents or biocides, it is useful to use fluorescent probes that reflect the type of inactivation that is likely to occur. For example, Korber *et al.* (1994a) applied fleroxacin (a fluoroquinolone DNA-gyrase inhibitor) to *P. fluorescens* biofilms that had been cultivated for 24 h. Fleroxacin inhibits the supercoiling of bacterial DNA, thereby altering nucleic acid conformation of affected cells and inhibiting cell replication (affected cells become elongated or pleomorphic). By applying acridine orange (AO), a nucleic acid probe whose wavelength of fluorescence (either green or red) is dependent on the conformation of DNA (Daley, 1979), fleroxacin-affected cells were identified using SCLM and dual-wavelength fluorescence detection. Use of AO was more appropriate than use of an indicator of redox potential, as cells not affected by fleroxacin initially sustained high metabolic activity. The conformation-specific nature of AO has also facilitated the determination of cellular growth rates from attached and planktonic populations of *Pseudomonas putida* strain KT2442 (Møller *et al.*, 1995).

5.2. Genetic Analyses

5.2.1. 16s rRNA

Ribosomal RNA probes and sequencing, PCR gene amplification, and hybrids of these methods can be applied to microbial communities. These techniques have already been used to index the genetic biodiversity that exists in natural microbial populations, to determine phylogenetic relationships, and to identify indicator strains from complex systems (Stahl *et al.*, 1984; Olsen *et al.*, 1986; Pace *et al.*, 1986; Woese, 1987; Amann *et al.*, 1990, 1992; Schmidt *et al.*, 1991; Hahn *et al.*, 1992; Muyzer *et al.*, 1993). Of these new methods, the development of phylogeny-specific molecular probes represents one of the most significant recent advances in determinative microbial ecology (Stahl *et al.*, 1984, 1988; DeLong *et al.*, 1989; Amann *et al.*, 1990, 1992). Based on the relatedness of slowly-evolving ribosomal RNA sequences (Woese, 1987), complimentary oligonucleotides to these conserved RNA regions permit phylogenetic analysis of organisms obtained from natural systems without the need for cell cultivation. Thus, the identity of community members or indicator organisms that are hard to culture and of unknown ecological significance may now be ascertained, in many cases, to the subspecies level (Stahl *et al.*, 1988; Amann *et al.*, 1990, 1992; Ward *et al.*, 1992).

Traditional approaches for examining the biodiversity of microbial communities involves culturing the inhabitants. Due to strong environmental stresses imposed during most enrichment and culture procedures, the diversity index for most *in situ* environments has been severely underestimated (Ward *et al.*, 1992; Wagner *et al.*, 1993). For example, Ward *et al.* (1990) determined that of 15 unique 16s rRNA sequences elucidated from a well-studied thermal microbial mat, none matched the rRNA sequences of organisms cultivated during previous studies, suggesting that a large number of unculturable organisms were present.

Methods for using phylogenetic probes have undergone steady refinement in terms of sensitivity towards target organisms, decreased background interference, and procedural refinements for *in situ* application in more complex systems. In a study examining antibiotic-mediated fluctuations in the microflora of the bovine rumen, Stahl et al. (1988) demonstrated that cultural enumeration techniques were less sensitive to changes in the relative abundance of organisms than were fluorescent rRNA probe conjugates. Other studies using fluor-conjugated phylogenetic probes have demonstrated single-mismatch specificity between target and nontarget sequences in complex intestinal systems (Amann et al., 1990). These improvements have also proven valuable for examining the composition of sludge-degrading and soil communities (Hahn et al., 1992; Wagner et al., 1993).

5.2.2. DGGE

Denaturing gradient gel electrophoresis (DGGE) analysis is another molecular approach applicable to community-level characterization of microbial populations (Muyzer et al., 1993). Based on the electrophoretic separation of PCR-amplified 16s rRNA gene fragments of the same length (PCR-amplified rDNA), DGGE may be used to profile microbial nucleic acids obtained from a range of environments both qualitatively and semiquantitatively. Upon examining *in situ* microbial communities from microbial mats and aerobic biofilms, Muyzer et al. (1993) demonstrated that between 5 to 10 different bands could be detected for each population, with some bands shared between different populations. The sensitivity of this approach was determined using mixed cultures of known composition. These studies showed that individuals accounting for less than 1% of the total community (*Desulfovibio* spp.) can be reliably detected. This approach was also used in combination with hybridization techniques to demonstrate the occurrence of sulfate-reducing bacteria (SRB) within a biofilm cultivated aerobically, demonstrating that anaerobic microniches formed as a result of metabolic oxygen consumption create suitable habitats for anaerobes.

5.2.3. RSGP

Reverse sample genome probing (RSGP) may also be used to determine the presence of indicator strains within environmental samples or microbial communities (Voordouw et al., 1991, 1992, 1993). This is done by denaturing bacterial DNA obtained from target organisms (the standards) and spotting the DNA onto a master filter. Environmental DNA is then labeled and hybridized with the filter to identify which of the bacterial genomes present on the filter is present in the environmental sample. The standards used during this procedure are available commercially or can be enriched and isolated from *in situ* environments. Voordouw et al. (1992) examined 56 Alberta oil field sites and demonstrated the potential for identifying constituent SRB using 35 different standards of SRBs that exhibited little or no cross-reactivity. Examination of these sites using genome probes demonstrated that salinity defined the SRB community, with two distinct SRB populations being evident. From the 35 SRB bacterial genomes utilized as standards, 10 were unique to fresh water, 18 to saline water, and 6 to both. In addition to providing qualitative data, this method may also be used quantitatively (provided that the cultivation step is not performed) by using scintillation counting to determine the amount of bound radioactivity for each specific standard genome. Voordouw et al. (1993) demonstrated the efficacy of this approach using samples obtained from oil field production sites from Alberta (water samples from water plants) and determined that sulfate-reducing bacteria were significantly more numerous in biofilms which developed on metal surfaces. While this technique does not normally establish the importance of community members that had not previously been isolated, it does permit the diagnostic detection of indicator or target organisms.

5.3. Fatty Acid Methyl Ester Profiles

Fatty acid analyses are useful as "fingerprints" of *in situ* communities from soils, groundwater, bioreactors, etc. (Nichols et al., 1987; Geesey and White, 1990). Phospholipid fatty acids (PLFA) represent a relatively stable percentage of bacterial biomass and the natural variations in phospholipids among microbial taxa are sufficient to allow the development of signature fatty acid profiles for several important groups. Phelps et al. (1991) used phospholipid analyses to determine microbial biomass in biofilm communities colonizing natural gas transmission pipelines, estimating microbial populations of $10^4 - 10^7$ cells per gram of separator and sludge catcher contents. They were thus able to make these determinations without culturing. Fatty acid profiles have also been used to determine the effect of organic contamination on the structure of native soil communities (Smith et al., 1986; White, 1986). McKinley et al. (1988) analyzed PFLPs associated with oil-field water injection wells, and determined that the proportion of fatty acid 18:1w7c was high in all samples, a feature correlated with the presence of anaerobes and the induction of the desaturase pathway in bacteria. They also found increasing proportions of cyclopropanes and increasing ratios of *trans-* to *cis-*monoenoic fatty acids that are characteristic of environmental stress (e.g., oxygen limitation, high temperatures, decreased pH) and starvation (Malmcrona-Friberg et al., 1986; Guckert et al., 1986).

While lab cultures cultivated under various conditions of temperature and medium composition have been shown to exhibit variable PLFA profiles, other organisms, including pathogens and those common to corrosive or degradative environments, have been reliably determined (Geesey and White, 1990). The validation and refinement of PLFA techniques is of considerable interest, thus, a number of studies have examined the reproducibility of these techniques. Recently, Haack et al. (1994) used a commercial gas chromatography-based fatty acid analysis technique, known as MIDI-FAME™ (microbial ID, Inc.–fatty acid methyl ester), to obtain fatty acid profiles for microbial isolates or mixed cultures. In an effort to evaluate the accuracy and reproducibility of this system, the authors examined two different mixed cultures in which the profiles and abundance of 10 common soil bacteria were known and compared to profiles obtained under various environmental conditions (temperature, growth medium, growth phase). The genus-specific fatty acid profiles were generally conserved.

5.4. Substrate Utilization Profiles

Substrate utilization profiles can be applied at the organismal or community level. Application at the organismal level requires that the members of a community be isolated and then tested for their ability to use a wide range of substrates. Application at the community level requires that the intact community be tested for its ability to use the same set of substrates. An expanded range of substrate utilization for the community as opposed to isolates indicates synergistic interactions. Traditional approaches for determining substrate utilization have relied upon pH indicators to detect substrate fermentations. However, this approach has been expanded to the utilization of substrates that are not necessarily fermented as well as to those which may not yield appreciable amounts of acid if they are fermented.

The BIOLOG™ system colorimetrically detects the utilization of various carbon substrates in 95-well microtitre plates via the reduction of tetrazolium violet to an insoluble, purple formazan crystal (Bochner, 1989). By comparing the metabolic fingerprint of isolates with those contained within an extensive computer database, medical and environmental isolates can be identified. A study by Klingler *et al.* (1992) examined the accuracy of the BIOLOG system and found that the system successfully identified (to the genus level) 98% of 39 ATCC reference strains, 76% of which were identified to the species level within 4–24 hrs. The BIOLOG system database and software have subsequently been updated to provide appropriate tests for identifying additional organisms, including those of environmental and agricultural significance. Garland and Mills (1991) compared the carbon-utilization patterns of microbial communities obtained from aquatic, soil, and rhizosphere environments. They determined that community-dependent patterns of carbon source utilization were identifiable. Spatial gradients in metabolic capabilities were found for soil and estuarine sites, suggesting that functional communities or assemblages of organisms remained distinct and operational within a larger system. Wolfaardt *et al.* (1994a) determined that when a diclofop methyl degrading community was cultivated using trypticase soy broth (TSB) rather than diclofop methyl, the community lost its ability to metabolize 43% of the carbon sources tested (most significantly, aromatic compounds such as p-hydroxyphenylacetic acid, inosine, and putrecine). A restoration of these metabolic capabilities was noted when biofilm communities cultivated using TSB were compared with biofilms cultivated using 4-(2,4-dichlorophenoxy) phenol, 2,4-dichlorophenol, and 1,3-dichlorobenzene as the sole substrates.

5.5. Other Technologies

FTIR, NMR, and X-ray microanalyses have been used to describe the chemical features within microbial biofilms and microcolonies. They have also been used to investigate the role of microbial communities as the causative agents in processes such as biocorrosion (Geesey and White, 1990; Bremer and Geesey, 1991). Microelectrodes have been useful in studying microbial communities (Alldredge and Cohen, 1987; Revsbech, 1989; Little et al., 1991; Lens et al., 1993; de Beer et al, 1994) and have been reviewed elsewhere (Revsbech and Jorgenson, 1988; Geesey and White, 1990).

6. Selected Examples of Community-Level Bacterial Associations

6.1. Degradative Microbial Consortia

For many environmental contaminants, the activity of more than one organism is required for complete mineralization (Federle and Pastwa, 1988; Federle and Schwab, 1989). Degradative networks, as opposed to isolated cell lines, may contribute to: 1) a larger, more flexible genetic pool from which new metabolic pathways and strategies for degradation and reproductive success may evolve, 2) chemical communication, interaction, and co-metabolism between different microbial species, 3) formation of permissive microniches for growth of stringent community members, and 4) protection within the biofilm-EPS matrix against extrinsic perturbations and maintenance of stable chemical and physical conditions (Fletcher, 1984; Bouwer, 1989; Lawrence et al., 1995).

Several community-level studies have focused on the breakdown of xenobiotic or recalcitrant pollutants (Slater, 1988; Tagger et al., 1990; Jiménez et al., 1991; Lang et al., 1992; Rozgaj and Glancer-Soljan, 1992). Most of these reveal a degradative hierarchy in which the primary users of a compound support other organisms by producing secondary substrates (Lappin et al., 1985; Tagger et al., 1990). Direct analysis of degradative communities using SCLM and molecular probes have provided insight into the spatial organization that characterizes these systems. Wolfaardt et al. (1993, 1994a–d, 1995) evaluated the importance of physical and chemical interactions between community members. They also demonstrated that from within a seemingly random mix of microorganisms, a spatial and temporal order arose in a highly reproducible way through the development of specific consortia.

Using SCLM and a set of fluorescent molecular probes, Wolfaardt et al.

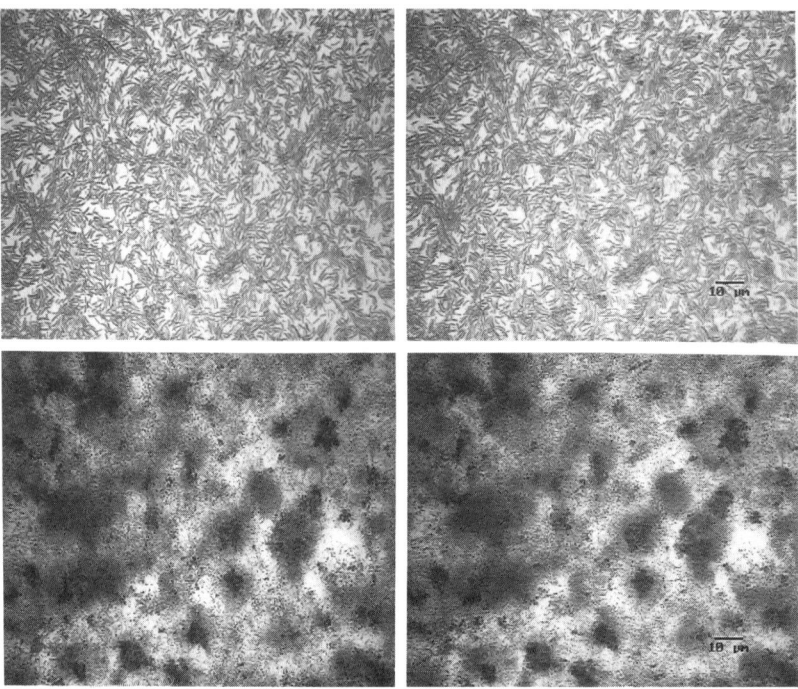

Figure 13. Three-dimensional stereo pairs (view with stereo glasses) of nature biofilms grown on 300 μg ml⁻¹ trypticase soy broth (TSB) (top pair) and 14 μg ml⁻¹ diclofop (bottom pair) showing the difference in biofilm architecture and reorganization of TSB-grown communities when cultivated using media containing diclofop (herbicide) as the sole source of carbon. Each stereo pair consists of a pair of through-view images projected side by side. Three-dimensional images were obtained by offsetting successive focal planes of the images on the right side of each pair by 1 μm. (Reprinted with permission, Wolfaardt et al. (1994.)

(1994a) demonstrated that the nature of the carbon source (e.g., diclofop methyl or an equivalent amount of a more labile carbon source) provided to organisms in continuous-flow slide culture affected the architecture of the resultant biofilm (Fig. 13). When diclofop methyl was the sole carbon source, the average thickness of the biofilm increased fourfold over biofilms cultivated using trypticase soy broth (23.6 μm vs. 4.4 μm). Consortia, including conical bacterial assemblages and grapelike clusters, were observed when the growth substrate was either diclofop methyl or other chlorinated ring compounds. These associations did not develop following growth on the labile carbon source. The architecture of the biofilms could also be controlled by switching between substrates. For example, a biofilm initially grown on 300 μg ml⁻¹ TSB for 21 days regained the characteristics of biofilms grown solely on the aromatic herbicide 2 days after switching to 14 μg ml⁻¹ diclofop methyl (Fig. 14).

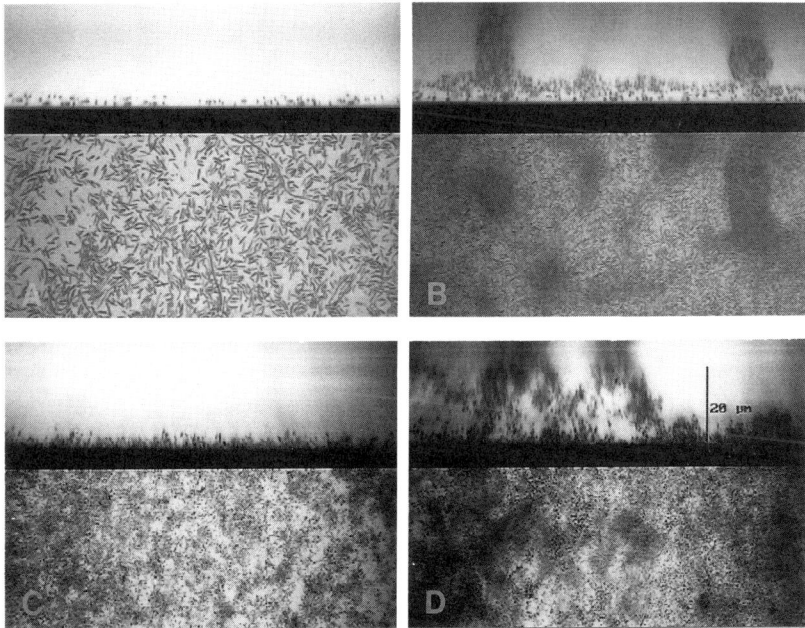

Figure 14. Laser micrographs (SCLM optical thin sections) of negatively stained biofilms illustrating the response of biofilms grown on TSB when switched to diclofop. Each image shows an *xy* through-view (demonstrating horizontal cellular distribution) with the corresponding *xz* image (showing vertical cellular distribution and biofilm thickness) directly above it. Biofilm (A) grown for 21 d on 300 μg ml^{-1} TSB, (B) grown for 21 d on 300 μg ml^{-1} TSB and then replaced with diclofop; 2 d after switch, (C) grown for 21 d on 18 μg ml^{-1} TSB, (D) grown for 21 days on 18 μg ml^{-1} TSB and then replaced with diclofop; 2 d after switch. (reprinted with permission, Wolfaardt et al. (1994.)

Wolfaardt *et al.* (1994c) also followed the fate (the accumulation and disappearance) of diclofop methyl within degradative biofilm communities using SCLM and diclofop fluorescence (Ex$_{max}$ of 420 nm and Em$_{max}$ of 504 nm). Fig. 15 shows negatively-stained (extracellular fluorescein solution) and positively-stained (using diclofop auto-fluorescence) SCLM optical thin sections of a 21 day biofilm accumulating diclofop methyl. Not all biofilm members bound diclofop (diclofop fluorescence was detected using SCLM) *in situ*. This was consistent with sole carbon source utilization data obtained using consortium isolates (Wolfaardt *et al.*, 1994a). After 14–21 days, specific consortium members accumulated disproportionate amounts of fluorescence (primarily on capsular EPS), whereas other immediately adjacent cells did not. EPS that bound diclofop methyl was relatively fucose-rich (Wolfaardt *et al.*, 1995) as indicated using a fucose-sensitive FITC-conjugated lectin. The binding of other lectins was spatially heterogeneous as well (Fig. 16).

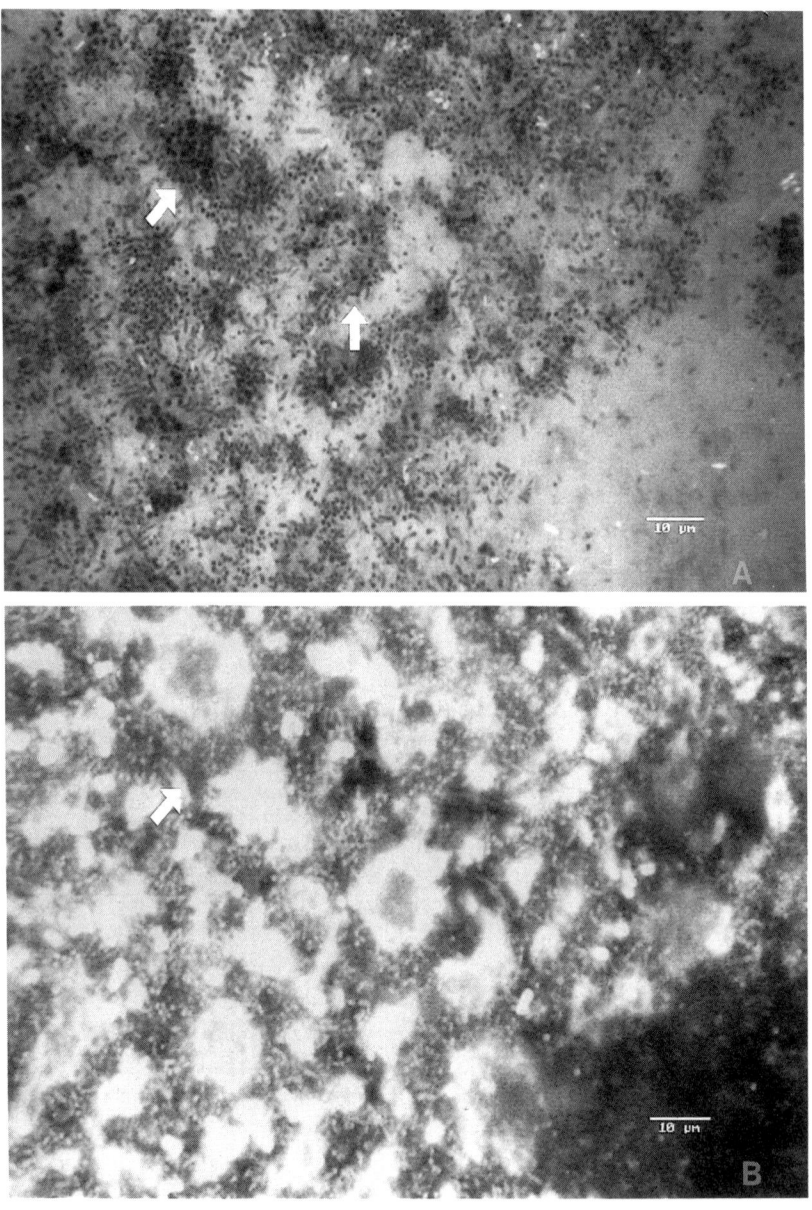

Figure 15. (A) An optical thin section, digitized at a depth of 3 μm from the base of a 21-d old, negatively stained biofilm showing cell aggregates (arrows). (B) An image of the same field showing accumulation of diclofop (bright areas of autofluorescence) within bacteria exopolymers encasing cells and cell aggregates (arrow). Laser intensity was set at a threshold value at which background fluorescence was negligible. (Reprinted with permission, Wolfaardt et al. (1994.)

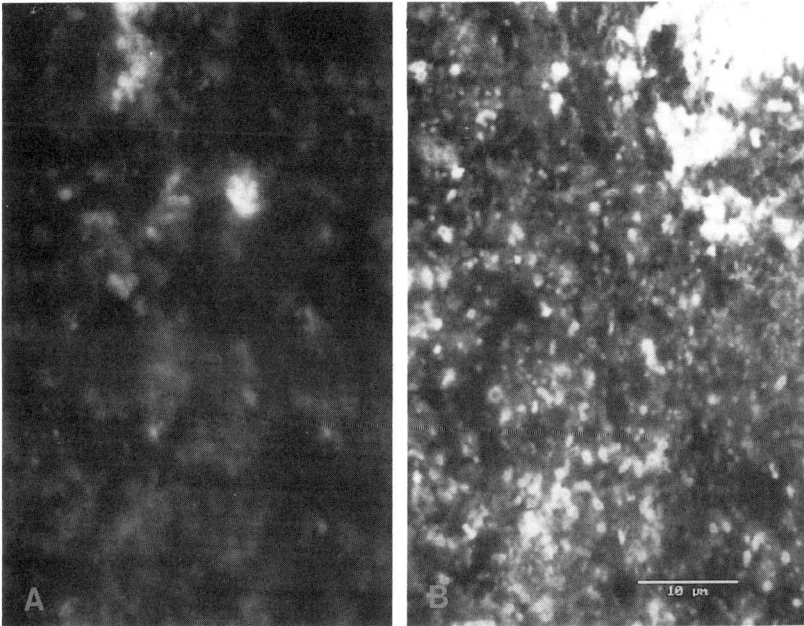

Figure 16. A confocal laser micrograph (optical thin section) obtained by utilizing dual channel imaging to illustrate binding and horizontal distribution in the same field, 1 μm from the glass surface of (A) TRITC-labeled *lycopersicon* (binds [D-glcNAc]$_3$) (red channel) and (B) FITC-labeled *concanavalin* A (binds a-D-mannose) (green channel).

The appearance and disappearance of diclofop, as well as its breakdown products, was quantified over time (Wolfaardt *et al.*, 1994c) based on diclofop fluorescence and the measurement of radiolabelled tracers. Figure 17 shows the accumulation of diclofop methyl, as determined by increased fluorescence, over a 21-day period. After 21 days, fluorescence intensity did not increase significantly following extended incubation in the presence of diclofop. Thus all binding sites present on specific, EPS-producing consortium members (not all biofilm members accumulated diclofop) may have became saturated. Extended incubation in the absence of diclofop methyl resulted in a gradual decrease in fluorescence to background levels. Adding more diclofop methyl once again resulted in an increase of fluorescence to maximum levels (Fig. 15). Studies using ^{14}C-labeled diclofop methyl previously demonstrated complete mineralization using the same consortium (Wolfaardt *et al.*, 1994b), therefore it is possible that bound aromatic compounds are partly degraded outside the cell by extracellular enzymes before being taken up by the cells for complete mineralization. In potassium-cyanide-inhibited biofilms, the intensity of fluorescence did not

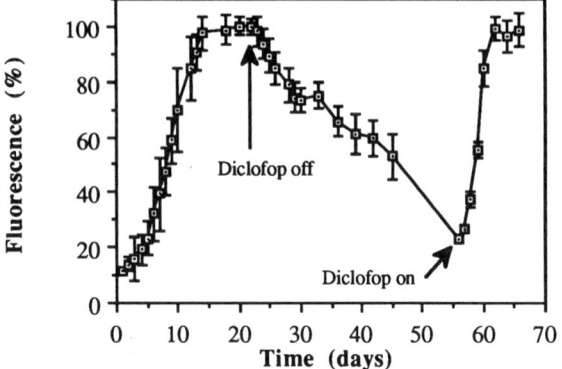

Figure 17. Accumulation of diclofop in the biofilm matrix resulted in an increase in fluorescence intensity with time. The gradual decrease in fluorescence after the diclofop supply was turned off suggests either desorption of bound diclofop or mineralization by the degradative consortium.

decrease significantly over time after diclofop methyl was omitted from the irrigation solution (Fig. 18), suggesting that microbial metabolism was responsible for the decrease in the amount of sorbed diclofop (Wolfaardt et al., 1994d). Only two of nine consortium isolates metabolized diclofop methyl in the presence of a labile carbon source. Thus the cells that bound diclofop methyl were required for the success of other consortium members.

6.2. Anaerobic Digestor Granules

Another important microbial community is the association of organisms that form bioactive granules in upflow anaerobic sludge bed (UASB) bioreactors. Central to the complete oxidation of dissolved organics to CO_2 and CH_4 are methanogenic bacteria that carry out the terminal steps in the process. A single bacterium, *Methanobacillus omelianskii*, was once thought to be responsible for the conversion of carbohydrates to methane. However, a range of compounds (e.g., ethanol, propionate and longer-chained fatty acids) are syntrophically catabolized by two bacteria, as first proposed by Bryant et al. (1967) following the observation that *M. omelianskii* consisted of a bacterium that produced acetate and H_2 from ethanol only when a H_2-consuming methanogen was present as a "contaminant" in the culture.

Anaerobic sludge communities include a spectrum of hydrolytic, fermentative, and methanogenic organisms. For example, formation of functional granules in whey-degrading reactors requires exopolysaccharide-producing organisms for initial sites of granule nucleation (Chartrain and Zeikus, 1986a,b). Enumerating the respective populations associated with lactose biomethanation,

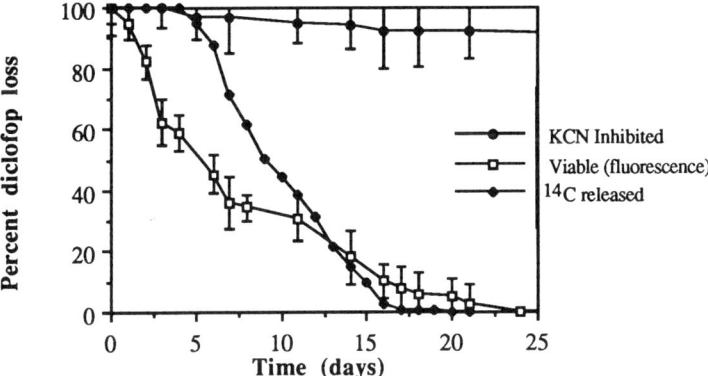

Figure 18. Degradation of sorbed diclofop. Biofilms were cultivated with diclofop as the sole carbon source for a minimum of 21 d to allow sorption of diclofop by EPS. The biofilms were subsequently irrigated with a minimal salts solution without a carbon source. Fluorescence was measured to determine the fate of the sorbed diclofop during this deprivation of carbon. A decrease in fluorescence in viable, but not in cyanide-inhibited biofilms, suggesting that microbial metabolism was responsible for a decrease in the amount of sorbed diclofop. Growth of the biofilm community on ^{14}C-labeled diclofop methyl followed by measurement of $^{14}CO_2$ evolved during irrigation with the carbon-free minimal salts solution confirmed mineralization of the sorbed diclofop.

Chartrain and Zeikus (1986b) determined (using MPN per ml) that the numbers of the three dominant bacterial trophic groups present varied with substrate composition: hydrolytic, 10^{10}; acetogenic, 10^7–10^{10}; and methanogenic, 10^6–10^9. Lactose-utilizing organisms identified included *Leuconostoc mesenteroides* and *Klebsiella oxytoca* (EPS-producers), whereas *Clostridium butyricum, Clostridium propionicum* and *Desulfovibrio vulgaris* were the main proton-reducing, acetogenic organisms identified. *Methanosarcina barkeri* and *Methanothrix soehngenii* were the most numerous acetate-utilizing methanogens, however, their MPN counts were low ($\sim 4 \times 10^6$) considering that acetate has been shown to be the primary precursor for methane production in sludge digestor systems (Jones *et al.*, 1984). *Methanobacterium formicicum* was the most common formate/hydrogen-utilizing methanogen in their system. Chartrain and Zeikus suggested that the difficulty in disrupting sludge granules may have caused underestimation of these community members. The enumeration techniques used by Chartrain and Zeikus (1986b) were based on the disruption of groups of organisms that likely required metabolic cooperation (interspecies hydrogen/formate transfer; IHT, IFT) for their growth. MPN enumeration would underestimate hydrogenotrophic methanogens and hydrogen-donating strains that have developed growth-interdependencies over evolutionary time, as well as other organisms with stringent or unknown growth requirements (Stams *et al.*, 1989). This underscores the importance of developing alternate methods for evaluating population architecture using non-destructive *in situ* methods. One of the most distinctive features of sludge-degrading communities is the fundamental role syntrophy plays in methanogen nutrition and thermodynamic equilibrium (Thiele *et al.*, 1988; Boone *et al.*, 1989; MacLeod *et al.*, 1990). Growth and fatty acid oxidation by acetogenic organisms are repressed by low concentrations of H_2. Failure to remove excess hydrogen represses the hydrogenase enzyme system, thus leaving reduced components (e.g., reduced ferridoxin, NADH) in a nonfunctional form (Ahring and Westerman, 1987). High partial pressures of H_2 (0.8 atm) completely inhibited growth of *Syntrophomonas wolfii* when grown in coculture with *Methanobacterium hungatei* (McInerney *et al.*, 1979). Removal of H_2 from *S. wolfei* pure cultures was not sufficient to allow *S. wolfii* to resume growth on butyrate; therefore, it appeared that the coexistence of a hydrogen-consuming organism was somehow obligatory. The low amounts of energy available from methane production (free energy change, $\Delta G_o'$) indicate that a balance between hydrogen (and/or formate) production and consumption must be maintained for methanogenesis to remain thermodynamically favorable (Boone *et al.*, 1989).

Mass balance data predict that IHT within granules would eliminate the need for hydrogen to equilibrate with the bulk phase pool, explaining methane production in excess of theoretical amounts based on bulk phase hydrogen concentrations (Conrad *et al.*, 1985; Thiele *et al.*, 1988). Each granule is composed of a complete community of organisms that carry out all the reactions required for hydrocarbon degradation (Stams *et al.*, 1989), thus chemical transfer of necessary equivalents (e.g., IHT or IFT) does not have to occur between two distinctly different granule communities. Ultrastructural studies have defined granule architecture in detail (Bochem *et al.*, 1982; Kinner *et al.*, 1983; Robinson *et al.*, 1984; Stams *et al.*, 1989; MacLeod *et al.*, 1990). Bochem *et al.* (1982) used electron microscopy to observe the morphology of sludge granule surfaces as well as the fine structure of thin sections. Bochem reported that granules consisted of three distinct layers: an outer layer which bore resemblance to a pseudoparenchyma, consisting of "macrocysts" and coccoid cells; a middle layer consisting of a built up layer of loosely-packed ovoid cells; and an inner layer where cavities containing rod-shaped organisms predominated. Methanogenic bacteria (resembling *Methanosarcina* and *Methanococcus* strains in ultrastructure) resided in both the outer pseudoparenchymal and middle granule layers (based on F_{420} U.V. irradiation). Although methane flowed through granule cavitations, the rod-shaped organisms that resided within cavities were not methanogens.

MacLeod *et al.* (1990) similarly identified a concentric arrangement of distinct bacterial layers in granules isolated from a sludge bed reactor. The outer layers of granules were covered by coccoid chains, long thin filaments, rods, and cocci, and included organisms resembling *Methanococcales* and *Methanothrix* spp. Freeze-fracture/EM examination of internal surfaces of sludge granules revealed bacterium-encased gas cavities surrounded by several distinct layers of bacteria, with cavities separating outer from inner layers. The middle layer was tightly packed, approximately 20 μm in thickness, and primarily composed of rods embedded in EPS. The third and innermost layer consisted almost exclusively of the angular-shaped rods characteristic of the Archaebacterial genus *Methanothrix* (viz. Patel, 1984). Based on these observations, MacLeod *et al.* (1990) formulated the following structural/functional model for sludge granule development. *Methanothrix* spp. function as condensation nuclei for syntrophobacteria, forming a core essential for formation of the granule. Hydrogen produced by the acetogenic bacteria is consumed by methanogens that then form methane and produce gas cavities. Associations of hydrolytic, fermentative bacteria form the outer layers of the granule, insulating the community against chemical perturbations (periodic increases in redox potential) and providing the acetogens with fatty acids.

Indirect support for the layered structure of the granule has also been obtained using microelectrode pH probes. Lens *et al.* (1993) obtained chemical profiles that suggest the activity of acetogenic bacteria in outer granule layers but not near the center of the granule where methanogenesis occurs. Direct microscopic verification for the positioning of methanogens opposite syntrophic proton-reducing bacteria from within granule systems has also been obtained by Stams *et al.* (1989). Using a propionate-fed sludge model reactor and immunospecific labeling, Stams *et al.* demonstrated that syntrophobacter spp.-type cells were commonly encircled (<0.5 μm distance) by *Methanobrevibacter* sp.

The role of community composition in the rate and pathways of granule formation remains largely speculative. However, it is clear that community structure is modulated by a range of physical and chemical factors including temperature, pH, nutrient concentration, and flow (Conrad *et al.*, 1985; ten Brummeler *et al.*, 1985; Chartrain and Zeikus, 1986a,b; Chartrain *et al.*, 1987; Goodwin *et al.*, 1988). *In situ*, community-based analytical techniques (described in the previous sections) would likely facilitate study of these complex systems. For example, 16S rRNA fluorescent oligonucleotide probes used in conjunction with SCLM may permit phylogenetic and spatial determinations of the principle community participants, whereas fluorescent probes specific for activity may provide indications of metabolic potential.

7. Criteria Used in the Isolation and Study of Communities

Communities are not isolated cell lines or mixtures of unrelated organisms. Consequently the criteria for obtaining and characterizing them differ from those used to obtain pure cultures, mixed cultures, and enrichments. Community cultures are defined by defining the environment of the culture as opposed to using aseptic technique. Each community is a network consisting of several members adapted to proliferate both as individuals and as part of one or more community associations. Community associations form through specific behavioral adaptations that allow physical, chemical, and biological interactions as well as correct spatially positioning (Lawrence *et al.*, 1995). These associations include consortia, biofilm communities, bioaggregates, microecosystems, and other networks of interacting microorganisms growing in association with one another. To confirm that a community culture has been obtained, it is necessary to demonstrate that the presumptive community meets criteria of communality, autopoiesis (self-organization), synergy, and homeostasis. These criteria have been discussed in more detail elsewhere (Caldwell *et al.*, 1997b) and are only briefly mentioned below.

The most important step in confirming the successful cultivation of a discrete community is meeting the criterion of communality. This requires that the community be cultivated along a spatial or temporal environmental gradient(s) to determine whether ecotones occur in response to environmental variation. If the organisms maintain themselves collectively through a network of interrelationships that are functional at the community level, then as the environment changes along the direction of the gradient there must be a transition from one set of relationships to another. This transition requires a coordinated response involving the links by which the members of the community interact. This normally includes chemoreceptors, exopolymers, metabolic exchange, intercellular communication and behavioral adaptations. Due to the difficulties coordinated response of a networked community, organisms appear and disappear as "sets" rather than as individuals, in response to environmental change. The presence of ecotones along gradients is thus indicative of a coordinated reorganization from one set of interrelationships to another. If there are no significant communal interrelationships then there should be a more gradual transition and the transition for each organism should be independent of the others.

Meeting the criterion of autopoiesis requires that the community be self organizing. This implies that the structure and architecture of the community arise spontaneously under septic conditions. The culture should also be stable, even when challenged by pure or mixed culture amendments. The criterion of synergy requires that the presumptive community proliferate and convert abiotic resources to biotic resources more effectively than its component members when cultivated individually. Synergy is indicated by a broader habitat range, more favorable growth constants (specific growth rate, half-saturation constant, cell yield), resistance to environmental stress, resilience following environmental stress, etc. The performance of the association should be better than that of isolated members of the association cultivated individually, and the performance of the community association should be at least partially regained by combining the isolates in mixed culture. Meeting the criterion of homeostasis requires that the community create a stable and favorable microenvironment within an unfavorable macroenvironment, and that the microenvironment be protected from periodically unfavorable conditions in the macroenvironment. Not all communities may be homeostatic and homeostasis should be considered as only one of many mechanisms of synergy.

8. Conclusions

In the absence of contrary information, it would be logical for human beings to assume that it is individuals like themselves (rather than genes, communities, or ecosystems) that are the primary mechanisms and endpoints of evolution. This bias must be set aside if human understanding of ecology and evolution is to move forward. As stated by Emerson (1841) more than a century ago:

> The method of nature: who could ever analyze it? The simultaneous life throughout the whole body, the equal serving of innumerable ends without the least emphasis or preference to any, but the steady degradation of each to the success of all, allows the understanding no place to work. Nature can only be conceived as existing to a universal and not to a particular end, to a universe of ends, and not to one . . . Nature knows neither palm nor oak, but only vegetable life, which sprouts into forest, and festoons the globe with a garland of grasses and vines. That no single end may be selected, and nature judged thereby, appears from this, that if man himself be considered as the end, and it be assumed that the final cause of the world is to make holy or wise or beautiful men, we see that it has not succeeded . . . The universal does not attract us until housed in an individual . . . A man's wisdom is to know that all ends are momentary, that the best end must be superseded by a better. But there is a mischievous tendency in him to transfer this thought from the life to the ends, to quit his agency and rest in his acts; the tools run away with the workman . . .

Making progress in the study of communities requires that life be viewed from this less biased perspective. Just as early astronomers mistakenly put the Earth at the center of the universe with the sun revolving around it, Darwin was mistaken when he placed the selection of individual races at the center of evolution and dismissed the occurrence of communities and ecosystems as indirect byproducts of competition. Natural selection must be reexamined and reconciled with the reality of scientific experimentation. In the case of astronomy, Ptolemy attempted to retain the concept of an Earth-centered universe by using a system of epicycles. Similarly, selection theorists attempt to maintain an egocentric view of evolution by explaining simple positive associations as a complex system of competitive interactions, which must always culminate in the success or failure of individual races of organisms (Keller and Lloyd, 1992). This perspective may serve the needs of organismal biologists (Mayr, 1994), but it does not adequately meet the needs of those who wish to develop testable hypotheses concerning the full range of biological complexity (Krassilov, 1994; Kauffman, 1993, 1995,

Kelly 1994) at all levels of biological organization (including the gene, transposon, community, ecosystem, and biosphere).

Evolution occurs through biological adaptations that arise as the result of somewhat random genetic variations due primarily to mutations (creation of new genes) and recombination (reorganization of genes in cells), as well as the emigration and immigration of organisms (reorganization of genomes in communities). The net result of these processes is the development of biological systems with multiple levels of organization and multiple levels of proliferation. Those organizations that optimize conversion of the abiotic to the biotic tend to proliferate more successfully, while those that don't tend to proliferate less successfully. This proliferation hypothesis appears to be the simplest, most economic, and most general of all conceivable explanations of ecology and evolution. Therefore, by the law of parsimony (Occam's razor, see footnote page 107) proliferation (self-organization) theory should be used in lieu of selection theory if the ecology and evolution of communities is to be understood through scientific experimentation. This theory is embodied by the following postulates of self organization.

Spontaneity of Information. Life is the self-organization (evolution) of information (order). It is thus a process of change rather than a set of static objects. Information (order) arises spontaneously through a process of self-organization in which self-replicating information systems proliferate more rapidly as the result of association and recombination among a diversity of self-replicating systems.

Recombination of Information. The evolution of life includes the mutation of genes, the recombination of genes, the recombination of organisms in communities (immigration and emigration), the spatial repositioning of organisms within geographic boundaries, the exchange and recombination of thoughts through conversation, the exchange and recombination of alphanumeric and digitally encoded information, as well as the development of all other forms of information.

Optimization of Information. Life optimizes conversion of the abiotic (non-proliferating information) to the biotic (proliferating information). Optimizing conversion of the abiotic to the biotic under continuously changing conditions requires an optimal balance of order (organization) and chaos (freedom, entropy). It does not reach a final endpoint or produce a final product.

Interrelation of Information. The evolution of life results in a nested hierarchy of interrelated systems, none of which can evolve independently of the rest. Reproductive strategies (proliferation strategies, propagation strategies) arise at all levels of organization within this hierarchy and are shared among all levels. Thus each living subsystem is the embodiment of an evolving strategy that includes not only the optimization of its own proliferation but also the optimization of proliferation at other levels of organization.

Life is the process by which useful information arises spontaneously, transcends any specific individual or group of individuals, and resides within and among all living things. This is a concept which can be tested through empirical experimentation. It is not an impenetrable mystery beyond any hope of comprehension. However, the information embodied by the diversity and universality of life is too complex to be comprehended in its entirety by any single individual.

9. Terms and Definitions

Tentative terms and definitions associated with the proliferation hypothesis are summarized below and represent a unified theory of ecology and evolution bridging gaps and inconsistencies between evolutionary ecology (Mayr, 1993; Krassilov, 1994; Kauffman, 1993, 1995), ecosystem ecology (Maynard-Smith, 1991; Loehle and Pechmann, 1988; Schulze and Mooney, 1993), microbial ecology (Caldwell, 1993; Margulis, 1990), and information theory (Rasmussen, 1988; Yockey, 1990, 1995; Kelly, 1994). They are in contrast to the dominant views of selection theorists as summarized by Keller and Lloyd (1992).

Association   A group of organisms normally interacting through periodic or continuous physical contact. Consortia, symbioses, biofilms, and bioaggregates are examples of microbial associations. In a broader sense, association also refers to groups occurring at any level of biological organization (molecules, macromolecules, cells, organisms, communities, and ecosystems) that may result in more effective proliferation through association.

Autopoiesis   The process by which new information is created through random recombination of existing information. The new information may lead to improved reproductive success or propagation through an increased rate of growth, an increased environmental range, increased resistance to periodic environmental stress, etc.

Biological Adaptation   Adaptation occurs through the modification and spatial rearrangement of information at all levels of biological organization (through emigration and immigration of plasmids and organisms as well as through genetic mutations and recombination) and sometimes results in more effective biological proliferation. It is not limited to the modification and rearrangement of genetic information within individual lineages. The displacement of one organism by another in a chemostat experiment is an example of biological adaptation through the rearrangement of information, as opposed to being the consequence of an environmental selection process. Genetic information is also reorganized among various species of bacteria by transformation, conjugation, and transduction.

Change   If spatial and temporal gradients of environmental change did not exist, then chaos, diversity, and other sources of variation would not be necessary. Thus without change, there would be no need to divert resources from proliferation and growth toward reorganization and innovation.

Chemostat   A continuous culture with a fixed volume of liquid as well as an inlet and outlet. The inlet and outlet provide control of the flux of nutrients needed to sustain interactive microbial associations in a quasi-steady state. It also results in continual removal of metabolic wastes. In a nutrient-limited chemostat, a pulse of additional nutrient produces a transient increase in biomass and is used to confirm that the culture is nutrient limited. However, the concentration of substrate can not be controlled directly and must be determined empirically. Communities grown in a chemostat culture have well-defined mathematical characteristics. For example, once a steady state has been obtained, growth becomes balanced. The doubling time for each subcellular biomolecule within each organism, as well as the doubling time for each cell line within a community, is equal to the natural log of 2 divided by the dilution rate (the dilution rate is equal to the rate of flow divided by the culture volume).

Clone   A group of cells produced from a single cell through asexual reproduction.

Communitization   The process by which communities arise through community-level proliferation and evolution (self-organization).

Community   An interactive network of associated organisms. To confirm that a community has been obtained as a laboratory culture it must meet criteria of communality, synergy, autopoiesis, and homeostasis, as discussed in section 7.

Community Culture   A community culture is defined by its environment rather than by isolation and aseptic technique. Adequately defining the environment requires that all of the relevant physicochemical conditions have been set and quantified (concentrations, fluxes, spatial variations, and temporal variations). The composition of the resulting community is controlled by environmental constraints. If the environment is not adequately defined, then the composition of the community cannot be adequately controlled or reproduced (Fig. 2). To confirm that a community has been obtained as a laboratory culture it should meet criteria of communality, synergy, autopoiesis, and homeostasis, as discussed in Section 7. (*Community cultures normally contain interacting populations as well as predators and parasites (multiple trophic levels). They are useful in determining the principles governing the behavior and formation of microbial communities. They are also useful in establishing communities as the causative agents in ecological and other processes.)

Competition   Competition is a misnomer when used in an ecological context. Human beings create competitions and designate winners and losers. However, bacteria either proliferate or they do not. Imagining that they must compete for survival is unnecessary. Imagining that they must collaborate to successfully compete for survival, is even more complex and unnecessary. There are positive and negative bacterial synergisms. Bacterial proliferation improves primarily through positive synergisms. Logic dictates (Ockham's razor) that this simpler and more comprehensive explanation of ecology and evolution should be accepted rather than more complicated alternatives.

Consortium   An association in which the members are in physical contact. The spatial arrangement of organisms is well defined and easily recognized by microscopy.

Continuous-Flow Slide Culture   A flow cell that can be continuously viewed during incubation and that is continuously irrigated with medium. Continuous flow slide culture allows control of the concentration, flux, and temporal variation of substrate. However, it does not allow control of spatial variations in substrate concentrations as does the microstat.

Ecogram (Environmental Response Surface)   A map of habitat range for a community or organism cultivated on the surface of two-dimensional diffusion gradients. Using fluorescent molecular probes and confocal laser microscopy, the community structure, viability, growth rate, metabolism, genetic expression, population interactions, biofilm architecture, and chemistry of the community or population is quantified. This data can be used to create an environmental response surface showing interaction effects between the responses to each of two perpendicular gradients (Caldwell, 1995). Ecograms of population structure reveal ecotones confirming that community level interactions occur and defining the boundaries between them.

Ecology The scientific investigation of principles concerning the mechanism by which evolution optimizes the conversion of abiotic resources (nonproliferating information) to biotic resources (proliferating information). This optimization occurs through a process of autopoiesis (self-organization). The contemporary definition of ecology (derived from the definition originated by Ernst Haeckel in 1866) is inadequate to test community-level theories. This definition envisions ecology as the study of relationships between organisms and their biotic and abiotic environments. The idea of ecology being the study of relationships between organisms (one level of biological complexity or organization) and both their biotic environment (another level of biological complexity or organization) as well as their physical environment, is a contradiction in terms that cannot serve as a logical basis in the formulation of community-level hypotheses and tests. Life is a hierarchy of nested systems that evolves both as a whole and as component subsystems.

Ecosystem A community or set of communities with well-defined boundaries, recycling C, N, S, and other elements between their oxidized and reduced forms, with multiple trophic levels, and a spatial organization, creating a favorable internal microenvironment within a relatively unfavorable macroenvironment, and in a dynamic quasi-steady-state.

Ecotone Ecotones are transitional areas between communities. Ecotone transitions represent the reorganization of communities from one set of interrelationships to another; they frequently form in response to both spatial and temporal environmental gradients. If all of the organisms in a community were unassociated, then the response of each organism to an environmental gradient would be independent. However, if communal synergies occur among sets of organisms within the community, then the response of the associated organisms to environmental variation must be coordinated. A sharp inflection in the community structure profile (percent of total population represented by each species plotted as a function of position along the gradient) for those organisms most directly involved in the synergy is indicative of the boundary between two distinct sets of communal relationships. For example, if two different communities are required to degrade two pesticides, then an ecotone forms at the boundary between the communities when they are cultured on the surface of a two-dimensional gradient of one pesticide versus the other (if the pesticide is the sole source of carbon and energy). If the same community is adequate to degrade both pesticides then no ecotone will form. In this way it is possible to discover the existence of specific communities and map their habitat range along defined environmental gradients.

Enrichment Culture A sample of organisms from an *in situ* environment is exposed to an environmental stress to enrich for individuals capable of a specific activity or process. An enrichment culture is normally used to isolate a cell line with specific properties or characteristics by making one organism numerically dominant through the use of specific environmental stresses. This simplifies its isolation in dilution tubes and on streak plates. The objective of enrichment culture is thus to reduce biological diversity, whereas the objective of community culture normally is to increase or sustain biological diversity. Enrichment methods have been reviewed recently by Holt and Kreig (1994).

Environmental Response Surface An ecogram showing interaction effects between the response of a population or community to variations in two environmental variables. Response surfaces often reveal ecotones and can thus be used to confirm the existence of specific communities and map their habitat range.

Environmental Selection (Group Selection, Individual Selection, Natural Selection, Selection) Although it is often said that the environment "selects," the term "environmental selection" is a misnomer. The environment is not capable of making a conscious decision to select one organism and reject another—it is not a referee. It is an arena providing the constraints within which biological evolution (self-organization) occurs. In most cases it is more accurate to use terms such as evolution, proliferation, propagation, displacement, predomination, development, enrichment, production, or environmental constraint instead of selection and/or environmental selection (see also *selection*).

Evolution Evolution is a process of self-organization that occurs directly at all levels of biological organization (including molecules, cells, organisms, communities, and ecosystems) capable of reproduction, propagation or proliferation. Evolution optimizes the conversion of abiotic (nonproliferating information) resources to biotic (proliferating information) resources (optimization being an increase in the quantity, quality, diversity, and utility of information available). This occurs through the rearrangement, proliferation, and attribution of genetic and other forms of information. It is the process by which the proliferation of life is optimized through biological self-organization and adaptation to environmental constraints. Optimization occurs through the more rapid or more effective proliferation of those information systems that make most effective use of environmental resources and habitat. Long-term optimization involves sustainable proliferation despite spatial and temporal variations in environmental conditions. New strategies arise not only through the organization (recombination) of genes within a genome, but through the modification and spatial redistribution of information at all levels of organization, including plasmids, viruses, chromosomes, bacterial cells, symbioses, consortia, protists, species, words, thoughts, computer code, communities, ecosystems, geographical boundaries, and the biosphere as a whole. This includes the reorganization of genes via recombination as well as the reorganization of entire genomes by immigration and emigration to and from communities. Evolution thus occurs simultaneously at all levels of biological organization and optimizes effective use of environmental resources. It is not restricted to individual organisms or species populations. Evolution normally occurs through mechanisms analogous to the self-organization and development of chemostat communities, in which some lineages (or other levels of organization) proliferate and some are lost by attrition, but none die in a "life and death" struggle (competition) between races or lineages. An example is the evolution of language through the continual creation of new words and the gradual attrition of older, less useful, words.

Habitat Range The range of environmental conditions within which a community occurs. The habitat range of an organism alone can be compared to its habitat range as a member of a community in order to determine if its range has been extended through the establishment of synergistic interactions with other organisms.

Individual (Organism)   One member of a community, normally a single member of a species population.

Information   Information includes any form of order, organization, knowledge, intelligence, understanding, meaning, or wisdom. All objects contain information in that every object has a spatial and temporal position with respect to all other objects. These spatial and temporal relationships make all objects part of a universal information system that evolves as a whole. All objects are thus part of a universal life force or process. Definitions in mathematical information theory use "order" and "information" in a somewhat different way (Yockey, 1990, 1995). According to conventional information theory, if there is a jar with 500 black beads on the bottom and 500 white beads on the top, the information in the system would be at a minimum and the order at a maximum. If the bottle were shaken until the beads were randomly distributed, then the order in the system would be at a minimum and the information would be at a maximum. Thus current mathematical thought defines information as the amount of computer memory required to record the relative position of the beads and interprets this memory requirement as information content. The lack of information content is defined as order. However, the position of the random beads represents "noise" rather than "useful information" (as "information" is used in normal conversation to equate information with order and intelligence). If a word were spelled out using white beads against a background of black beads, the information content of the jar would be less than that of the random beads (from the mathematical point of view). Life occurs specifically at this interface between extremes of order and chaos. Useful information related to life differs from information that is "noise" in that it is capable of propagating or proliferating. However, the noise due to chaos provides the spontaneous variation (innovation) necessary for continued propagation despite changing conditions. In this sense, it might be considered important information that is necessary for life. There is no apparent message in it until the "potential" or "inactive" information is recombined with "useful" or "active" information and leads to successful proliferation. This simple process, which requires both chaos and order, is the essence of life. Thus to understand the role of information as a part of life it is necessary to discriminate between "inactive" and "active" forms of information, just as we discriminate between potential and kinetic forms of energy. It is thus necessary to think of chaos as potentially useful information, rather than as disorder. It is also necessary to imagine a universe which consists not just of matter and energy but of matter/energy (objects occupying space/time) and information/life (spatial and temporal relationships between objects).

Isolated Cell Line (Clone)   A group of cells produced from one parental cell through asexual reproduction and maintained in isolation by use of aseptic technique.

Life   Life is a self-optimizing process whereby the proliferation of self-replicating information systems is optimized through recombination. Recombination of information often involves the spatial reshuffling of atoms, molecules, genes, plasmids, ideas, organisms, words, computer code, etc. Chaotic recombination is the only mechanism by which useful information (information that optimizes its own proliferation) can be created spontaneously from potential information (information that does not proliferate or lead directly to proliferation). Recombination of diverse information resources is necessary for optimization due to the lack of an infinite preexisting source of useful information, capable of optimizing life under all possible environmental conditions. Living systems optimize the conversion of the abiotic (nonproliferating information resources) to the biotic (proliferating information resources) through any number of processes that must be both partially ordered (to allow the accrual of information) and partially chaotic (to allow the recombination of information). Note that life forms need not consist of cells, contain DNA, or contain carbon atoms to fulfill these criteria. However, they must proliferate (although they may be dependent on other forms of life for their own proliferation) and they must manifest an optimal balance between order and chaos. In this sense a community has a life of its own, as does an organism, language, gene, consortium, book, biosphere, plasmid, ecosystem, prion, or virus.

Microstat The microstat is a flow cell in which two-dimensional environmental gradients are used to study the response of bacterial populations and communities to environmental variations. It consists of a two-dimensional steady-state diffusion gradient within a gel or porous medium. A biofilm community forms on the surface of the diffusion matrix in response to environmental gradients created by diffusion. The response of microbial communities to environmental gradients within the microstat is visualized as an environmental response surface (Caldwell, 1995). The microstat allows control of the concentration, flux, spatial variation, and temporal variation of substrate.

Microecosystem An ecosystem consisting of microorganisms.

Mixed Culture A mixture of isolated cell lines maintained by using aseptic technique to prevent contamination by unwanted organisms.

Natural Selection A misnomer for "environmental selection" or "selection." To say that a particular type of selection is natural or artificial is arbitrary. (see also *environmental selection*)

Nutristat A substrate concentration-controlled continuous culture (Rutgers *et al.*, 1993). It allows control of the concentration, flux, and temporal variation of substrate. The chemostat permits control only of the flux of substrate.

Population A lineage of genetically related individuals (a clone or species population).

Pure Culture A misnomer for "isolated cell line," "isolate," "strain," or "clone." Bacterial cells cannot be purified in the same sense that atoms or molecules are purified. They must be cloned during the isolation process and are not immutable. Cloning results in spontaneous mutation and cellular differentiation rather than a completely homogeneous population of cells.

Quasi-Steady State This is the condition of most microecosystems that are in a dynamic state. Nutrient cycling is relatively rapid compared to the rate of change in total biomass and other state variables.

Reproduction The process by which a genetic element, organism, or community propagates itself and thus proliferates.

Reproductive Success The long-term development and proliferation of information contained within a genetic element, cell, organism, community, or other level of order and organization.

Selection (Natural Selection, Environmental Selection) The term "selection" is a misnomer. Neither the environment nor the ecosystem is capable of consciously selecting one organism over another. The environment merely provides physical resources and space necessary for proliferation. Self-organizing biological systems then optimize the conversion of abiotic environmental resources to biological resources by proliferating through adaptation, association, and interaction (see also *environmental selection*).

Self-Organization   See *autopoiesis*

Speciation   The process by which species arise through species-level proliferation and evolution.

Species (Species Population)   A lineage sharing a common gene pool but isolated from the gene pool of other species by elaborate sexual reproductive mechanisms. These reproductive mechanisms facilitate the reshuffling of genes within species (chaos) while preventing the reshuffling of genes between species (order). This provides the essential balance between chaos and order, which is a prerequisite for the adaptation of life.

Universe   The universe is a unified information system which evolves as a whole. Every object in the universe contains information in that it has a spatial and temporal position with respect to all other objects. The movement of one object reorders all of the other objects. The universe consists of matter/energy (objects occupying space/time) and information/life (spatial and temporal relationships between objects). The spatial and temporal relationships between all objects make them part of a universal information system, or life force, which evolves as a whole.

ACKNOWLEDGMENTS. We are grateful to Lynn Margulis for her encouragement and for pointing us in the direction of non-reductive, community-level thinking. Nearly two decades ago she infected us with an idea about the origin of life that seemed implausible from conventional perspectives. Today this thinking is central to our understanding of microbial communities and life as a whole.

We also thank Jim Shapiro and Martin Dworkin for encouraging us to write this and for helping us to think of bacteria (and bacterial communities) as multicellular organisms. We thank Bill Costerton for helping us to appreciate the importance of biofilms and for supporting us in our endeavors. We are grateful to Lynn Margulis, Tom Schmidt, Paul Kolenbrander, Joseph Schmutz, Dirk Schmid, Marty Dworkin, and Jim Shapiro, and others for comments on the manuscript. We acknowledge Anita Choi for providing photomicrographs of salt-stressed bacteria.

The authors acknowledge the U. S. Department of Energy, Environment Canada (National Hydrology Research Institute and GASReP), the Saskatchewan Agricultural Development Fund, and the National Science and Engineering Research Council of Canada for financial support.

References

Aftring, R. P., Chalker, B. E., and Taylor, B. F., 1981, Degradation of phthalic acids by denitrifying mixed cultures of bacteria, *Appl. Environ. Microbiol.* **41**:1177–1183.

Ahmadjian, V., and Hale, M. E., 1973, *The Lichens*, Academic Press, London.

Ahring, B. K., and Westermann, P., 1987, Thermophilic anaerobic degradation of butyrate by a butyrate-utilizing bacterium in coculture and triculture with methanogenic bacteria, *Appl. Environ. Microbiol.* **53**:429–433.

Allard, A.-S., Hynning, P.-A., Remberger, M., and Neilson, A. H., 1992, Role of sulfate concentration in dechlorination of 3,4,5-trichlorocatechol by stable enrichment cultures grown with coumarin and flavanone glycones and aglycones, *Appl. Environ. Microbiol.* **58**:961–968.

Alldredge, A. L., and Cohen, Y., 1987, Can microscale chemical patches persist at sea? Microelectrode study of marine snow, fecal pellets, *Science* **235**:689–91.

Allison, D. G., and Sutherland, I. W., 1987, The role of expolysaccharides in adhesion of freshwater bacteria, *J. Gen. Microbiol.* **133**:1319–1327.

Amann, R. I., Krumholz, L., and Stahl, D. A., 1990, Fluorescent-oligonucleotide probing of whole cells for determinative, phylogenetic, and environmental studies in microbiology, *J. Bacteriol.* **172**:762–770.

Amann, R. I., Springer, N., Ludwig, W., Gortz, H. D., and Schleifer, K. H., 1991, Identification and phylogeny of uncultured bacterial endosymbionts, *Nature* (London) **351**:161–165.

Amann, R. I., Stromley, J., Devereux, R., Key, R., and Stahl, D. A., 1992, Molecular and microscopic identification of sulfate-reducing bacteria in multispecies biofilms, *Appl. Environ. Microbiol.* **58**:614–623.

Amann, R. I., Ludwig, W., and Schleifer, K. H., 1994, Identification of uncultured bacteria: a challenging task for molecular taxonomists, *ASM News* **60**:360–365.

Aoki, K., 1982, Additive polygenic formulation of Hamilton's model of kin selection, *Heredity* **49**:163–170.

Aoki, K., 1986, Stable polymorphic equilibria in a toy model of group selection, *Japan. J. Genet.* **61**:481–490.

Aviles, L. 1986, Sex-ratio bias and possible group selection in the social spider, *Amer. Nat.* **128**:1–12.

Back, J. P., and Kroll, R. G., 1991, The differential fluorescence of bacteria stained with acridine orange and the effects of heat, *J. Appl. Bacteriol.* **71**:51–58.

Bagley, D. M., and Gossett, J. M., 1995, Chloroform degradation in methanogenic methanol enrichment cultures and by *Methanosarcina barkeri* 227, *Appl. Environ. Microbiol.* **61**:3195–3201.

Batra, S. W. T., and Batra, L. R., 1967, The fungus gardens of insects, *Sci. Amer.* **217**:112–120.

Beijerinck, M. W., 1889, Auxanography, a method useful in microbiological research, involving diffusion in gelatin, *Archives Neerlandaises des Sciences Exactes et Naturelles Haarlem* **23**:367–372.

Beijerinck, M. W., 1901, Enrichment culture studies with urea bacteria, *Centralblatt f. Bakteriologie Part II* **7**:33–61.

Betts, R. P., Bankes, P., and Banks, J. G., 1989, Rapid enumerations of viable micro-organisms by staining and direct microscopy, *Lett. Appl. Microbiol.* **9**:199–202.

Beurskens, J. E. M., Dekker, C. G. C., Van Den Heuvel, H., Swart, M., De Wolf, J., 1994, Dechlorination of chlorinated benzenes by an anaerobic microbial consortium that selectively mediates the thermodynamically most favorable reactions, *Environ. Sci. Technol.* **28**:701–706.

Bhatnagar, L., and Fathepure, B. Z., 1991, Mixed cultures in detoxification of hazardous waste, in: *Mixed Cultures in Biotechnology* (J. G. Zeikus, and E. A. Johnson, eds.) McGraw-Hill, New York, pp. 293–340.

Bochem, H. P., Schoberth, S. M., Sprey, B., and Wengler, P., 1982, Thermophilic biomethanation of acetic acid: morphology and ultrastructure of a granular consortium, *Can. J. Microbiol.* **28**:500–510.

Bochner, B., 1989, "Breathprints" at the microbial level, *ASM News* **55**:536–539.

Boone, D. R., Johnson, R. L., and Liu, Y., 1989, Diffusion of the interspecies electron carriers H_2 and formate in methanogenic ecosystems and its implications in the measurement of K_m for H_2 and formate uptake, *Appl. Environ. Microbiol.* **55:**1735–1741.

Bottomley, P. J., and Maggard, S. P., 1990, Determination of viability within serotypes of a soil population of *Rhizobium leguminosarum* bv. *trifolii*, *Appl. Environ. Microbiol.* **56:**533–540.

Bouwer, H., 1989, Transformations of xenobiotics in biofilms, in: *Structure And Function Of Biofilms* (W. G. Characklis and P. H. Wilderer, eds.), John Wiley and Sons, Toronto, pp. 251–267.

Bradshaw, D. J., McKee, A. S., and Marsh, P. D., 1989, Effects of carbohydrate pulses and pH on population shifts within oral microbial communities in vitro, *J. Dent. Res.* **68:**1298–1302.

Brakenhoff, G. J., van der Voort, H. T. M., Baarslag, M. W., Mans, B., Oud, J. L., Zwart, R., and van Driel, R., 1988, Visualization and analysis techniques for three dimensional information acquired by confocal microscopy, *Scanning Microsc.* **2:**1831–1838.

Brannan, D. K., 1995, Cosmetic preservation, *J. Soc. Cosmet. Chem.* **46:**199–220.

Brayton, P. R., Tamplin, M. L., Huq, A., and Colwell, R. R., 1987, Enumeration of *Vibrio cholerae* 01 in Bangladesh waters by fluorescent-antibody direct viable count, *Appl. Environ. Microbiol.* **53:**2862–2865.

Brefeld, O., 1881, *Botanische Untersuchungen uber Schimmelpilze: Culturemethoden*, Leipzig.

Bremer, P. J., and Geesey, G. G., 1991, Laboratory-based model of microbiologically induced corrosion of copper, *Appl. Environ. Microbiol.* **57:**1956–1962.

Brock, T. D., and Madigan, M., 1988, *The Biology of Microorganisms*, Prentice-Hall, New Jersey.

Brown, M. J., and Lester, J. N., 1982, Role of bacterial extracellular polymers in metal uptake in pure bacterial culture and activated sludge-I, *Water Res.* **16:**1539–1548.

Brown, M. R. W., Allison, D. G., and Gilbert, G., 1988, Resistance of bacterial biofilms to antibiotics: a growth-rate related effect?, *J. Antimicrob. Chemother.* **22:**777–780.

Brown, S. W., and Oliver, S. G., 1982, Isolation of ethanol-tolerant mutants of yeast by continuous culture selection, *Eur. J. Appl. Microbiol. Biotechnol.* **16:**119–122.

Bryant, M. P., Wolin, E. A., Wolin, M. J., and Wolfe, R. S., 1967, *Methanobacillus omelianskii*, a symbiotic association of two species of bacteria, *Arch. Mikrobiol.* **59:**20–31.

Bryers, J. D., 1993, The biotechnology of interfaces, *J. Appl. Bact. Symp. Suppl.* **74:**98S–109S.

Bungay, H. R., 1995, A challenge for modelling mutualism, *Binary: Computing in Microbiology.* **7:**100–102.

Busscher, H. J., Bellon-Fontaine, M.-N., Mozes, N., Van Der Mei, H. C., Sjollema, J., Cerf, O., and Rouxhet, P. G., 1990, Deposition of *Leuconostoc mesenteroides* and *Streptococcus thermophilus* to solid substrata in a parallel plate flow cell, *Biofouling* **2:**55–63.

Caldwell, D. E., 1993, The microstat: Steady-state microenvironments for subculture of steady-state consortia, communities, and microecosystems, in: *Trends in Microbial Ecology* (R. Guerrero and C. Pedros-Alio, eds.), Spanish Society for Microbiology, Barcelona, pp. 123–128.

Caldwell, D. E., 1995, *Cultivation and Study of Biofilm Communities*. in: *Microbial Biofilms*, (H. M. Lappin-Scott and J. W. Costerton, eds), Cambridge University Press, Cambridge, pp. 64–79.

Caldwell, D. E., and Costerton, J. W., 1996, Are bacterial biofilms constrained to Darwin's concept of evolution through natural selection? *Microbiologia SEM* **12:**347–358.

Caldwell, D. E., and Germida, J. J., 1985, Evaluation of difference imagery for visualizing and quantitating microbial growth, *Canad. J. Microbiol.* **31:**35–44.

Caldwell, D. E., and Hirsch, P., 1973, Growth of microorganisms in two-dimensional steady-state diffusion gradients, *Can. J. Microbiol.* **19:**53–58.

Caldwell, D. E., and Lawrence, J. R., 1986, Growth kinetics of *Pseudomonas fluorescens* microcolonies within the hydrodynamic boundary layers of surface microenvironments, *Microb. Ecol.* **2:**299–312.

Caaldwell, D. E., and Lawrence, J. R., 1988, Study of attached cells in continuous-flow slide

culture, in: *CRC Handbook of Laboratory Model Systems for Microbial Ecology Research*, Vol. 1 (J. W. T. Wimpenny, ed.), CRC Press, Boca Raton, pp. 117–138.
Caldwell, D. E., and Lawrence, J. R., 1989, Microbial growth and behavior within surface microenvironments, in: *Proceedings of ISME-5* (T. Hattori, Y. Ishida, Y. Maruyama, R. Y. Morita, and A. Uchida, eds.), JSS Press, Tokyo, pp. 140–145.
Caldwell, D. E., Lai, S. H., and Tiedje, J. M., 1973, A two-dimensional steady-state diffusion gradient for ecological studies, in: *Modern Methods in Microbial Ecology* (Thomas Rosswall, ed.), Bull. Ecol. Res. Comm. (Stockholm) **17:**151–158.
Caldwell, D. E., Caldwell, S. J., and Tiedje, J. M., 1975, An ecological study of sulfur-oxidizing bacteria from the littoral zone of a Michigan lake and a sulfur spring in Florida, *Plant and Soil* **43:**101–114.
Caldwell, D. E., Brierley, J. A., and Brierley, C. L., 1985, *Planetary Ecology*, Van Nostrand Reinhold, New York.
Caldwell, D. E., Korber, D. R. and Lawrence, J. R., 1992a, Confocal Laser Microscopy and Computer Image Analysis, in: *Advances in Microbial Ecology* Vol. 12 (K. C. Marshall, ed.), Plenum Press, New York, pp. 1–67.
Caldwell, D. E., Korber, D. R., and Lawrence, J. R., 1992b, Imaging of bacterial cells by fluorescence exclusion using scanning confocal laser microscopy, *J. Microbiol. Methods* **15:**249–261.
Caldwell, D. E., Korber, D. R., and Lawrence, J. R., 1993, Analysis of Biofilm Formation Using 2-D Versus 3-D Digital Imaging, in: *Microbial Cell Envelopes: Interactions and Biofilms* (L. B. Quesnel, P. Gilbert, and P. S. Handley, eds), Blackwell Scientific, Oxford, pp. 52–66S.
Caldwell, D. E., Atuku, E., Wilkie, D. C., Wivcharuk, K. P., Karthikeyan, S., Korber, D. R., Schmid, D. R., and Wolfaardt, G. M., 1997a, Germ theory versus community theory in understanding and controlling the proliferation of biofilms, *Adv. Dental Res.* **11:**4–13.
Caldwell, D. E., Wolfaardt, G. M., Korber, D. R., and Lawrence, J. R., 1997b, Cultivation of microbial consortia and communities, in: *Manual of Environmental Microbiology* (C. J. Hurst, G. R. Knudsen, M. J. McInerney, L. D. Stetzenbach, M. V. Walter, American Society of Microbiology Press, Washington, D.C., pp. 79–90.
Characklis, W. G., 1988, Model biofilm reactors, in: *CRC Handbook of Laboratory Model Systems for Microbial Ecology Research*, Vol. 1 (J. W. Wimpenny, ed.), CRC Press, Boca Raton, pp. 155–174.
Characklis, W. G., McFeters, G. A. and Marshall, K. C., 1990, Physicological ecology in biofilm systems, in: *Biofilms* (W. G. Characklis and K. C. Marshall, eds.), J. Wiley and Sons, New York, pp. 341–393.
Chartrain, M., and Zeikus, J. G., 1986a, Microbial ecophysiology of whey biomethanation: Intermediary metabolism of lactose degradation in continuous culture, *Appl. Environ. Microbiol.* **51:**180–187.
Chartrain, M., and Zeikus, J. G., 1986b, Microbial ecophysiology of whey biomethanation: Characterization of bacterial trophic populations and prevalent species in continuous culture, *Appl. Environ. Microbiol.* **51:**188–196.
Chartrain, M., L. Bhatnagar, and Zeikus, J. G. 1987, Microbial ecophysiology of whey biomethanation: Comparison of carbon transformation parameters, species composition, and starter culture performance in continuous culture, *Appl. Environ. Microbiol.* **53:**1147–1156.
Christersson, C. E., Glantz, P-O. J., and Baier, R. E., 1988, Role of temperature and shear forces on microbial detachment, *Scand. J. Dent. Res.* **96:**91–98.
Claasen, P. A. M., Korstee, G. J. J., Ossterveld Van Vliet, W. M., and Van Neerven, A. R. W., 1986, Colonial heterogeneity of Thiobacillus, *J. Bacteriol.* **168:**791–794.
Conrad, R., T. J., Phelps, and Zeikus, J. G., 1985, Gas metabolism evidence in support of the juxtaposition of hydrogen-producing and methanogenic bacteria in sewage sludge and lake sediments, *Appl. Environ. Microbiol.* **50:**595–601.

Costerton, J. W., Geesey, G. G., and Cheng, K.-J., 1978, How bacteria stick, *Sci. Am.* **238:**86–95.
Costerton, J. W., Lewandowski, Z., DeBeer, D., Caldwell, D. E., Korber, D. R., and James, G. A., 1994, Biofilms: the customized microniche, *J. Bacteriol.* **176:**2137–2142.
Daley, R. J., 1979, Direct epifluorescence enumeration of native aquatic bacteria: Uses, limitations, and comparative accuracy, in: *Native Aquatic Bacteria: Enumeration, Activity, and Ecology, STP 695* (J. W. Costerton and R. R. Colwell, eds.), American Society for Testing and Materials, New York, pp. 29–45.
Damuth, J. 1985, Selection among species. A formulation in terms of natural functional units, *Evolution* **39:**1132–1146.
Darwin, C., 1859, *The Origin of Species By Means of Natural Selection or the Preservation of Favoured Races in the Struggle for Life,* New American Library, New York.
Darwin, C., 1868, *The Variation of Animals and Plants Under Domestication. Vol. 2,* Organe Judd, New York, p. 204.
Dawson, K. A., Allison, M. J., and Hartman, P. A. 1980, Characteristics of anaerobic oxalate degrading enrichment cultures from the rumen, *Appl. Environ. Microbiol.* **40:**840–846.
de Beer, D., Stoodley, P., Roe, F., and Lewandowski, Z., 1994, Effect of biofilm structures on oxygen distribution and mass transport, *Biotechnol. Bioeng.* **43:**1131–1138.
DeLong, E. F., Wickham, G. S., and Pace, N. R., 1989, Phylogenetic stains: ribosomal RNA-based probes for the identification of single cells, *Science* **243:**1360–1363.
Dietrich, G., and Winter, J. 1990, Anaerobic degradation of chlorophenol by an enrichment culture, *Appl. Environ. Microbiol.* **34:**253–258.
Distefano, T. D., Gossett, J. M., and Zinder, S. H., 1991, Reductive dechlorination of high concentrations of tetrachloroethene to ethene by an anaerobic enrichment culture in the absence of methanogenesis, *Appl. Environ. Microbiol.* **57:**2287–2292.
Dolfing, J., and Beurskens, J. E. M., 1995, The microbial logic and environmental significance of reductive dehalogenation, *Adv. Microb. Ecol.* **14:**188.
Doolittle, W. F., and Sapienza, C., 1980, Selfish genes, the phenotype paradigm, and genome evolution, *Nature* (London) **284:**601–603.
Drake, J. W., 1970, *The Molecular Basis of Mutation,* Holden-Day, San Francisco, pp. 39–62.
Dunbar, M. J., 1971, Higher levels of organization, the evolution of stability in marine environments: natural selection at the level of the ecosystem, in: *Group Selection* (G. C. Williams, ed.), Aldine Atherton, Chicago, pp. 120–139.
Dworkin, M., 1985, The myxobacteria, in: *Developmental Biology of the Bacteria* (M. Dworkin, ed.), Benjamin/Cummings, Menlo Park, CA, pp. 105–149.
Dworkin, M., and Kaiser, D., 1985, Cell interactions in myxobacterial growth and development, *Science* **230:**18–24.
Emerson, D., Worden, R. M., and Breznak, J. A., 1994, A diffusion gradient chamber for studying microbial behavior and separating organisms, *Appl. Environ. Microbiol.* **60:**1269–1278.
Emerson, R. W., 1841, *The Method of Nature: An Oration Delivered Before the Society of the Adelphi In Waterville College, Maine, August 11, 1841,* Books on Line, http://www-cgi.cs.cmu.edu.
Eng, R. H. K., Padberg, F. T., Smith, S. M., Tan, E. N., and Cherubin, C. E., 1991, Bactericidal effects of antibiotics on slowly growing and nongrowing bacteria, *Antimicrob. Agents Chemother.* **35:**1824–1828.
Evenboom, W., Van Der Does, J. Bruning, K., and Mur, L. M., 1981, A non-heterocystous mutant of *Aphanizomenon flos-aquae,* selected by competition in light-limited continuous culture, *FEMS Microbiol. Lett.* **10:**11–16.
Fairbairn, B., 1994, History from the ecological perspective: gaia theory and the problem of cooperatives in turn-of-the-century Germany, *Amer. Historic. Rev.* **99:**1203–1239.
Farrar, J. F., 1976. The lichen as an ecosystem: observation and experiment, in: *Lichenology:*

Progress and Problems (D. H. Brown, D. L. Hawksworth, and R. H. Bailey, eds.), Academic Press, New York, pp. 19–46.

Federle, T. W. and Pastwa, G. M., 1988, Biodegradation of surfactants in saturated subsurface sediments: a field study, *Groundwater* **26:**761–70.

Federle, T. W., and Schwab, B. S., 1989, Mineralization of surfactants by microbiota of aquatic plants, *Appl. Environ. Microbiol.* **55:**2092–2113.

Fix, A. G., 1984, Kin groups and trait groups population structure and epidemic disease, *Amer. J. Phys. Anth.* **65:**201–212.

Fletcher, M., 1984, Comparative physiology of attached and free-living bacteria, in: *Microbial Adhesion And Aggregation* (K. C. Marshall, ed.), Springer-Verlag, New York, pp. 223–232.

Foster, P. L., 1993, Adaptive mutation: the uses of adversity, *Annu. Rev. Microbiol.* **47:**467–504.

Fulthorpe, R. R., and Wyndham, R. C., 1989, Survival and activity of a 3-chlorobenzoate-catabolic genotype in a natural system, *Appl. Environ. Microbiol.* **55:**1584–1590.

Fulthorpe, R. R., and Wyndham, R. C., 1991, Transfer and expression of the catabolic plasmic pBRC60 in wild bacterial recipients in a freshwater ecosystem, *Appl. Environ. Microbiol.* **57:**1546–1553.

Fulthorpe, R. R., and Wyndham, R. C., 1992, Involvement of a chlorobenzoate-catabolic transposon, Tn5271, in community adaptation to chlorobiphenyl, chloroaniline, and 2,4-dichlorophenoxyacetic acid in a freshwater ecosystem, *Appl. Environ. Microbiol.* **58:**314–325.

Fulthorpe, R. R., McGowan, C., Maltseva, O. V., Holben, W. E., and Tiedje, J. M., 1995, 2, 4-Dichlorophenoxyacetic acid-degrading bacteria contain mosaics of catabolic genes, *Appl. Environ. Microbiol.* **61:**3274–3281.

Garland, J. L., and Mills, A. L., 1991, Classification and characterization of heterotrophic microbial communities on the basis of patterns of community-level sole-carbon-source utilization, *Appl. Environ. Microbiol.* **57:**2351–2359.

Geesey, G. G., and White, D. C., 1990, Determination of bacterial growth and activity at solid-liquid interfaces, *Ann. Rev. Microbiol.* **44:**579–602.

Geesey, G. G., Mutch, R., Costerton, J. W., and Green, R. B., 1978, Sessile bacteria: an important component of the microbial population in small mountain streams, *Limnol. Oceanogr.* **23:**1214–1223.

Gest, H., 1993, Bacterial growth and reproduction in nature and in the laboratory, *ASM News* **59:**542–543.

Ghosal, D., You, I.-S., Chatterjee, D. K., and Chakrabarty, A. M., 1985, Microbial degradation of halogenated compounds, *Science* **228:**135–142.

Gilbert, P., Collier, P. J., and Brown, M. R. W., 1990, Influence of growth rate on susceptibility to antimicrobial agents: biofilms, cell cycle, dormancy, and stringent response, *Antimicrob. Agents Chemother.* **34:**1856–1868.

Goodnight, C. J., 1990a, Experimental studies of community evolution I. The ecological basis of the response to community selection, *Evolution* **44:**1614–1624.

Goodnight, C. J., 1990b, Experimental studies of community evolution II. The ecological basis of the response to community selection, *Evolution* **44:**1625–1636.

Goodnight, C. J., Schwartz, J. M., and Stevens, S. L., 1992, Contextual analysis of models of group selection, soft selection, hard selection, and the evolution of altruism, *American Naturalist* **140:**743–761.

Goodwin, S., Conrad, R., and Zeikus, J. G., 1988, Influence of pH on microbial hydrogen metabolism in diverse sedimentary ecosystems, *Appl. Environ. Microbiol.* **54:**590–593.

Gottschal, J. C., and Dijkhuizen, L., 1988, The place of continuous culture in ecological research, in: *CRC Handbook of Laboratory Model Systems for Microbial Ecology Research*, Vol. 1 (J. W. T. Wimpenny, ed.), CRC Press, Boca Raton, pp. 19–49.

Guckert, J. B., Hood, M. A., and White, D. C., 1986, Phospholipid ester-linked fatty acid profile changes during nutrient deprivation of *Vibrio cholerae:* increases in the trans/cis ratio and proportions of cyclopropyl fatty acids, *Appl. Environ. Microbiol.* **52:**794–801.

Guede, H., 1979, Grazing by protozoa as selection factor for activated sludge bacteria, *Microb. Ecol.* **5:**225–238.

Haack, S. K., Garchow, H., Odelson, D. A., Forney, L. J., and Klug, M. J. 1994, Accuracy, reproducibility, and interpretation of fatty acid methyl ester profiles of model bacterial communities, *Appl. Environ. Microbiol.* **60:**2483–2493.

Haeckel, E., 1866, *Generelle Morphologie der Organism,* Reimer, Berlin.

Haefele, D. M., and Lindow, S. E., 1987, Flagellar motility confers epiphytic fitness advantages upon *Pseudomonas syringae, Appl. Environ. Microbiol.* **53:**2528–2533.

Hahn, D., Amann, R. I., Ludwig, W., Akkermans, A. D. L., and Schleifer, K.-H., 1992, Detection of micro-organisms in soil after *in situ* hybridization with rRNA-targeted, fluorescently labelled oligonucleotides, *J. Gen. Microbiol.* **138:**879–887.

Harder, W., and Veldkamp, H., 1971, Competition of marine phychrophilic bacteria at low temperatures, *Antonie van Leeuwenhoek* **37:**51–63.

Harder, W., Kuenen, J. G., and Matin, A , 1977, Microbial selection in continuous culture, *J. Appl. Bacteriol.* **43:**1–24.

Haugland, R. P., 1992, *Molecular Probes: Handbook of Fluorescent Probes and Research Chemicals,* Molecular Probes Inc., Eugene, OR.

Hawksworth, D. L., 1982, Secondary fungi in lichen symbioses: parasites, saprophytes and parasymbionts, *J. Hattori Botan. Lab.* **52:**357–366.

Hawksworth, D. L., 1988, The variety of fungal-algal symbioses, their evolutionary significance and the nature of lichens, *Botan. J. Linnean Soc.* **96:**3–20.

Herbert, R. A., 1988, Bidirectional compound chemostats: applications of compound diffusion-linked chemostats in microbial ecology, in: *CRC Handbook of Laboratory Model Systems for Microbial Ecology Research,* Vol. 1 (J. W. T Wimpenny, ed.), CRC Press, Boca Raton, pp. 99–115.

Herdman, M., 1977, The cyanelle: chloroplast or endosymbiotic prokaryote? *FEMS Microbiol. Lett.* **1:**7–12.

Hernandez-Cruz, A., Sala, F., and Adams, P. R., 1990, Subcellular calcium transients visualized by confocal microscopy in a voltage-clamped vertebrate neuron, *Science* **247:**858–862.

Hickey, D. A., 1982, Selfish DNA: A sexually-transmitted nuclear parasite, *Genetics* **106:**519–531.

Hirsch, P., 1980, Some thoughts on and examples of microbial interactions in the natural environment, in: *Aquatic Microbial Ecology* (R. R. Colwell and A. J. Foster, eds.), University of Maryland, College Park, pp. 36–54.

Hirsch, P., 1984, Microcolony formation and consortia, in: *Microbial Adhesion and Aggregation* (K. C. Marshall, ed.), Springer Verlag, New York, pp. 373–393.

Hoff, K. A., 1988, Rapid and simple method for double staining of bacteria with 4', 6-Diamidino-2-phenylindole and fluorescein isothiocyanate-labeled antibodies, *Appl. Environ. Microbiol.* **54:**2949–2952.

Holt, J. G., and Krieg, N. R., 1994, Enrichment and isolation, in: *Methods for General and Molecular Bacteriology* (P. Gerhardt, ed.), American Society for Microbiology, Washington, D.C., pp. 179–204.

James, G. A., Caldwell, D. E., and Costerton, J. W., 1993, Spatial relationships between bacterial species within biofilms, *Proceedings of The CSM/SIM Annual Meeting* (abstract) Toronto, Canada.

Jannasch, H. W., 1967, Enrichment of aquatic bacteria in continuous culture, *Arch. Mikrobiol.* **59:**165–173.

Jensen, R. H., and Woolfolk, C. A., 1985, Formation of filaments by *Pseudomonas putida, Appl. Environ. Microbiol.* **50:**364–372.

Jeon, K. W., 1972, Development of cellular dependence in infective organisms; microsurgical studies in amoebas, *Science* **176**:1122–1123.

Jeon, K. W., and Ahn, T. I., 1978, Temperature sensitivity: A cell character determined by obligate endosymbionts in amoebas. *Science* **202**:635–637.

Jeon, K. W., and Jeon, M. S., 1976, Endosymbiosis in amoebae: Recently established endosymbionts have become required cytoplasmic components, *J. Cell. Physiol.* **89**:337–344.

Jiménez, L., Breen, A., Thomas, N., Federle, T. W., and Sayler, G. S., 1991, Mineralization of linear alkylbenzene sulfonate by a four-member aerobic bacterial consortium, *Appl. Environ. Microbiol.* **57**, 1566–1569.

Jones, W. J., Guyot, J.-P., and Wolfe, R. S., 1984, Methanogenesis from sucrose by defined immobilized consortia, *Appl. Environ. Microbiol.* **47**:1–6.

Kauffman, S., 1995, *At Home In The Universe: The Search for Laws of Self-Organization and Complexity*, Oxford University Press, New York.

Kauffman, S. A., 1993, *The Origins of Order: Self-Organization and Selection in Evolution*, Oxford University Press, New York.

Keller, E. F., and Lloyd, E. A., 1992, *Keywords in Evolutionary Biology*, Harvard University Press, Cambridge, Massachusetts.

Kelly, K., 1994, *Out of Control—The New Biology of Machines, Social Systems and the Economic World*, Addison-Wesley, New York.

Kendrick, B., 1991, Fungal symbioses and evolutionary innovations, in: *Symbiosis as a Source of Evolutionary Innovation* (L. Margulis and R. Fester, eds.), MIT Press, Cambridge, Massachusetts.

Kieft, T. L., and Caldwell, D. E., 1984, Chemostat and in-situ colonization kinetics of *Thermothrix thiopara* on calcite and pyrite surfaces, *Geomicrobiol. J.* **3**:217–229.

Kinner, N. E., Balkwill, D. L., and Bishop, P. L., 1983, Light and electron microscope studies of microorganisms growing in rotating biological contactor biofilms, *Appl. Environ. Microbiol.* **45**:1659–1669.

Kjelleberg, S., 1984, Effects of interfaces on survival mechanisms of copiotrophic bacteria in low-nutrient habitats, in: *Current Perspectives in Microbial Ecology* (M. J. Klug and C. A. Reddy, eds.), Wiley, New York, pp. 151–159.

Klinger, J. M., Stowe, R. P., Obenhuber, D. C., Groves, T. O., Mishra, S. K., and Pierson, D. L., 1992, Evaluation of the Biolog automated microbial identification system, *Appl. Environ. Microbiol.* **58**:2089–2092.

Koch, A. I., 1991, Diffusion: The crucial process in many aspects of the biology of bacteria, in: *Advances in Microbial Ecology*, Vol. 11 (K. C. Marshall, ed.), Plenum Press, New York, pp. 37–70.

Koch, R., 1881, Methods for the study of pathogenic organisms, *Mittheilungen aus dem Kaiserlichen Gesundheitsamte* **1**:1–48.

Koch, R., 1884, The etiology of tuberculosis, *Mitthelungen aus dem Kaiserlichen Gesundheitsamte* **2**:1–88.

Korber, D. R., Lawrence, J. R., Sutton, B., and Caldwell, D. E., 1989, Effects of laminar flow velocity on the kinetics of surface recolonization by mot$^+$ and mot$^-$ *Pseudomonas fluorescens*, *Microb. Ecol.* **18**:1–19.

Korber, D. R., Lawrence, J. R., Zhang, L., and Caldwell, D. E., 1990, Effect of gravity on bacterial deposition and orientation in laminar flow environments, *Biofouling* **2**:335–50.

Korber, D. R., Lawrence, J. R., Hendry, M. J., and Caldwell, D. E. 1992, Programs for determining statistically representative areas of microbial biofilms, *Binary,* **4**:204–210.

Korber, D. R., Lawrence, J. R., Hendry, M. J., and Caldwell, D. E., 1993, Analysis of spatial variability within mot$^+$ and mot$^-$ *Pseudomonas fluorescens* biofilms using representative elements, *Biofouling* **7**:339–358.

Korber, D. R., James, G. A., and Costerton, J. W., 1994a, Evaluation of fleroxacin activity against established *Pseudomonas fluorescens* biofilms, *Appl. Environ. Microbiol.* **60**:1663–1669.
Korber, D. R., Caldwell, D. E., and Costerton, J. W., 1994b, Structural analysis of native and pure-culture biofilms using scanning confocal laser microscopy, *Abstracts of the National Association of Corrosion Engineers (NACE) Canadian Region Western Conference*, Calgary, Alberta.
Korber, D. R., Lawrence, J. R., and Caldwell, D. E., 1994c, Effect of motility on surface colonization and reproductive success of *Pseudomonas fluorescens* in dual-dilution continuous culture and batch culture systems, *Appl. Environ. Microbiol.* **60**:1421–1429.
Korber, D. R., Lawrence, J. R., Lappin-Scott, H. M., and Costerton, J. W., 1995, Growth of microorganisms on surfaces, in: *Bacterial Biofilms* (H. M. Lappin-Scott and J. W. Costerton, eds.), Cambridge University Press, Cambridge, U.K., pp. 15–45.
Korber, D. R., Choi, A., and Caldwell, D. E., 1996, Bacterial plasmolysis as a physical indicator of viability, *Appl. Environ. Microbiol.* **62**:3939–3947.
Krassilov, V. A., 1994, Evolutionary synthesis, *Trends Ecol. Evol.* **9**:149.
Labatiuk, C. W., Schaefer III, F. W., Finch, G. R., and Belosevic, M., 1991, Comparison of animal infectivity, excystation, and fluorogenic dye as measures of *Giardia muris* cyst inactivation by ozone, *Appl. Environ. Microbiol.* **57**:3187–3192.
Lang, E., Viedt, H., Egestorff, J., and Hanert, H. H., 1992, Reaction of the soil microflora after contamination with chlorinated aromatic compounds and HCH, *FEMS Microbiol. Ecol.* **86**:275–282.
Lappin, H. M., Greaves, M. P., and Slater, J. H., 1985, Degradation of the herbicide mecoprop 2-2 methyl-4-chlorophenoxypropionic-acid by a synergistic microbial community, *Appl. Environ. Microbiol.* **49**:429–433.
Lattanzio, Jr., F. A., 1990, The effects of pH and temperature on fluorescent calcium indicators as determined with chelex-100 and EDTA buffer systems, *Biochem. Biophys. Res. Comm.* **171**:102–108.
Lattanzio, Jr., F. A., and Bartschat, D. K., 1991, The effect of pH on rate constants, ion selectivity and thermodynamic properties of fluorescent calcium and magnesium indicators, *Biochem. Biophys. Res. Comm.* **177**:184–191.
Lawrence, J. R., and Caldwell, D. E., 1987, Behavior of bacterial stream populations within the hydrodynamic boundary layers of surface microenvironments, *Microb. Ecol.* **14**:15–27.
Lawrence, J. R., and Korber, D. R., 1994, Aspects of microbial surface colonization behavior, in: *Trends in Microbial Ecology*, (R. Guerrero and C. Pedros-Alio, eds.), Spanish Society for Microbiology, Barcelona, pp. 113–118.
Lawrence, J. R., Delaquis, P. J., Korber, D. R., and Caldwell, D. E., 1987, Behavior of *Pseudomonas fluorescens* within the hydrodynamic boundary layers of surface microenvironments, *Microb. Ecol.* **14**:1–14.
Lawrence, J. R., Korber, D. R., Hoyle, B. D., Costerton, J. W., and Caldwell, D. E., 1991, Optical sectioning of microbial biofilms, *J. Bacteriol.* **173**:6558–6567.
Lawrence, J. R., Korber, D. R., and Caldwell, D. E., 1992, Behavioral analysis of *Vibrio parahaemolyticus* variants in high and low viscosity microenvironments using digital image processing, *J. Bacteriol.* **174**:5732–5739.
Lawrence, J. R., Wolfaardt, G. M., and Korber, D. R., 1994, Monitoring diffusion in biofilm matrices using confocal laser microscopy, *Appl. Environ. Microbiol.* **60**:1166–1173.
Lawrence, J. R., Korber, D. R., and Wolfaardt, G. M., Caldwell, D. E., 1995, Behavioral strategies of surface-colonizing bacteria, in: *Advances in Microbial Ecology*, Vol. 14 (J. G. Jones, ed.), Plenum Press, New York, pp. 1–75.
Leigh, E. G., Jr., 1983, When does the good of the group override the advantage of the individual? *Proc. Nat. Acad. Sci. USA* **80**:2985–2989.
Lens, P. N. L., De Beer, D., Cronenberg, C. C. H., Houwen, F. P., Ottengraf, S. P. P., and

Verstraete, W. H., 1993, Heterogeneous distribution of microbial activity in methanogenic aggregates: pH and glucose microprofiles, *Appl. Environ. Microbiol.* **59**:3803–3815.

Lenski, R. E., and Travisano, M., 1994, Dynamics of adaptation and diversification: a 10,000-generation experiment with bacterial populations, *Proc. Natl. Acad. Sci. USA* **91**:6808–6814.

Little, B., Ray, R., Wagner, P., Lewandowski, Z., Lee, W. C., Characklis, W. G., and Mansfeld, F., 1991, Impact of biofouling on the electrochemical behavior of 304 stainless steel in natural seawater, *Biofouling* **3**, 45–49.

Loehle, C., and Pechmann, J. H. K., 1988, Evolution: The missing ingredient in systems ecology, *American Naturalist* **132**:884–899.

Lomnicki, A., 1978, Adventures of ecologists and evolutionists in the land of super-organisms, *Wiadomosci Ekologiczne* **24**:249–260.

Lovelock, J. E., 1979, *Gaia: A New Look at Life on Earth*, Oxford University Press, Oxford, UK.

Lovelock, J. E., 1988, *Ages of Gaia: A Biography of Our Living Earth*, Norton, New York.

Lovelock, J. E., and Margulis, L., 1974, Atmospheric homeostasis by and for the biosphere: the Gaia hypothesis, *Tellus* **26**:2–10.

Lovitt, R. W., and Wimpenny, J. W. T., 1981, Physiological behavior of *Escherichia coli* grown in opposing gradients of oxidant and reductant in the gradostat, *J. Gen. Microbiol.* **127**:269.

Luby-Phelps, K., Lanni, F., and Taylor, D. L., 1988, The submicroscopic properties of cytoplasm as a determinant of cellular function, *Ann. Rev. Biophys. Chem.* **17**:369–396.

Mackie, R. I., Krecek, R. C., Els, H. J., van Niekerk, J. P., Kirschner, L. M., and Baecker, A. A. W., 1989, Characterization of the microbial community colonizing the anal and vulvar pores of helminths from the hindgut of zebras, *Appl. Environ. Microbiol.* **55**:1178–1186.

MacLeod, F. A., Guiot, S. R., and Costerton, J. W., 1990, Layered structure of bacterial aggregates produced in an upflow anaerobic sludge bed reactor, *Appl. Environ. Microbiol.* **56**:1598–1607.

Madsen, T., and Aamand, J., 1992, Anaerobic transformation and toxicity of trichlorophenols in a stable enrichment culture, *Appl. Environ. Microbiol.* **58**:557–561.

Maenhaut-Michel, G., and Shapiro, J. A., 1994, The roles of selection and starvation in the emergence of araB-lacZ fusion clones, *EMBO J.* **13**:5229–5239.

Maigetter, R. Z., and Pfister, R. M., 1974, A mixed bacterial population in a continuous culture with and without kaolinite, *Can. J. Microbiol.* **21**:173–180.

Malmcrona-Friberg, K., Tunlid, A., Marden, P., Kjelleberg, S., and Odham, G., 1986, Chemical changes in cell envelope and poly-β-hydroxybutyrate during short-term starvation of a marine bacterial isolate, *Arch. Microbiol.* **144**:340–245.

Margulis, L. 1981, *Symbiosis in Cell Evolution: Life and Its Environment on the Early Earth*, W. H. Freeman, San Francisco.

Margulis, L., 1990, Introduction, in: *Handbook of Protoctista* (L. Margulis, J. O. Corliss, M. Melkonian, D. J. Chapman, and H. I. Mckhann, eds.), Jones and Bartlett, Boston.

Margulis, L., 1992, *Symbiosis in Cell Evolution: Microbial Communities in the Archean and Proterozoic Eons*, W.H. Freeman, Salt Lake City.

Margulis, M., 1993, Microbial communities as units of selection, in: *Trends in Microbial Ecology* (R. Guerrero and C. Pedros-Alio, eds.), Spanish Society of Microbiology of Barcelona, pp. 349–352.

Margulis, L., 1995, From kefir to death, in: *How Things Are*. (J. Brochmer, Ed.), William Morrow, New York, pp. 69–78.

Margulis, L., and Fester, R., 1991, *Symbiosis as a Source of Evolutionary Innovation*, MIT Press, Cambridge, Massachusetts.

Margulis, L., and Guerrero, R., 1991, Two plus three equal one: individuals emerge from bacterial communities, in: *Gaia 2. Emergence: The New Science of Becoming*, Lindisfarne Press, New York, pp. 60–67.

Margulis, M., and West, O., 1993, Gaia and the colonization of Mars, *GSA Today*, **3**:277–291.

Marshall, K. C., 1994, Microbial ecology: wither goest thou? in: *Trends in Microbial Ecology* (R. Guerrero and C. Pedros-Alio, eds.), Spanish Society for Microbiology, Barcelona, pp. 5–8.

Marshall, P. A., Loeb, G. I., Cowan, M. M., and Fletcher, M., 1989, Response of microbial adhesives and biofilm matrix polymers to chemical treatments as determined by interference reflection microscopy and light section microscopy, *Appl. Environ. Microbiol.* **55:**2827–2831.

Marxsen, J., 1988, Investigations into the number of respiring bacteria in *Groundwater* from sandy and gravelly deposits, *Microb. Ecol.* **16:**65–72.

Maynard-Smith, J., 1976, Group selection, *Quart. Rev. Biol.* **51:**277–283.

Maynard-Smith, J., 1991, A darwinian view of symbiosis, in: *Symbiosis as a Source of Evolutionary Innovation* (L. Margulis and R. Fester, eds.), MIT Press, Cambridge, Massachusetts, pp. 83–92.

Mayr, E., 1993, What was the evolutionary synthesis? *Trends Eco. Evol.* **8:**31–34.

McInerney, M. J., Bryant, M. P., and Pfennig, N., 1979, Anaerobic bacterium that degrades fatty acids in syntrophic association with methanogens, *Arch. Microbiol.* **122:**129–135.

McKinley, V. L., Costerton, J. W., and White, D. C., 1988, Microbial biomass, activity, and community structure of water and particulates retrieved by backflow from a waterflood injection well, *Appl. Environ. Microbiol.* **54:**1383–1393.

Mitchell, J., Pearson, G. L., Dillon, S., and Kantalis, K., 1995, Natural assemblages of marine bacteria exhibiting high-speed motility and large accelerations, *Appl. Environ. Microbiol.* **61:**4436–4440.

Møller, S., Kristensen, C. S., Poulsen, L. K., Carstensen, J. M., and Molin, S. 1995, Bacterial growth on surfaces: Automated image analysis for quantification of growth rate-related parameters, *Appl. Environ. Microbiol.* **61:**741–748.

Monod, J., 1942, *Recherches sur la Croissance des Cultures Bacteriénnes*, Hermann, Paris.

Monod, J., 1949, The growth of bacterial cultures, *Ann. Rev. Microbiol.* **3:**371–394.

Muyzer, G., De Waal, E. C., and Uitterlinden, A. G., 1993, Profiling of complex microbial populations by denaturing gradient gel electrophoresis analysis of polymerase chain reaction-amplified genes coding for 16s rRNA, *Appl. Environ. Microbiol.* **59:**695–700.

Neilson, A. H., Allard, A.-S., Hynning, P.-A., and Remberger, M., 1988, Transformations of halogenated aromatic aldehydes by metabolically stable anaerobic enrichment cultures, *Appl. Environ. Microbiol.* **54:**2226–2236.

Neu, T. R., and Marshall, K. C., 1991, Microbial "footprints"—a new approach to adhesive polymers, *Biofouling* **3:**101–112.

Ney, U., Schoberth, S. M., Sahm, H., 1991, Anaerobic degradation of sulfite evaporator condensate in a fixed-bed loop reactor by a defined bacterial consortium, *Appl. Microbiol. Biotechnol.* **34:**818–822.

Nichols, P. D., Henson, J. M., Antworth, C. P., Parsons, J., Wilson, J. T., and White, D. C., 1987, Detection of a microbial consortium, including type II methanotrophs, by use of phospholipid fatty acids in an anaerobic halogenated hydrocarbon-degrading soil column enriched with natural gas, *Environ. Toxicol. Chem.* **5:**89–97.

Nix, P. G., and Daykin, M. M., 1992, Resazurin reduction tests as an estimate of coliform and heterotrophic bacterial numbers in environmental samples, *Bull. Environ. Contam. Toxicol.* **49:**354–360.

Noack, D., 1986, Directed selection of differentiation mutants of *Streptomyces noursei* using chemostat cultivation, *J. Basic Microbiol.* **26:**231–239.

Novick, A., and Silard, L., 1950, Experiments with the chemostat on mutations of bacteria, *Proc. Nat. Acad. Sci. USA* **36:**708–719.

Nunney, L., 1985, Group selection, altruism, and structured deme models, *Am. Naturalist* **126:**212–230.

Olsen, G. J., Lane, D. L., Giovannoni, S. J., and Pace, N. R., 1986, Microbial ecology and evolution: a ribosomal RNA approach, *Ann. Rev. of Microbiol.* **40:**337–365.

Pace, N. R., Stahl, D. A., Lane, D. L., and Olsen, G. J., 1986, The analysis of natural microbial populations by rRNA sequences, *Adv. Microb. Ecol.* **9:**1–55.
Palleroni, N. J., 1994, Some reflections on bacterial diversity, *ASM News* **60:**537–540.
Parkes, R. J., and Senior, E., 1988, Multistate chemostats and other models for studying anoxic ecosystems, in: *Handbook of Laboratory Model Systems for Microbial Ecosystems*, Vol. 1 (J. W. T. Wimpenny, ed.), CRC Press, Boca Raton, pp. 51–71.
Patel, G. B., 1984, Characterization and nutritional properties of *Methanothrix concilii* sp. nov., a mesophilic, aceticlastic methanogen, *Can. J. Microbiol.* **30:**1383–1396.
Peck, J. R., 1992, Group selection, individual selection, and the evolution of genetic drift, *J. Theor. Biol.* **159:**163–187.
Peters, A. C., 1990, Using image analysis to map bacterial growth on solid media, *Binary* **2:**73–75.
Petrini, O., Hake, U., and Dreyfuss, M. M., 1990, An analysis of fungal communities isolated from fruticose lichens, *Mycologia* **82:**444–451.
Phelps, T. J., Schram, R. M., Ringelberg, D., Dowling, N. J., and White, D. C., 1991, Anaerobic microbial activities including hydrogen mediated acetogenesis within natural gas transmission lines, *Biofouling* **3:**265–276.
Pringsheim, E. G., 1946, The biphasic or soil-water culture method for growing algae and flagellata, *J. Ecol.* **33:**193–204.
Prosser, J. I., 1989, Modeling nutrient flux through biofilm communities, in: *Structure and Function of Biofilms* (W. G. Characklis and P. A. Wilderer, eds.), John Wiley and Sons, Toronto, pp. 239–250.
Rasmussen, S. 1988, Toward a quantitative theory of the origin of life, in: *Artificial Life, SFI Studies in the Sciences of Complexity* (C. Langton, ed.) Addison-Wesley, New York, pp. 79–104.
Rasmussen, S. 1991, Aspects of information, life, reality and physics, in: *Artificial Life II, SFI Studies in the Sciences of Complexity*, Vol. X, (C. G. Langton, C. Taylor, J. D. Farmer, and S. Rasmussen, eds.), Addison-Wesley, New York, pp. 767–773.
Rasmussen, S., Knudsen, C., Feldberg, R., Hindsholm, M., 1990, The coreworld: emergence and evolution of cooperative structures in a computational chemistry, *Physica D* **42:**111–134.
Rajogopal, B. S., Brahmaprakash, G. P., and Sethunanthan, N., 1984, Degradation of carbofuran by enrichment cultures and pure cultures of bacteria from flooded soils, *Environmental Pollution Series A, Ecological and Biological* **36:**61–74.
Ratnam, D. A., Pavlou, S., and Fredrickson, A. G., 1982, Effects of attachment of bacteria to chemostat walls in a microbial predator-prey relationship, *Biotechnol. Bioeng.* **24:**2675–2694.
Revsbech, N. P., 1989, Diffusion characteristics of microbial communities determined by use of oxygen microsensors, *J. Microbiol. Methods* **49:**111–122.
Revsbech, N. P., and Jorgenson, B. B., 1988, Microelectrodes: their use in microbial ecology, in: *Advances in Microbial Ecology*, Vol. 9 (K. C. Marshall, ed.), Plenum Press, New York, pp. 293–352.
Robarts, R. D., and Zohary, T., 1993, Fact or fiction—Bacterial growth rates and production as determined by [^{3}H-methyl]thymidine, in: *Advances in Microbial Ecology*, Vol. 13 (G. F. Jones, ed.), Plenum Press, New York, pp. 371–418.
Robinson, R. W., Akin, D. E., Nordstedt, R. A., Thomas, M. V., and Aldrich, H. C., 1984, Light and electron microscopic examinations of methane-producing biofilms from anaerobic fixed-bed reactors, *Appl. Environ. Microbiol.* **48:**127–136.
Rodriguez, G. G., Phipps, D., Ishiguro, K, and Ridgway, H. F., 1992, Use of a fluorescent redox probe for direct visualization of actively respiring bacteria, *Appl. Environ. Microbiol.* **58,** 1801–1808.
Rosenberg, E., 1984, *Myxobacteria: Development and Cell Interactions*, Springer-Verlag, New York.
Rothmel, R. K., Haugland, R. A., Coco, W. M., Sangodkar, U. M. X., and Chakrabarty, A. M. 1989, Natural and directed evolution: microbial degradation of synthetic chlorinated com-

pounds, in: *Recent Advances in Microbial Ecology* (T. Hattori, Y. Ishida, Y. Maruyama, R. Y. Morita, and A. Uchida, eds.), JSS Press, Tokyo, pp. 605–610.

Rozgaj, R., and Glancer-Soljan, M., 1992, Total degradation of 6-aminonaphthalene-2-sulphonic acid by a mixed culture consisting of different bacterial genera, *FEMS Microbiol. Ecol.* **86**:229–235.

Rutgers, M., Bogte, J. J., Breure, A. M., and van Andel, J. G., 1993, Growth and enrichment of pentachlorophenol-degrading microorganisms in the nutristat, a substrate concentration-controlled continuous culture, *Appl. Environ. Microbiol.* **59**:3373–3377.

Saavedra-Molina, A., Uribe, S., and Devlin, T. M., 1990, Control of mitochondrial matrix calcium: studies using fluo-3 as a fluorescent calcium indicator, *Biochem. Biophys. Res. Comm.* **167**:148–153.

Sauch, J. F., Flanigan, D., Galvin, M. L., Berman, D., and Jakubowski, W., 1991, Propidium iodide as an indicator of *Giardia* cyst viability, *Appl. Environ. Microbiol.* **57**:3243–3247.

Schiefer, G. E., and Caldwell, D. E., 1982, Synergistic interaction between *Anabaena* and *Zoogloea* spp. in carbon dioxide limited continuous cultures, *Appl. Environ. Microbiol.* **44**:84–87.

Schmidt, E. L., 1972, Fluorescent antibody techniques for the study of microbial ecology, in: *Modern Methods in the Study of Microbial Ecology*, Vol. 17 (T. Rosswall, ed.), Swedish Natural Science Research Council, Stockholm, pp. 67–76.

Schmidt, T. M., Delong, E. F., and Pace, N. R., 1991, Analysis of a marine picoplankton community by 16S rRNA gene cloning and sequencing, *J. Bacteriol.* **173**:4371–4378.

Schulze, E. D., and Mooney, H. A., 1993, *Biodiversity and Ecosystem Function Ecological Studies 99*, Springer-Verlag, New York.

Schwemmler, W., 1989, *Symbiogenesis: A Macro-mechanism of Evolution*, Walter de Gruyter, Berlin.

Scott, O. T., 1985, Petridoglycan envelope in the cyanelles of *Glaucocystis nostochinearum*, in: *Planetary Ecology* (D. E. Caldwell, J. A. Brierley, and C. L. Brierley, eds.), Van Nostrand Reinhold Co., New York, pp. 27–40.

Senior, E., Bull, A. T., and Slater, J. H., 1976, Enzyme evolution in a microbial community growing on the herbicide Dalapon, *Nature* **263**:476–479.

Shapiro, J. A., 1984, The use of Mudlac transposons as tools for vital staining to visualize clonal and non-clonal patterns of organization in bacterial growth on agar surfaces, *J. Gen. Microbiol.* **130**:1169–1181.

Shapiro, J. A., 1985a, Photographing bacterial colonies, *ASM News* **51**:62–69.

Shapiro, J. A., 1985b, Mechanisms of DNA reorganization in bacteria, *Int. Rev. Cytol.* **93**:25–56.

Shapiro, J. A., 1988, Bacteria as multicellular organisms, *Sci. Am.* **256**:82–89.

Shapiro, J. A., 1992, Differential action and differential expression of *E. coli* DNA polymerase I during colony development, *J. Bacteriol.* **174**:7262–7272.

Shapiro, J. A., and Higgins, N. P., 1988, Variation of B-galactosidase expression from Mudlac elements during the development of *E. coli* colonies, *Annales de l'Institut Pasteur* **139**:79–103.

Shapiro, J. A., and Trubatch, D., 1991, Sequential events in bacterial colony morphogenesis, in: *Waves and Patterns in Chemical and Biological media* (H. L. Swinney and V. I. Krinski eds.), Elsevier Science, Amsterdam, pp. 214–223.

Shen, C. F., Kosaric, N., and Blaszczyk, R., 1993, The effect of heavy metals (Ni, Co and Fe) on anaerobic granules and their extracellular substance, *Water Res.* **27**:25–33.

Shotton, D. M., 1989, Confocal scanning optical microscopy and its applications for biological specimens, *J. Cell Sci.* **94**:175–206.

Sissons, C. H., Wong, L., Cutress, T. W., 1995, Patterns and rates of growth of microcosm dental plaque biofilms, *Oral Microbiol. Immunol.* **10**:160–167.

Sjollema, J., Busscher, H. J., and Weerkamp, A. H., 1989a, Experimental approaches for studying adhesion of microorganisms to solid substrata: Applications and mass transport, *J. Microbiol. Meth.* **9**:79–90.

Sjollema, J., Busscher, H. J., and Weerkamp, A. H., 1989b, Real-time enumeration of adhering microorganisms in a parallel plate flow cell using automated image analysis, *J. Microbiol. Meth.* **9**:73–78.

Sjollema, J., Van der Mei, H. C., Uyen, H. M. W., and Busscher, H. J., 1990, The influence of collector and bacterial cell surface properties on the deposition of oral streptococci in a parallel plate flow cell, *J. Adhesion Sci. Technol.* **4**:765–777.

Skryabin, G. K., Golovleva, L. A., Golovlev, E. L., Pertsova, R. N., and Zyakun, A. M., 1978, Degradation of DDT and its analogs by soil microflora, *Izvestiya Akademii Nauk Sssr Seriya Biologicheskaya* **3**:352–365.

Slater, J. H., 1988, Microbial population and community dynamics, in: *Micro-organisms in Action: Concepts* (J. M. Lynch and J. E. Hobbie, eds.), Blackwell Scientific, Palo Alto, CA, pp. 51–74.

Slater, J. H., and Hartman, D. J., 1982, Microbial ecology in the laboratory: experimental systems, in: *Experimental Microbial Ecology* (R. G. Burns and J. H. Slater, eds.), Blackwell Scientific, Oxford, pp. 255–274.

Smith, G. A., Nickels, J. S., Kerger, B. D., Davis, J. D., Collins, S. P., Wilson, J. T., McNabb, J. F., and White, D. C., 1986, Quantitative characterization of microbial biomass and community structure in subsurface material: a prokaryotic consortium responsive to organic contamination, *Can. J. Microbiol.* **32**:104–111.

Sober, E., 1984, *The Nature Of Selection: Evolutionary Theory in Philosophical Terms*, Bradford, Cambridge, MA.

Sonea, S., 1991, Bacterial evolution without speciation, in: *Symbiosis as a Source of Evolutionary Innovation* (L. Margulis and R. Fester, eds.), MIT Press, Cambridge, Massachusetts, pp. 95–105.

Sonea, S., and Panisset, M., 1983, *A New Bacteriology*, Jones and Bartlett, Boston, MA.

Stahl, D. A., Lane, D. J., Olson, G. J., and Pace, N. R., 1984, Analysis of hydrothermal vent-associated symbionts by ribosomal RNA sequences, *Science* **224**:409–411.

Stahl, D. A., Flesher, B., Mansfield, H. R., and Montgomery, L., 1988, Use of phylogenetically based hybridization probes for studies of ruminal microbial ecology, *Appl. Environ. Microbiol.* **54**:1079–1084.

Stams, A. J. M., Grotenhuis, J. T. C., and Zehnder, A. J. B. 1989, Structure-function relationship in granular sludge, in: *Recent Advances in Microbial Ecology* (T. Hattori, Y. Ishida, Y. Maruyama, R. Y. Morita, and A. Uchida, eds.), Japan Scientific Societies, Tokyo, pp. 440–445.

Starmer, W. T., Ganter, P., Aberdeen, B., Lachance, M. A., and Phaff, H. J., 1987, The ecological role of killer yeasts in natural communities of yeasts, *Can. J. Microbiol.* **33**:783–796.

Starmer, W. T., Ganter, P. F., and Aberdeen, B., 1992, Geographic distribution and genetics of killer phenotypes for the yeast *Pichia kluyveri* across the United States, *Appl. Env. Microbiol.* **58**:990–997.

Stevens, T. O., and Holbert, B. S., 1990, Density-dependent growth patterns exhibited by bacteria from terrestrial subsurface environments, *Abstracts of the Conference on Multicellular and Interactive Behavior of Bacteria*, American Society of Microbiology, Marine Biological Laboratory, Woods Hole, Massachusetts, p. 20.

Stewart, P. S., Peyton, B. M., Drury, W. J., and Murga, R., 1993, Quantitative observations of heterogeneities in *Pseudomonas aeruginosa* biofilms, *Appl. Environ. Microbiol.* **59**:327–329.

Szathmary, E. and Demeter, L., 1987, Group selection of early replicators and the origin of life, *J. Theoret. Biol.* **128**:463–486.

Szybalski, W. 1952, Gradient plates for the study of microbial resistance to antibiotics, *Bacteriol. Proc.* **36**.

Szybalski, W., and Bryson, V., 1953, Genetic studies on microbial cross-resistance to toxic agents. I. Cross resistance of *Escherichia coli* to fifteen antibiotics, *J. Bacteriol.* **64**:489–499.

Tabor, P. S., and Neihof, R. A., 1982, Improved method for determination of respiring individual microorganisms in natural waters, *Appl. Environ. Microbiol.* **43**:1249–1255.

Tagger, S., Truffaut, N., and Le Petit, J., 1990, Preliminary study on relationships among strains forming a bacterial community selected on naphthalene from a marine sediment, *Can. J. Microbiol.* **36**:676–681.
Takana, H., Matsumura, M., and Veliky, I. A., 1984, Diffusion characteristics of substrates in Ca-alginate gel beads, *Biotechnol. Bioeng.* **26**:53–58.
ten Brummeler, E., Hulshoff Pol, L. W., Dolfing, J., Lettinga, G., and Zehnder, A. J. B., 1985, Methanogenesis in an upflow anaerobic sludge blanket reactor at pH 6 on an acetate-propionate mixture, *Appl. Environ. Microbiol.* **49**:1472–1477.
Thiele, J. H., Chartrain, M., and Zeikus, J. G., 1988, Control of interspecies electron flow during anaerobic digestion: role of floc formation in syntrophic methanogenesis, *Appl. Environ. Microbiol.* **54**:10–19.
Troy, F. A., 1979, The chemistry and biosynthesis of selected bacterial capsular polymers, *Ann. Rev. Microbiol.* **33**:519–560.
Trulear, M. G., and Characklis, W. G., 1982, Dynamics of biofilm processes, *J. Wat. Poll. Control Fed.* **54**:1288–1301.
Tsien, R. Y., 1989, Fluorescent indicators of ion concentrations, *Meth. Cell Biol.* **30**:127–156.
Tsien, R. Y., and Waggoner, A., 1990, Fluorophores for confocal microscopy: photophysics and photochemistry, in: *Handbook of Confocal Microscopy* (J. B. Pawley, ed.), Plenum Press, New York, pp. 169–178.
Uhlinger, D. J., and White, D. C., 1983, Relationship between physiological status and formation of extracellular polysaccharide glycocalyx in *Pseudomonas atlantica*, *Appl. Environ. Microbiol.* **45**:64–70.
Upton, A. D., Nedwell, D. B., Wynn-Williams, D. D., 1990, The selection of microbial communities by constant or fluctuating temperatures, *FEMS Microb. Ecol.* **74**:243–252.
Veldkamp, H., 1977, Ecological studies with the chemostat, in: *Advances in Microbial Ecology*, Vol. 1 (M. Alexander, ed.), Plenum Press, New York, pp. 59–94.
Veldkamp, H., and Jannasch, H. W., 1972, Mixed culture studies with the chemostat, *J. Appl. Chem. Biotechnol.* **22**:105–123.
Voordouw, G., Shen, Y., Harrington, C. S., Telang, A. J., Jack, T. R., and Westlake, D. W. S., 1993, Quantitative reverse sample genome probing of microbial communities and its application to oil field production waters, *Appl. Environ. Microbiol.* **59**:4101–4114.
Voordouw, G., Voordouw, J. K., Jack, T. R., Foght, J., Fedorak, P. M., and Westlake, D. W. S., 1991, Reverse sample genome probing, a new technique for identification of bacteria in environmental samples by DNA hybridization, and its application oto the identification of sulfate-reducing bacteria in oil field samples, *Appl. Environ. Microbiol.* **57**:3070–3078.
Voordouw, G., Voordouw, J. K., Jack, T. R., Foght, J., Fedorak, P. M., and Westlake, D. W. S., 1992, Identification of distinct communities of sulfate-reducing bacteria in oil fields by reverse sampling genome probing, *Appl. Environ. Microbiol.* **58**:3542–3552.
Wade, M. J., 1978, *O. Rev. Biol.* **53**:101–114.
Wagner, M., Amann, R., Lemmer, H., and Schleifer, K.-H., 1993, Probing activated sludge with oligonucleotides specific for proteobacteria: Inadequacy of culture-dependent methods for describing microbial community structure, *Appl. Environ. Microbiol.* **59**:1520–1525.
Ward, D. M., Bateson, M. M., Weller, R., and Ruff-Roberts, A. L., 1992, Ribosomal RNA analysis of microorganisms as they occur in nature, in: *Advances in Microbial Ecology*, Vol. 12 (K. C. Marshall, ed.), Plenum Press, pp. 219–286.
Ward, D. M., Weller, R., and Bateson, M. M., 1990, 16s rRNA sequences reveal uncultured inhabitants of a well-studied thermal community, *Nature* **345**:63–65.
Weber, N. A., 1966, Fungus-growing ants, *Science* **153**:587–604.
Weber, N. A., 1972, The fungus-culturing behavior of ants, *American Zoologist* **12**:577–587.
Weinberg, E. D., 1957, Double-gradient agar plates, *Science* **125**:196.

White, D. C., 1986, Environmental effects testing with quantitative microbial analysis: chemical signatures correlated with *in situ* biofilm analysis by FT/IR, *Toxicity Assessment* **1:**315–338.

White, J. G., Amos, W. B., and Fordham, M., 1987, An evaluation of confocal versus conventional imaging of biological structure by fluorescence light microscopy, *J. Cell Biol.* **105:**41–48.

Wilkinson, T. G., Topiwala, H. H., and Hamer, G., 1974, Interactions in a mixed bacterial population growing on methane in continuous culture, *Biotechnol. Bioeng.* **16:**41–59.

Wilson, D. S., 1980, *The Natural Selection of Populations and Communities*, Benjamin-Cummings, Menlo Park, CA.

Wilson, D. S., 1987, Altruism in mendelian populations derived from sibling groups. The Haystack model revisited, *Evolution* **41:**1059–1070.

Wilson, D. S., 1992, Complex interactions in metacommunities with implications for biodiversity and higher levels of selection, *Ecology* **73:**1984–2000.

Wilson, J. B., 1987, Group selection in plant populations, *Theoret. Appl. Genet.* **74:**493–502.

Wimpenny, J. W. T., 1988, Bidirectionally linked continuous culture: the gradostat, in: *CRC Handbook of Laboratory Model Systems for Microbial Ecology Research*, Vol. 1 (J. W. T. Wimpenny, ed.), CRC Press, Boca Raton, pp. 73–98.

Wimpenny, J. W. T., 1992, Microbial systems: Patterns in space and time, in: *Advances in Microbial Ecology*, Vol. 12 (K. C. Marshall, ed.), Plenum Press, New York, pp. 469–522.

Wimpenny, J. W. T., and Waters, P., 1984, Growth of microorganisms in gel-stabilized two-dimensional diffusion gradient systems, *J. Gen. Microbiol.* **130:**2921–2936.

Wimpenny, J. W. T., Waters, P., and Peters, A., 1988, Gel-plate methods in microbiology, in: *CRC Handbook of Laboratory Model Systems for Microbial Ecology Research*, Vol. 1 (J. W. T. Wimpenny, ed.), CRC Press, Boca Raton, pp. 229–251.

Wimpenny, J. W. T., Gest, H., and Favinger, J. L., 1986, The use of two-dimensional gradient plates in determining the responses in non-sulphur purple bacteria to pH and NaCl concentration, *FEMS Microbiol. Lett.* **37:**367–371.

Wirsen, C. O., and Jannasch, H. W., 1970, Growth response of *Spirosoma* sp. to temperature shifts in continuous culture, *Bacteriological Proceedings* **G118:**32.

Woese, C. R., 1987, Bacterial evolution, *Microb. Rev.* **51:**221–271.

Wolfaardt, G. M., Lawrence, J. R., Hendry, M. J., Robarts, R. D., and Caldwell, D. E., 1993, Development of steady-state diffusion gradients for the cultivation of degradative microbial consortia, *Appl. Environ. Microb.* **59:**2388–2396.

Wolfaardt, G. M., Lawrence, J. R., Robarts, R. D., Caldwell, S. J., and Caldwell, D. E., 1994a, Multicellular organization in a degradative biofilm community, *Appl. Environ. Microbiol.* **60:**434–446.

Wolfaardt, G. M., Lawrence, J. R., Robarts, R. D., and Caldwell, D. E., 1994b, The role of interactions, sessile growth and nutrient amendment on the degradative efficiency of a bacterial consortium, *Can. J. Microbiol.* **40:**331–340.

Wolfaardt, G. M., Lawrence, J. R., Headley, J. V., Robarts, R. D., and Caldwell, D. E., 1994c, Microbial exopolymers provide a mechanism for bioaccumulation of contaminants, *Microbial Ecology* **27:**279–291.

Wolfaardt, G. M., Lawrence, J. R., Robarts, R. D., and Caldwell, D. E., 1994d, Bioaccumulation of the herbicide diclofop in extracellular polymers and its utilization by a biofilm community during starvation, *Appl. Environ. Microbiol.* (in press).

Wolfaardt, G. M., Lawrence, J. R., Robarts, R. D., and Caldwell, D. E., 1995, *In situ* characterization of biofilm exopolymers involved in the accumulation of chlorinated organics (submitted).

Wright, J. B., Costerton, J. W., and McCoy, W. F. 1988, Filamentous growth of *Pseudomonas aeruginosa*, *J. Indust. Microbiol.* **3:**139–146.

Yockey, H. P., 1990, When is random random? *Nature* **344:**823.

Yockey, H. P., 1995, Information in bits and bytes: reply to Lifson's review of "Information Theory and Molecular Biology", *Bioessays* **17**:85–88.
Yu, F. P. and McFeters, G. A., 1994, Physiological response of bacteria in biofilms to disinfection, *Appl. Environ. Microbiol.* **60**:2462–2466.
Zahavi, A., 1981, Some comments on sociobiology, *Auk* **98**:412–415.
Zahavi, A., and Ralt, D., 1984, Social adaptations in myxobacteria, in: *Myxobacteria: Development and Cell Interactions* (E. Rosenberg, ed.), Springer-Verlag, New York, pp. 216–245.
Zeikus, J. G., and Johnson, E. A., 1991, *Mixed Cultures in Biotechnology,* McGraw-Hill, New York.

5

Ecology of Terrestrial Fungal Entomopathogens

ANN E. HAJEK

1. Introduction

Fungal pathogens are capable of causing sensational levels of mortality in insect populations. As early as about 1000 AD, sericulturists in Asia reported *Beauveria bassiana* infections in silkworms (Steinhaus, 1956). The "germ theory of disease," the concept that microbes can cause disease, was first experimentally proven by Agostino Bassi in 1834 working with *B. bassiana* and silkworms. From the late 1800s through 1925, research on the potential use of fungi for insect control was conducted. In recent years, largely due to our present knowledge of the hazards and inefficiencies of dependence on synthetic chemical pesticides for insect control, interest in developing fungal pathogens for control purposes has increased dramatically (Roberts and Hajek, 1992; Vandenberg, 1993). Although abundant research on use of entomopathogenic fungi for control has been conducted, there are major gaps in our understanding of the basic ecology of these fungal species, in part due to the complexity of host/pathogen/environment interactions and the diversity of host/pathogen systems to be studied. Lack of successful control in some systems and difficulties in adapting some species for typical control practices has promoted a shift in research emphasis. At present, some research efforts are directed toward understanding the factors leading to fungal infection in insects in order to investigate the potential for manipulating these systems to enhance levels of infection and promote development of epizootics (disease outbreaks).

The ability to infect insects has evolved many times, as evidenced by the

Table I. Some Commonly Studied
Genera of Entomopathogenic Fungi[a]

Phylum Zygomycota
  Class Zygomycetes
    Order Entomophthorales
      Conidiobolus
      Entomophthora
      Entomophaga
      Erynia
      Furia[b]
      Neozygites
      Pandora[b]
      Strongwellsea
      Zoophthora
Phylum Ascomycota
  Order Ascosphaerales
    Ascosphaera
  Order Hypocreales
    Cordyceps
Phylum Deuteromycota ("Fungi Imperfecti")
  Class Hyphomycetes
    Aschersonia
    Beauveria
    Hirsutella
    Metarhizium
    Nomuraea
    Paecilomyces
    Verticillium

[a] Genera including those mentioned in this chapter. Classification scheme follows Alexopoulos et al., 1996 and R. A. Humber (personal communication). Deuteromycota and categories within this group are artificial groupings of taxa from which sexual states are unknown.
[b] Controversy exists regarding this generic name. This genus is considered under *Erynia* by Keller (1991) and *Erynia* is therefore used instead of *Pandora* or *Furia* by some authors.

occurrence of entomopathogens in several major fungal lineages (McCoy et al., 1988). However, of the approximately 750 species of fungi infecting insects, the majority belong to the Deuteromycetes and Entomophthorales (Table I). Hosts occur in most orders of insects and almost all entomopathogenic fungi infect more than one species of insect. Insect species hosting fungal pathogens live in a great diversity of habitats and species of fungi infecting insects have conformed to the requirements for living in these environments. Fungal entomopathogens are known in terrestrial ecosystems from arid to moist regions (Carruthers and Soper, 1987), where hosts range from aerial to epigeal (dwelling on the surface of the ground) to subterranean. Fungal pathogens infecting insects also occur in fresh-water aquatic ecosystems.

Diverse aspects of entomopathogenic fungi have been reviewed over the past ten years. For general information on the biology of entomopathogenic fungi, readers are referred to recent reviews by McCoy *et al.* (1988) and Samson *et al.* (1988). Fungal pathogens of subterranean insects (Keller and Zimmermann, 1989), epigeal and aerial insects (Evans, 1989), and several specific insect/fungus systems (Glare and Milner, 1991) have recently been reviewed, as well as interactions between insects and entomopathogenic fungi (Hajek and St. Leger, 1994). Studies of fungal attachment (Boucias and Pendland, 1991), infection processes (Dillon and Charnley, 1991; Charnley and St. Leger, 1991; St. Leger, 1991, 1993, 1995; Clarkson and Charnley, 1995), and physiology and genetics (Khachatourians, 1991) have also been summarized. The epizootiology of a diversity of fungal entomopathogens inhabiting differing habitats has been reviewed by Carruthers and Soper (1987). Several reviews have focused on use of entomopathogenic fungi for insect control, including Roberts and Hajek (1992), Wainwright (1992), Roberts *et al.* (1991), Hajek (1993), and Ferron *et al.* (1991). In addition, a manual presenting techniques for working with Entomophthorales (Papierok and Hajek, 1997) and Deuteromycetes (Goettel and Inglis, 1997) in both the laboratory and field has recently been published. Whereas only some of the more general reviews from the last 10 years are cited above, due to the importance of abiotic conditions to fungal activity an older summation by Roberts and Campbell (1977) specifically on this subject must be mentioned.

The aim of this review will be to synthesize our knowledge of the ecology of entomopathogenic fungi. The majority of this information will, necessarily, be drawn from systems of outbreak insects, frequently with fungal diseases that cause epizootics, although the lesser known enzootic host/pathogen relationships will be included when possible. Due to the breadth of fungal/insect relationships, this review will exclude aquatic host/pathogen systems. A wealth of information exists regarding entomopathogenic fungi and the scope of this chapter cannot include all relevant papers, but will highlight trends. When appropriate, examples will be drawn from *Entomophaga maimaiga* and its gypsy moth host due to the author's familiarity with this system.

2. Life Histories of Insect/Fungus Systems

Entomopathogenic fungi kill their hosts and must, therefore, subsequently disperse, locate and recognize a new host, and successfully cause infection (Fig.

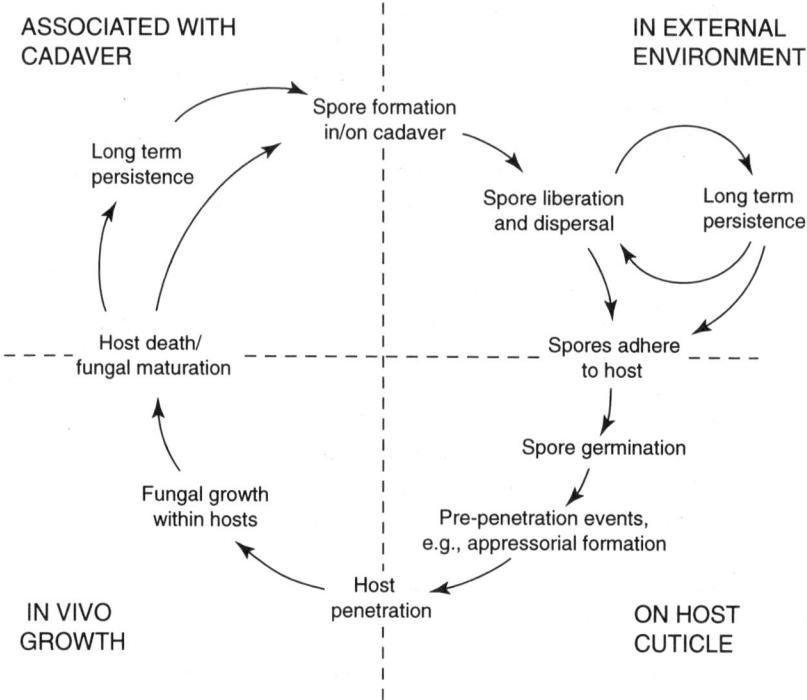

Figure 1. Schematic representation of the generalized structure of a successful cycle of infection for one generation of an entomopathogenic fungus. (Drawn by Frances L. Fawcett.)

1). The strategies of the two major groups of entomopathogenic fungi, Deuteromycetes and Entomophthorales, differ fundamentally.

Deuteromycetes, or Fungi Imperfecti, routinely lack the teleomorphic (sexual) stage of their life cycle and exist only as anamorphic (asexual) stages. Some deuteromycetes have now been recognized as anamorphs of ascomycete species but, since teleomorphic associations are not known for most species, anamorphic designations will be used. After infection, entomopathogenic deuteromycetes usually grow throughout the host and subsequently produce naked conidia (not within a fruiting structure) externally on cadavers (Fig. 2). Conidia are not actively discharged from cadavers although they can be wind dispersed; they adhere to new hosts after hosts directly contact conidia borne on cadavers or contaminating the environment.

Species of Entomophthorales, by contrast, typically produce two types of spores. An example of an entomophthoralean life cycle is presented in Fig. 3. After fungal cells have colonized hosts, conidia (asexual spores) are produced

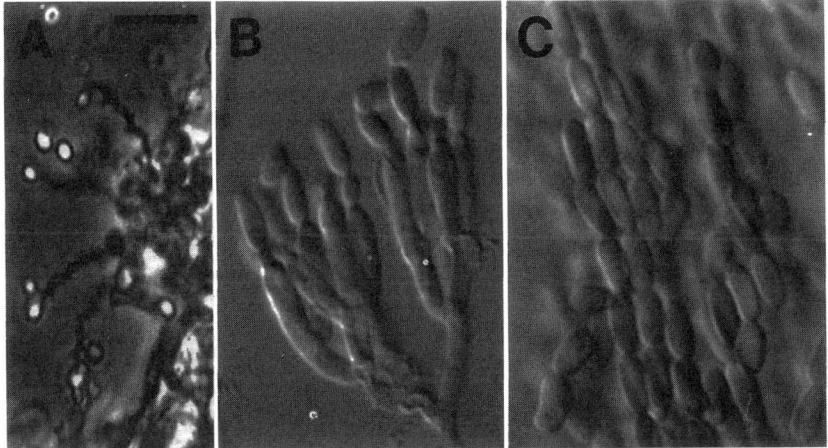

Figure 2. Fruiting structures and conidia of two Deuteromycetes. (A) Rachis of *Beauveria bassiana* bearing conidia, (B) conidia of *Metarhizium anisopliae* produced by conidiogenous cells, (C) conidia of *M. anisopliae* produced in chains. (A–C: bar = 10 μm) (Part A courtesy of R. A. Humber, Parts B and C reproduced with permission of Humber, 1996.)

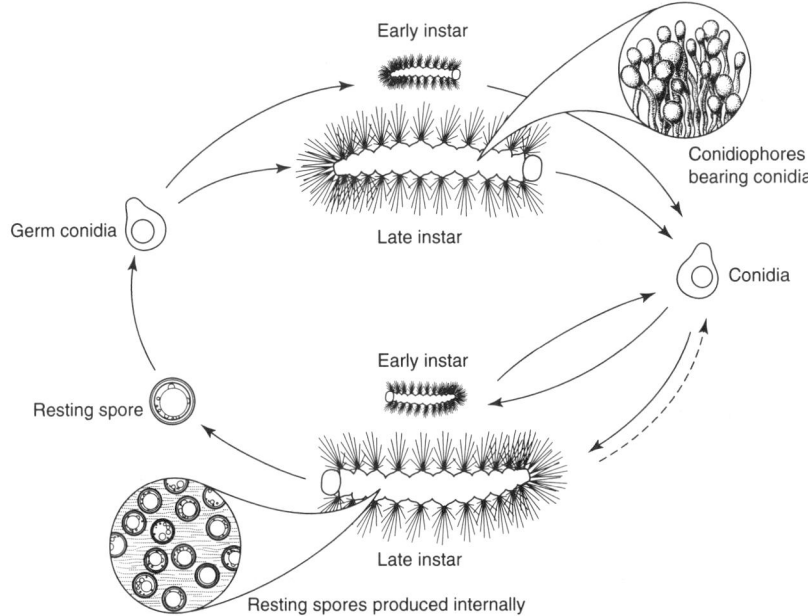

Figure 3. Life cycle of the gypsy moth fungal pathogen *Entomophaga maimaiga*. Conidia not landing on a host can germinate to produce actively ejected secondary conidia (not shown). Secondary conidia can germinate to produce tertiary conidia. While production of supernumerary conidia has not been specifically studied in this host/pathogen system, other species of Entomophthorales are known to produce quaternary conidia (Drawn by Frances L. Fawcett.)

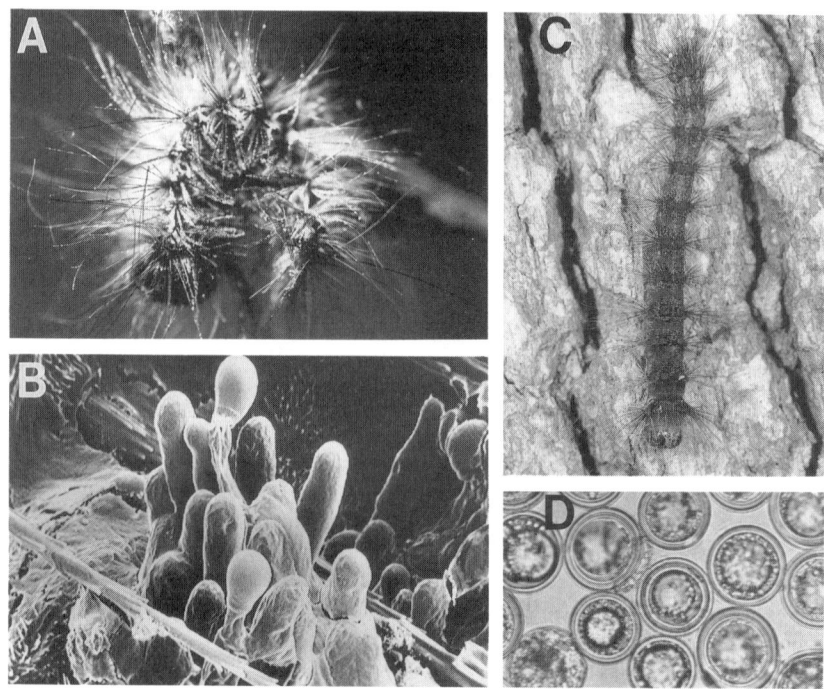

Figure 4. *E. maimaiga* and gypsy moth. (A) Cadaver from which conidia have been discharged, some attaching to setae, (B) conidiophores bearing conidia, (C) Cadaver of a late instar internally containing resting spores, (D) Resting spores (diameter ≈30 μm). (Parts A and C courtesy of D. Specker, Part B courtesy of T. Butt).

externally on cadavers (Fig. 4 A,B) (or, in a few species, on the surfaces of living hosts), and are actively discharged from the cadaver surface by hydrostatic pressure. If primary conidia discharged from cadavers do not land on a host, these spores can germinate to actively eject secondary conidia, which are often morphologically similar but slightly smaller. If secondary conidia do not land on a host, these spores can actively eject tertiary conidia and this process can be repeated for quaternary conidial production. Mainly in *Zoophthora* and *Neozygites*, primary conidia can germinate to produce "adhesive" capilliconidia on stalks, elevating spores up to 250 μm above the substrate (King and Humber, 1981); capilliconidia are not dislodged by winds but adhere to insects coming into direct contact with them. Resting spores, formed either as azygospores or zygospores, are generally produced within cadavers (Fig. 4 C, D), providing a resistant stage for survival through periods when hosts are not present or when environmental conditions are unfavorable. In some species resting spores are never formed (e.g., *Pandora neoaphidis*), while in others resting spores and conidia are always produced in different cadavers (e.g., *Massospora* spp.), or both resting spores and conidia can be formed in the same cadaver (e.g., *E. maimaiga*). When resting spores germinate, they produce one to several germ conidia that can be morphologically similar to conidia. However, germ conidia have long been thought to differ from primary or secondary conidia. Our laboratory has shown that when germ conidia of *E. maimaiga* infect gypsy moth larvae, infections always only result in production of conidia, whereas infections initiated by conidia produced externally on cadavers can result in either conidia or resting spores (Fig. 3; Hajek, 1997). Whether other entomophthoralean species exhibit this differential reproduction from infections initiated by germ conidia or conidia has not yet been investigated.

Besides obvious differences in life cycles, entomophthoralean entomopathogens are generally thought to differ from deuteromycetes in environmental relations. Entomophthoralean fungi are frequently considered temperate organisms with temperature optima generally around 20°C or below (Hall and Papierok, 1982). However, studies on entomophthoraleans, such as those documenting *Neozygites fresenii* causing epizootics in cotton aphids in hot Arkansas cotton fields (Steinkraus *et al.*, 1991), *Neozygites* cf. *floridana* causing epizootics on cassava green mites in northeastern Brazil (Delalibera *et al.*, 1992), and optimal temperatures for *Z. radicans* conidial germination of 26–32°C (see Magalhaes *et al.*, 1991a), challenge the assumption regarding restriction of Entomophthorales to cooler, temperate climates. Laboratory studies have demonstrated that temperature optima for many Deuteromycete species range between 20–25°C for species of *Beauveria*, *Verticillium*, *Paecilomyces*, and *Nomuraea* while 25–30°C are optimal for *Hirsutella thompsonii* and *Metarhizium anisopliae* (Hall and Papierok, 1982). Conversely, Tasmanian strains of *M. anisopliae* active against soil-dwelling scarabs to as low as 5°C have been identified (Rath *et al.*, 1995b); therefore, Deuteromycetes with lower temperature optima also occur.

3. Survival of Entomopathogenic Fungi during Unfavorable Periods

During periods that are unfavorable for associations with hosts (e.g., winter, dry seasons, host diapause, low density host populations), survival of pathogens is critical to their persistence in the environment. Stages of entomopathogenic fungi that can survive these conditions, whether in an active or inactive state, provide a reservoir for resumption of activity during favorable conditions. Survival is important within one year, e.g., gypsy moth larvae are only present each year from May through June in northeastern North America, so the host-specific *E. maimaiga* survives without hosts for about 10 months out of every 12. More

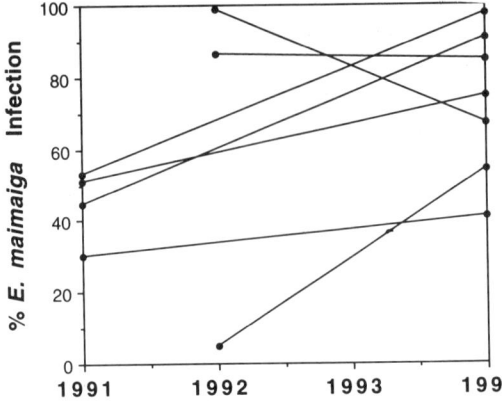

Figure 5. Persistence of *Entomophaga maimaiga* for 2–3 years after inoculative release at locations in Virginia, U.S.A. (Study reported in Hajek, 1995).

importantly, entomopathogenic fungal populations also frequently persist for over one year (Keller, 1983; Shimazu *et al.*, 1993; Rath *et al.*, 1995a) (Fig. 5).

Entomopathogenic fungi range from facultative to obligate pathogens. Species exemplifying the extremes of this continuum can persist during periods without hosts using two trends in strategies. More obligate pathogens persist as stages not requiring growth whereas facultative pathogens can grow saprophytically until once more living as pathogens. While many entomophthoralean species are obligate pathogens, entomopathogenic fungi in the Deuteromycetes range from obligate to facultative pathogens, although many of the better known species predominantly act as pathogens in nature (R. A. Humber, personal communication).

3.1. Inactive Resting Stages

3.1.1. Entomophthoralean Resting Spores

Zygospores or azygospores are produced by the majority of species in the Entomophthorales after host death. For *E. maimaiga,* only infections initiated by conidia, and not germ conidia, result in resting spore formation (Hajek, 1997). For *E. maimaiga* conidial infections, the single factor most important in promoting production of resting spores is host age; early instars predominantly produce conidia and older instars produce resting spores, although frequently producing conidia as well (Hajek and Shimazu, 1996). In addition, resting spore production by *E. maimaiga* is positively associated with temperature, humidity, and fungal dose as well as extensive serial passage, and it is negatively associated with molting and varies by fungal isolate (Shimazu, 1987; Hajek and Shimazu, 1996). For other host/pathogen systems, increased resting spore production is frequently positively associated with host and age negatively associated with temperature (see Hajek and Shimazu, 1996). The cues initiating resting spore production often signal the end of the presence of hosts; although several species produce resting spores under colder temperatures as fall approaches, *E. maimaiga* produces resting spores under warmer conditions because the susceptible stage of its univoltine host is complete in July. In other host/pathogen systems, resting spore production has also been associated with lack of light, decreased or increased humidity, specific fungal isolates or host species, as well as increased dose or serial passage. *Zoophthora radicans* infecting spotted alfalfa aphid produced resting spores more frequently when specific isolates were used in combination for infections (Glare *et al.*, 1989).

After insects die from entomophthoralean infections, cadavers containing resting spores frequently fall to the ground where resting spores are leached into the soil. Resting spores of the grasshopper pathogen *Entomophaga grylli* and the aphid pathogen *Conidiobolus obscurus* have been recovered from soil at 38–120 and 320 resting spores/g dry soil, respectively, in areas where epizootics had occurred that same season (MacDonald and Spokes, 1981; Li *et al.* 1988). Using a regression to correct for losses during soil extraction and density gradient centrifugation, densities of *E. maimaiga* resting spores at the bases of tree trunks that had been covered with cadavers of gypsy moth larvae the previous season ranged from 1202–3895 resting spores/g dry soil (Hajek and Wheeler, 1994). These high densities of *E. maimaiga* resting spores are restricted to the O_e soil layer (the duff layer) at the tree base with densities decreasing precipitously with increasing soil depth and distance from the tree base (A.E.H., unpublished data). *E. maimaiga* resting spores can also be recovered when bark from trees covered with cadavers from the previous season is sonicated (Hajek and Roberts, 1991). *E. maimaiga* resting spores overwintering in/on bark can survive a winter in the northeastern United States (Hajek and Roberts, 1991) and Japan (Shimazu *et al.*, 1987).

Resting spores have a double wall that is thought to enhance their environmental resistance. Resting spores of most species undergo an obligate constitutive dormancy after maturation. Research efforts have concentrated on the minimum length of time resting spores must remain dormant and what cues trigger their release from dormancy. Five of the eight species for which requirements have been identified break dormancy after exposure to cold under moist conditions for several months (Table II). Alternatively, *Zoophthora canadensis* resting spores germinate readily when incubated under a photophase of 12 hours or more (Wallace *et al.*, 1976). Uncharacteristic of this order, *Conidiobolus thromboides* resting spores do not require a dormant period. However, exposure of *C. thromboides* resting spores to aromatic compounds or fungal extracts increased germination significantly (Soper *et al.*, 1975).

In many instances, when resting spore germination is first possible, few spores germinate. However, this percentage increases with increasing time and,

Table II. Entomophthoralean Zygospore and Azygospore Dormancy Requirements

Fungal species	Conditions necessary for initiation of germination	References
Conidiobolus obscurus	3–7°C for 3 months at ≥95% RH	Perry and Latgé, 1982
Conidiobolus thromboides	No dormancy required	Soper et al., 1975
Furia crustosa	4–12°C for 3 months in moist soil	Perry and Fleming, 1989b
Neozygites fresenii	5–14°C for 14–15 days at high RH	Ben-Ze'ev et al., 1990
Pandora bullata	≤20°C for 2 months	Perry, 1988
Zoophthora canadensis	>12 hours light every 24 hours after 32 days at 4°C	Wallace et al., 1976
Zoophthora phytonomi	4°C for 2.5 months in moist soil or at 100% RH	Perry, unpublished
Zoophthora radicans	4°C for 2 months in moist soil or at 100% RH	Perry and Fleming, 1989a

for several species, maximal germination is observed only with spores older than 6 months (Perry, 1988). In addition, after 6 months, resting spores of several species will germinate across a broader range of temperatures than when they are younger, e.g., for Z. radicans 8–20°C at 3 months compared with 4–32°C at 8 months (Perry and Fleming, 1989a). Thus, conditions allowing germination of older resting spores are more flexible and might help to synchronize resting spore activity with activity of host insects in temperate climates. Bioassays conducted in the field demonstrate that resting spore activity is indeed closely associated with the presence of hosts; infections due to resting spores of the gypsy moth larval pathogen E. maimaiga begin in spring, about 2 weeks prior to the onset of host egg hatch and cease when late instars or pupae are present (Hajek and Roberts, 1991). Reasons for asynchrony in germination are not known. Curiously, when E. maimaiga resting spores were extracted from cadavers collected in the spring and placed in the laboratory at 15°C, germination began after 4 days and ceased within 2 weeks, although resting spore germination occurs over > 2 months in the field (A.E.H., unpublished data).

Resting spore germination has been studied in detail for few species due to the requirements for satisfying dormancy prior to germination and difficulties with saprophytic contamination of resting spores overwintered in the environment. One major problem with studies of resting spore germination is that methods for determining whether non-germinating resting spores are viable have not been developed; thus, lack of germination could be due to dormancy or inviability. The dynamics of resting spore germination differs strikingly from conidial germination; after dormancy requirements have been satisfied for E. maimaiga, resting spore germination only begins after about 4 days at 20°C

(A.E.H., unpublished data) while conidia begin to germinate several hours after production (Hajek et al., 1990a). For Z. radicans, maintenance of resting spores at 16–20°C for 6–10 months yielded the fastest time to germination but total percent germination was greater at temperatures from 8–16°C (Perry and Fleming, 1989a). At 25°C and a photophase of 16 hours, the percentage of E. grylli resting spores germinating increased rapidly, attaining maximum germination of approximately 33% by 9 days; these spores required an aerobic environment for germination (Stoy et al., 1988). For the aphid pathogen C. obscurus, resting spores germinated more readily at 15°C than at 10°C or 20°C (Latteur et al., 1982). C. obscurus resting spores desiccated for 12 days were not affected, but 24–48 days of desiccation reduced production of germ conidia by 35%. Using in-vitro-produced spores, production of germ conidia by E. aulicae resting spores occurred at pH 6.5, while only hyphae were produced from resting spores under more alkaline conditions (Nolan et al., 1976).

Even after extended periods, resting spore bioassays usually document germination of far less than 100% (e.g., Krejzova, 1971; Soper et al., 1975; Perry 1988; A.E. Hajek, unpublished data). This germination can range from 15–77% for different isolates of Z. radicans (Perry and Fleming, 1989a). Entomophthoralean resting spores not only allow disease persistence over one winter but may allow persistence for longer periods. It appears that many entomophthoralean resting spores may not germinate only after one year after production, e.g., azygospores of E. maimaiga in the field are known to germinate up to six years after their production (Weseloh and Andreadis, 1997). Similarly, zygospores of C. thromboides survived for five years, although during this time they remained at 7–14°C and did not experience warmer temperatures as would occur in the field (Krejzova, 1971).

3.1.2. Vegetative Fungal Stages

If an infected insect dies and humidity is not high enough for outgrowth of conidiophores that would subsequently bear conidia, the fungus in cadavers of some species can retain the ability to develop and reproduce once favorable conditions occur. Mummies, containing desiccated hyphal stages, thereby appear to provide survival for some species of Entomophthorales. For example, Z. radicans (Pell and Wilding, 1992), P. neoaphidis, C. obscurus (Wilding, 1973), E. aulicae (Tyrrell, 1988), Neozygites floridana (Kenneth et al., 1972) and Pandora sp. (Newman and Carner, 1975) can all survive as hyphal stages within dried cadavers. Survival of Z. radicans in mummified cadavers was documented for up to 34 weeks at 4°C (Pell and Wilding, 1992). Pandora neoaphidis mummies can survive 6–8 months when held near 0°C and 15% relative humidity (RH), although mummies did not survive < 0°C, humidities near 0%, or lyophilization (see Wilding and Latteur, 1987). However, many entomophthoralean species probably survive in mummified cadavers over the shorter term, e.g., during transient cool or dry periods. This ability of mycelia of entomophthoralean fungi to survive in a desiccated state has been exploited for several species by researchers searching for an optimal entomophthoralean stage for storage after mass production for biological control use (e.g., Li et al., 1993).

For entomophthoralean species, conidia can survive for over one month in some instances (see Section 4.6). However, entomophthoralean conidia are generally not thought to persist over long periods of unfavorable conditions (i.e., winter), although this dogma should be reexamined due to recent evidence demonstrating survival of *P. neoaphidis* conidia for over 9 months at 3°C on soil (Morgan et al., 1992).

Among the Deuteromycetes, entomopathogenic fungi generally survive as mummified cadavers or conidia. In the field, *Nomuraea rileyi* in mummified cadavers of corn earworm larvae survived and produced spores after at least 281 days on the soil surface and at least 194 days buried 10 cm in the soil (Sprenkel and Brooks, 1977). *N. rileyi* in mummified cadavers of the velvetbean caterpillar survived for at least 12 weeks when stored dry at room temperature; environmentally resistant structures were observed in overwintering cadavers, including externally produced, thick-walled chlamydospores (Pendland, 1982). Many studies of conidial survival have been conducted due to the production of deuteromycete conidia for insect control (reviewed in section 4.6).

Species of the deuteromycete genus *Sorosporella* persist as thick-walled vegetative chlamydospores in the environment (Steinhaus, 1949). After death of infected grasshoppers, the host cuticle becomes pale and fragile, resulting in ready release of the internally produced chlamydospores (Wellings et al., 1995). These chlamydospores do not necessarily require a period of rest, can be viable after dry storage for 14 months, and can survive the low temperatures experienced during northeastern North American winters.

3.2. Persistence in Soil and Facultative Saprophytic Growth

Soil provides the matrix for maintenance of a natural reservoir of many entomopathogenic fungi. Soils can be inoculated with entomopathogenic fungi either by an infected insect entering into the soil and subsequently dying or by deposition of spores on the soil surface. For some species, the soil environment also provides a medium for growth and potential dispersal. For pathogens also able to grow as saprophytes, fungal growth can extend far beyond cadavers in the soil environment, e.g., hyphae of *B. bassiana* spread throughout the surrounding soil with colonies radiating from pecan weevil cadavers (Gottwald and Tedders, 1984) or armyworm cadavers (Studdert and Kaya, 1990b), as far as 8.5 and 4.9 cm in diameter respectively.

B. bassiana conidia have been isolated from soil at the base of elm trees that died from Dutch elm disease as well as from the bark (Doberski and Tribe, 1980). It was suggested that these conidia had originated from insects, predominantly under the bark, that had died from *B. bassiana* infections. *M. anisopliae* and *Paecilomyces farinosus* were rarely isolated from soil at elm tree bases. An analysis of soil surveys for fungi suggests that neither *B. bassiana* nor *M. anisopliae* are frequently found except in locations that are known to harbor insect populations (Doberski and Tribe, 1980). Conidia of entomopathogenic fungi deposited on the surface of soil become washed into the soil to varying degrees depending on soil type, but tend to remain very close to the soil surface. *N. rileyi* conidia percolated readily through sand, while 90% of conidia remained at < 2 cm depth in silt-loam soil (Ignoffo *et al.*, 1977). In sandy clay-loam soil, the majority of *B. bassiana* conidia remained in the top 5 cm (Storey and Gardner, 1987; Storey *et al.*, 1989).

Location of conidia within the soil can enhance fungal survival; *B. bassiana* (Gaugler *et al.*, 1989) and *N. rileyi* (Ignoffo *et al.*, 1978) conidia below the soil surface survived better than conidia applied to the surface. However, many abiotic and biotic conditions in the soil influence survival and activity of fungi, including temperature, soil moisture, aeration, and soil characteristics (McCoy *et al.*, 1988). High temperature and high moisture levels adversely affect survival of conidia of *B. bassiana* (Studdert *et al.*, 1990; Krueger *et al.*, 1991a) and dry mycelium of *M. anisopliae* and *Beauveria brongniartii* (Krueger *et al.*, 1991b) applied to soils. However, these studies evaluated conidial survival but not infectivity. A study of *B. bassiana* infection of beet armyworm pupae buried in soil demonstrated that conidia in soil at -37 and -200 bars water potential (equal to less water than that for the permanent wilting point for many plants) resulted in high infection rates as temperature was increased from 8°C to 25°C (Studdert and Kaya, 1990a). Curiously, although the greatest infection occurred at -200 bars, this is a water potential at which conidial half-lives were very low (Studdert and Kaya, 1990a). The authors suggest that the high dosage of 3.2×10^7 or 1×10^8 conidia/cm^3 allowed enhanced infection even at low moisture levels redundancy because soil microbial activity would have been negligible, resulting in less interference with *B. bassiana* conidia.

Researchers have found increases in the number of *B. bassiana* colony forming units over time after application of conidia to soil (see Storey *et al.*, 1989), supposedly due to saprophytic growth. However, there are probably more examples of fungistatic effects of soils on *B. bassiana*, *M. anisopliae*, and *N. rileyi* (see Groden and Lockwood, 1991; McCoy *et al.*, 1988). Fungistasis has been attributed to intense competition within the microbial community for limited resources in the energy-poor soil environment. *M. anisopliae* is known to be a poor competitor when growing saprophytically in the soil; *M. anisopliae* only successfully invaded cadavers of wax moth larvae in sterile soil where competitors were absent (Keller and Zimmermann, 1989). *B. bassiana* and *Paecilomyces* sp. conidia can remain viable in soil for 2–3 years, and it has been suggested that in this situation the fungistatic influence of soil may be beneficial, acting to preserve conidial activity (see McCoy et al., 1988).

Some fungal species are only rarely pathogenic to insects. It has been hypothesized that these species may commonly survive as dormant epiphytes, opportunistically invading leaves wounded by herbivorous insects and only occasionally becoming abundant enough to infect these insects and grow as pathogens (Carroll, 1991).

4. Activity of Entomopathogenic Fungi during Favorable Periods

For infection to occur once a conidium lands on a host, it must adhere to the cuticle, germinate, possibly differentiate, and penetrate through the host cuticle (Fig. 1). After cuticular penetration, fungi can encounter defensive host reactions and/or produce toxins, supposedly to facilitate their invasion of hosts. These highly specialized processes are completely dissimilar from other events in the fungal life cycle. Conidia not landing on hosts can either not germinate, produce a non-penetrant germ tube, or form conidiophores to produce microconidia, secondary conidia, or adhesive conidia (e.g., capilliconidia) (Boucias and Pendland, 1991). Of these options, only the last can result in infection.

The most common mode of infection is through the insect integument, although there have been reports of infection through spiracles (Charnley and St. Leger, 1991) or the digestive tract (for reports for *B. bassiana* see Feng et al., 1994). However, Allee et al. (1990) demonstrated that *B. bassiana* conidia passed through the gut of Colorado potato beetles without infecting, and excreted conidia in frass subsequently infected beetles externally. While the locust gut pH is optimal for growth of *M. anisopliae*, germination there is inhibited, potentially due to fungitoxic compounds produced by bacteria inhabiting the locust hindgut (reviewed in Dillon and Charnley, 1991).

4.1. Conidial Adhesion

When a conidium lands on a potential host, firm attachment of the pathogen to the host surface is necessary for prepenetration development and infection processes. Attachment is mediated by chemical components of the outer layers of the spore and the host epicuticle. Adhesion processes differ for hydrophobic (e.g., some deuteromycete species) and hydrophilic (many entomophthoralean species, *Verticillium, Aschersonia, Hirsutella*) conidia (Boucias and Pendland, 1991). The hydrophilic conidia of many Entomophthoraleans and some Deuteromycetes are covered by an extracellular mucus that aids in adherence to substrates (Fig. 6). In addition, it has been suggested that this mucilage modifies epicuticular waxes (Wraight et al., 1990), facilitates splash dispersal of some

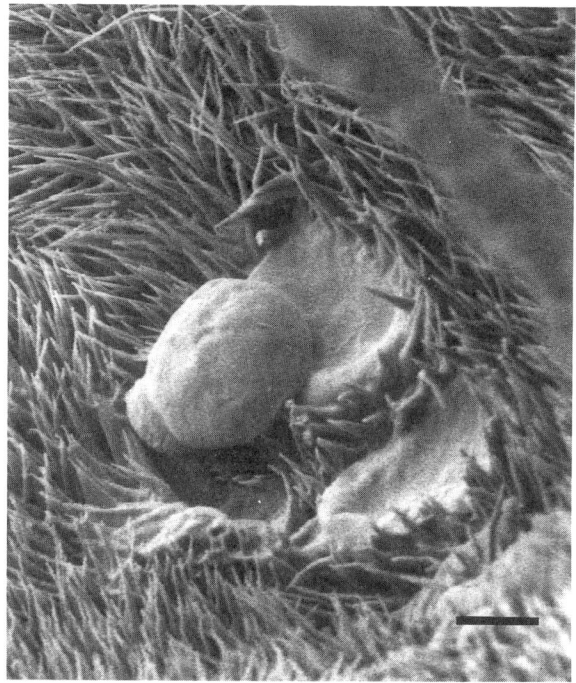

Figure 6. Conidium of *Entomophaga maimaiga* on gypsy moth cuticle, slightly dislodged so that mucus used for adhesion is visible (bar = 10.0 μm).

species (see Section 4.5), and may protect conidia from sunlight and desiccation (Dickinson, 1986). Mucilaginous coats on conidia are generally not actively secreted after dispersal but are constitutive components of conidia at the time of their production. Dry, hydrophobic conidia of *B. bassiana, M. anisopliae, N. rileyi,* and *Paecilomyces fumosoroseus* are covered by a rodlet layer composed of protein, lipoprotein, glycoprotein, or polysaccharide (Boucias and Pendland, 1991). This rodlet layer passively attaches conidia to host cuticles by strong binding forces involving hydrophobic and, to a minor extent, electrostatic interactions. Surface hydrophobicity is also thought to influence adhesion of fungal pathogens; greater numbers of *B. bassiana, M. anisopliae,* and *N. rileyi* conidia bind to neonate larval cuticle than to less hydrophobic fourth instar larval cuticle (Boucias and Pendland, 1991).

The majority of adhesion is non-specific with equivalent binding to host and non-host surfaces (Boucias and Pendland, 1991). In only a few cases have entomopathogenic conidia been shown to adhere specifically, e.g., *M. anisopliae* conidia adhere to cuticle of only some scarabs (Fargues, 1984) and *Z. radicans* conidia adhere only to certain leafhopper species (McGuire, 1985). Conidial adhesion may be enhanced by cuticular topography and chemical properties of the epicuticle. While conidia of *B. bassiana, M. anisopliae*, and *N. rileyi* are capable of attaching to all body regions, more conidia adhered to cuticular areas with short spines and conidia were frequently found tightly bound to these spines (Boucias *et al.*, 1988). *M. anisopliae* conidia were more easily removed from smooth, exposed sclerite epicuticle but remained firmly attached to epicuticle in intersegmental folds (see Boucias and Pendland, 1991).

4.2. Conidial Germination

Conidial germination frequently occurs within several hours under optimal conditions, e.g., 93% of *N. fresenii* capilliconidia germinated by 6 hours after discharge at 25°C and 100% RH (Steinkraus and Slaymaker, 1994). Abiotic optima for conidial germination are frequently very similar to conditions necessary for sporulation. For example, for *B. bassiana* and *M. anisopliae*, optimal temperatures (25–30°C) and relative humidity (100% RH) were equivalent for conidial germination, mycelial growth, and sporulation (Walstad *et al.*, 1970). The lowest moisture level for both outgrowth of *Zoophthora phalloides* from cadavers and conidial germination was 98% RH (Glare *et al.*, 1986). However, *E. maimaiga* requires free water for germination but only high humidities are necessary for fungal outgrowth and conidial discharge (Hajek *et al.*, 1990a). Although many entomopathogenic fungi require high moisture levels for germination, *E. muscae* (Kramer, 1980) and *Entomophthora planchoniana* (Holdom, 1984) are capable of germination at low humidities. Curiously, *E. muscae* does not require high humidity for sporulation while *E. planchoniana* only sporulates at RH > 95%. While research emphasis frequently focuses on moisture levels as critical to conidial germination, interactions between temperature and humidity can govern response to moisture, e.g., if humidity is high but temperatures are too low for a high percentage of spores to germinate, infections will not result (Hall and Papierok, 1982).

Temperature also can affect the types of secondary conidia produced when primary conidia do not initially land on suitable hosts. At ≤ 10°C, the majority of secondary conidia produced by primary conidia of *P. neoaphidis* were lemon-shaped, while at > 20°C, a rounded secondary conidium was formed (Morgan *et al.*, 1995). Primary conidia of *C. obscurus* produced a secondary conidium with a small papilla at 5°C and 10°C, while at 20°C secondary conidia bore large round papillae (Latteur, 1980). Unfortunately, the ecological significance of these morphologically different secondary conidia has not yet been determined.

Most entomopathogenic fungi are thought to produce spores and infect at night. It follows that 50% of *E. maimaiga* conidia germinated within 3.4 hours in the dark, while 9 hours were required before 50% germination in the light (Hajek *et al.*, 1990a). Similarly, high rates of germination of *N*. cf. *floridana* primary conidia occurred in dark, while light completely inhibited germination (Oduor, 1995). Infective *N*. cf. *floridana* capilliconidia produced by primary conidia could germinate in light (20.3%), although more capilliconidia germinated in darkness (44.93%) (Oduor, 1995). In fact, infective capilliconidia germinate over a wider range of temperatures, light regimes, and moisture levels than primary conidia, allowing more flexibility for conditions under which infection can occur.

Entomophthoralean conidia are generally not known to require an exogenous source of nutrients to germinate, e.g., most studies of germination are conducted with conidia showered onto coverslips or water agar. In contrast, deuteromycete conidia tend to require external sources of nutrients for germination, e.g., to conduct germination studies of Deuteromycetes, conidia are often showered onto nutrient media (e.g., Walstad *et al.*, 1970). *B. bassiana* conidia required carbon and nitrogen sources for germination and germ tube growth (Smith and Grula, 1981) and *N. rileyi* requires even more specific nutrients (Boucias and Pendland, 1984). In contrast, other researchers have found that *B. bassiana* and *P. fumosoroseus* conidia do not always require nutrients and can germinate on water agar (J. D. Vandenberg, personal communication).

Fungistatic compounds (fatty acids) on surfaces can terminate fungal development as seen with *B. bassiana* conidia on three insect hosts (Smith and Grula, 1982; Saito and Aoki, 1983). However, the presence of nutrients can alter the effect of these compounds (Smith and Grula, 1982). Conversely, lipids associated with insect cuticle can stimulate infection, e.g., cuticular extracts regulated conidial germination in several systems (Kerwin, 1984; Boucias and Pendland, 1984; Latgé *et al.*, 1987; Bidochka and Khachatourians, 1992; Nadeau *et al.*, 1996) although this effect can be non-specific (Boucias and Latgé, 1988).

4.3. Infection and Growth *in Vivo*

Different types of entomophthoralean conidia can differ extensively in infectivity. Bellini *et al.* (1992) found that secondary conidia of *E. muscae* were approximately 10 times more infective for house flies than primary conidia (primary conidial LC50 = 67 conidia/mm^2; secondary conidial LC50 = 0.36 conidia/mm^2). For species in the genera *Neozygites* and *Zoophthora* producing capilliconidia secondarily from primary conidia, the capilliconidia are infective, while primary conidia and those secondary conidia morphologically similar to primaries are considered dispersive forms (see Morgan *et al.*, 1995).

4.3.1. Pre-Penetration Events

After conidia adhere to host cuticle and begin to germinate, resulting germ tubes must orient and attach to the epicuticular surface before penetration can

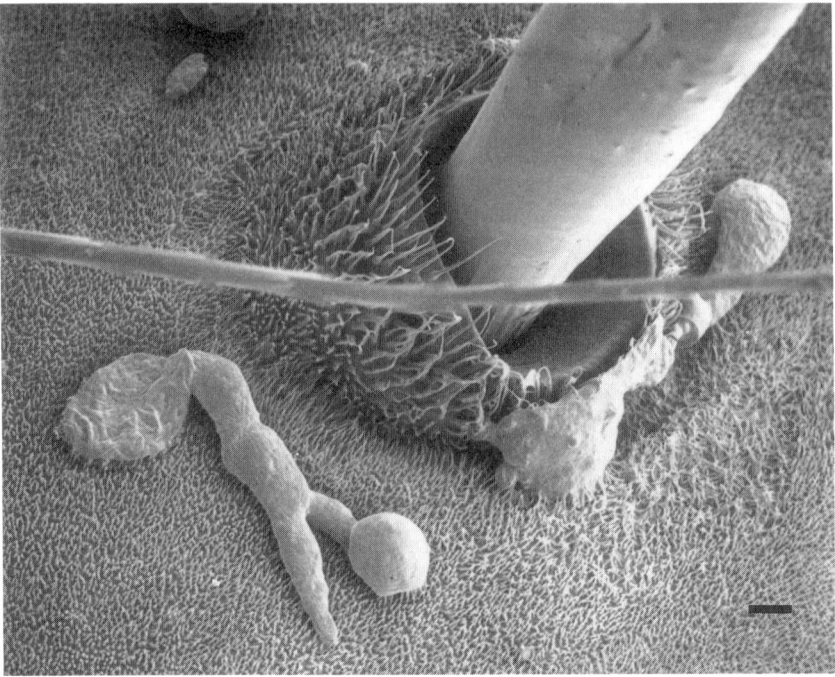

Figure 7. Conidia of *Entomophaga maimaiga* (conidium on left is collapsed) that have produced germ tubes at the end of which appressoria have formed around the base of a secondary seta on gypsy moth cuticle (bar = 10.0 μm).

proceed. Germ tubes, appressoria growing from mucus-coated conidia, and hydrophobic conidia all may actively produce mucus for adhesion (see Boucias and Pendland, 1991).

The penetration of insect cuticle often requires the formation of specialized infection structures such as appressoria (Fig. 7). Broadly defined, appressoria are structures adhering to host surfaces to achieve penetration (Emmett and Parbery, 1975). They frequently are the swollen end of a germ tube or hypha and are usually separated from the germ tube by a septum. While appressoria are well known from many insect pathogenic systems, infection without production of distinct appressoria has also been documented (Fargues, 1984), especially for soft-bodied insects such as bagworms (Berisford and Tsao, 1975) and aphids (Butt *et al.*, 1995). This supports the hypothesis that only hard surfaces stimulate appressorial production, although appressoria can also be produced in the relatively soft intersegmental membranes of waxmoth larvae (Butt *et al.*, 1995), so cuticular rigidity is not the only factor inducing appressorial formation. Appressorial formation can be induced by low levels of complex nitrogenous compounds but can be repressed by catabolites (St. Leger *et al.*, 1989, 1992; Magalhaes *et al.*, 1991a).

Appressorial formation can also be triggered by thigmotropic stimuli. In tobacco hornworm, appressoria of *M. anisopliae* were formed more readily on the flat cuticle of fifth-day fifth instars than the convoluted cuticle of first-day fifth instars (St. Leger *et al.*, 1991), and appressoria were preferentially formed at the bases of setal sockets on both native cuticle and plastic replicas of cuticle (St. Leger *et al.*, 1988). Wraight *et al.* (1990) found that modes of germination of *Z. radicans* were different based on the part of the leafhopper onto which conidia were showered; conidia deposited on sclerotized cuticle showed strong directional growth parallel to the axis of the leafhopper body toward intersegmental membranes where appressoria were formed.

St. Leger *et al.* (1991) propose that differentiation of germ tubes is initiated after a contact-induced change in the potential of germ tube membranes, resulting in disruption of the apical Ca^{+2} gradient necessary for polar growth. In agreement, *Z. radicans* requires external Ca^{+2} for formation of appressoria (Magalhaes *et al.*, 1991b). In addition, a significant increase in levels of cAMP occurs at the onset of appressorial formation (St. Leger *et al.*, 1990). Molecular-level regulation of genes associated with appressorial formation by *M. anisopliae* is presently being investigated (R. J. St. Leger, personal communication; see Clarkson and Charnley, 1995).

4.3.2. Penetration

Insect pathogenic fungi penetrate the multilayered insect cuticle using different combinations of enzymes and mechanical pressure (St. Leger, 1993). A penetration peg is produced either from a germ tube or appressorium. This penetration peg must initially invade the epicuticle, a thin layer lacking chitin but containing phenol-stabilized protein and covered by a waxy layer containing fatty acids, lipids, and sterols (Anderson, 1979). Beneath the epicuticle, the much thicker procuticle contains chitin fibrils embedded in a protein matrix with lipids and quinones. Approximately 70% of the cuticle is composed of protein.

A number of extracellular enzymes are produced by *M. anisopliae* conidia grown *in vitro* on insect cuticle (see Clarkson and Charnley, 1995; St. Leger, 1995). The majority of these enzymes are proteases and peptidases with only one chitinase, as might be expected because cuticle is composed of such a high percentage of protein. Among these enzymes, the serine protease PR1 is the major protein synthesized during appressorium development of *M. anisopliae* (St. Leger *et al.*, 1989). In fact, PR1-like proteases have been found in all of the entomopathogenic deuteromycete and ascomycete species tested (St. Leger *et al.*, 1987). In *M. anisopliae*, penetration of the epicuticle apparently occurs through enzymic degradation because PR1 has been detected in abundance at the point of penetration, with no mechanical deformation (Goettel *et al.*, 1989). The biochemical and molecular level investigations of PR1 and associated enzymes have been recently reviewed (St. Leger, 1993; Clarkson and Charnley, 1995). In contrast, penetration of the procuticle involves mechanical processes as well as enzymic degradation, as detected by deformation of procuticular lamellae. Unfortunately, studies of penetration processes of entomophthorales have not been conducted for comparison with deuteromycetes.

4.3.3. Survival and Growth within Hosts

In general, several hours are required from conidial inoculation through successful cuticular penetration, e.g., at 25°C *E. maimaiga* requires 8.7 hours for successful infection of 50% of gypsy moth larvae inoculated (Hajek *et al.*, 1993). After penetration into the host, the environment surrounding fungal cells changes drastically as they are now surrounded by a sea of nutrients. Many fungal species grow vegetatively within the insect hemocoel as yeastlike structures (blastospores) or hyphal bodies, while some species of Entomophthorales grow within the hemolymph as protoplasts lacking a cell wall. These growth forms cannot survive naturally outside of hosts and are thought to provide fungi with increased surface area for acquisition of nutrients, the ability to readily disperse and colonize the hemocoel through circulation in the hemolymph, and possible evasion and dissipation of the host cellular and/or humoral immune response. Many fungi do not penetrate internal organs before host death. For example, early in infection *E. grylli* cells adhere to fat body and penetrate neural tissue, although skeletal muscle is not penetrated until after host death (Funk *et al.*, 1993). For many fungi, host death is attributed to nutrient depletion. In agreement, by the time *E. maimaiga* kills gypsy moth larval hosts, samples taken from the hemolymph are completely filled with fungal cells and virtually no hemocytes remain. That entomophthoralean pathogens do not kill hosts before all of the available nutrients have been utilized is consistent with the highly obligate nature of their relationships with hosts. In contrast, some insects infected by deuteromycetes die after only limited vegetative growth of fungi within their bodies, due to activity of lethal mycotoxins (see below).

Not all fungal cells that successfully penetrate the insect cuticle grow and reproduce. Hosts mount humoral and cellular defense responses and, in a few instances, behavioral alterations of hosts can cure fungal infections. The possibility that hemolymph and tissues of some insect species do not provide the correct and adequate nutrients for growth of some fungi has not been investigated.

4.3.3.a. Non-Self Recognition. A fundamental basis for host specificity is not only the ability to penetrate insect cuticle but also the ability to survive within hosts and escape host defense responses. Mechanisms used by hosts to detect entomopathogenic fungi are poorly understood. Sugars in cell walls are thought to be crucial for host detection of fungal cells; therefore, the surface components of potentially pathogenic fungi may determine whether invading fungi are recognized as non-self by host hemocytes. Principal components of fungal cell walls are β-glucans and chitin. One major fungal mechanism for evading host immune responses is to cease producing the sugar-rich cell wall. Entomophthoralean protoplasts lack cell walls and can thereby escape host recognition; while wall-less protoplasts bearing few surface sugars are not detected by host defenses, walled hyphal bodies become encapsulated by hemocytes (Beauvais *et al.*, 1989; T. M. Butt, unpublished data). A similar evasive strategy has also been reported from *N. rileyi*; hyphal bodies having cell walls shed a galactomannan coat that can be recognized by defensive hemocytes (Pendland *et al.*, 1993). Some studies have shown differential recognition of fungal cells growing *in vivo* versus *in vitro*. *P. farinosus* blastospores produced *in vitro* and injected into beet armyworm larvae were recognized by hemocytes, while *in vivo*-grown blastospores lacked a galactomannan surface layer and were not recognized (Pendland *et al.*, 1995). An induced galactose-binding lectin in beet armyworm hemolymph assists in recognition of *in vitro*-produced *N. rileyi* blastospores by hemocytes, but *in vivo*-grown hyphal bodies are not recognized (Pendland *et al.*, 1988).

However, nonrecognition of cell wall sugars alone cannot completely explain fungal survival *in vivo*. Both *E. grylli* and *E. maimaiga* produce protoplasts *in vivo*, yet *E. grylli* (a grasshopper pathogen) is encapsulated by gypsy moth hemocytes while the gypsy moth-specific *E. maimaiga* is not (Butt and Humber, 1989). *E. maimaiga* and *E. aulicae* belong to the same species complex, both infecting Lepidoptera, but *E. maimaiga* cannot infect some hosts of *E. aulicae* and vice versa (Hajek *et al.*, 1991b; Bidochka and Hajek, 1996b). Major differences in mannose glycosylation have been observed in proteins of protoplast plasma membranes of these species; it is possible that these glycoproteins in protoplast plasma membranes may be important in fungal recognition (Bidochka and Hajek, 1996a). Treatment of hemolymph of *E. maimaiga* hosts with *E. aulicae* protoplast plasma membranes resulted in higher polyphenoloxidase activity compared with treatment with *E. maimaiga* protoplast plasma membranes (Bidochka and Hajek, 1996b), demonstrating recognition of the non-pathogen and suggesting involvement of humoral host defenses in regulation of specificity in this host/pathogen system.

4.3.3.b. Immune Responses. Insect immune responses are primitive compared to those of higher eukaryotes; they consist of both cellular and humoral components but demonstrate little "immunologic memory" (Boucias and Latgé, 1988). The primary reaction to invading fungal cells is generally hemocyte aggregation around invaders with subsequent melanization (encapsulation). Cellular defenses can be readily activated by soluble, cell-bound glucan molecules on invading fungal cells. Fungal cells penetrating cuticle are frequently recognized and melanized, leaving darkened cuticular lesions (Butt et al., 1988). However, it is not known whether this is a defense response or wound repair and, furthermore, this response did not prevent penetration of leafhopper cuticle by *Z. radicans* (Butt et al., 1988).

Once fungi are within the host body, encapsulation of fungal cells may only provide protection for host insects against weakly virulent pathogens. *B. bassiana* conidia injected into migratory grasshoppers can germinate within nodules formed around them by hosts and subsequently reinvade the hemolymph (Bidochka and Khachatourians, 1987). With a hypervirulent strain of *B. bassiana*, either hosts cannot form typical defensive capsules around invaders or fungi overcome encapsulation (Hung et al., 1993). In addition, after infection of beet armyworm by *B. bassiana*, the number of granulocytes declines precipitously after 3 days (Hung and Boucias, 1992). It has been hypothesized that the toxic metabolites produced by *B. bassiana* target host hemocytes and thereby dampen the host cellular defenses (Hung et al., 1993). These examples suggest that *B. bassiana* restrains the cellular response at the same time as overcoming it. For insect hosts with few hemocytes, e.g., Diptera and some Homoptera, humoral defense responses have been adequate defenses in some instances, but were unable to contain growth of *Z. radicans* in potato leafhoppers (Butt et al., 1988).

As another method of host defense, protease inhibitors in the hemolymph of insects resistant to *N. rileyi* have been shown to inhibit germination of *N. rileyi* conidia and growth of germ tubes (Boucias and Pendland, 1987). An antifungal protein has been identified from the hemolymph of a flesh fly (Iijima et al., 1993). However, these findings of antifungal chemicals in insect hemolymph are unique, and the activity of these proteins has yet to be investigated fully.

4.3.3.c. Host Behavioral Response to Infection. In a few systems, use of host behavior to induce fever and, thereby, to control fungal infection has been demonstrated. Both grasshoppers and flies normally bask in the sun to increase body temperatures before flight or other activities. Grasshoppers infected with *E. grylli* and allowed to bask at 25–30°C almost totally eliminated infection, in comparison with infected controls exposed to diffuse light (Carruthers et al., 1992). House flies infected with *E. muscae* and exposed to temperatures of 40°C early in the infection period (1–3 days) eliminated infections, while most flies exposed to high temperatures on days 4–6 after infection did not recover (Watson et al., 1993). Flies given free choice of temperature exhibited behavioral fevers on days 2–3 after infection but not on days 1, 4, or 5 and, by day 5, dying flies moved to the coolest part of the temperature gradient. As cooler temperatures are beneficial to *E. muscae,* it seems that by day 5, the fungal pathogen might be influencing fly behavior. While basking behavior is not known from gypsy moth larvae, constant exposure to a temperature of 30°C after infection with *E. maimaiga* cured larvae of infections (Hajek et al., 1990a), and similar cures with exposure to temperatures lethal to fungal pathogens are known from other entomopathogenic fungi.

4.3.3.d. Fungal Toxin Production. Most of the biologically active mycotoxins that have been studied are produced by Deuteromycetes. Our knowledge of these toxins is predominantly based on studies using toxins extracted from filtrates of *in vitro* cultures of entomopathogenic fungi or from mycelium (Clarkson and Charnley, 1995). Some of these compounds have been found in infected insects but few studies have been conducted establishing the impact of these toxins in disease development. Khachatourians (1991) presents a review of relatively recent studies of toxins. Extensive research has been conducted with the cyclodepsipeptide destruxins produced by *M. anisopliae*. These compounds are produced in amounts that correlate with toxicosis and differential virulence of isolates (Samuels *et al.*, 1988a, 1988b). Recently, a high-molecular-mass insecticidal protein toxin (>10 kDa) has been extracted from beet armyworm larvae infected with *B. bassiana* (Mazet *et al.*, 1994).

Hosts infected by toxin-producing fungi are frequently killed fairly rapidly, before the fungus has entirely colonized the cadaver. After host death, the fungus completely colonizes the cadaver and then sporulates. Therefore there can be a lag of several days between host death and fungal outgrowth resulting in sporulation. Survival of fungi in cadavers of insects killed by toxigenic species has not been investigated; fungi are growing in these cadavers for a period of time prior to sporulation without the assistance of the insect immune system, and, during this period, must compete with other microbes.

While the majority of species producing toxins belong to the Deuteromycetes, a few toxins have been obtained from entomophthoralean cultures grown *in vitro*. However, the activity of these substances *in vivo* has not been confirmed (Wilding, 1981). A highly labile toxin produced by *E. aulicae* has been identified that is produced only for a brief interval precisely at the time of host death (D. Tyrrell, unpublished data).

4.4. Host Death and Production of Spores for Dispersal

The length of time from conidial inoculation until host death varies both within and between species as well as by fungal dose, temperature, and size and stage of host. For entomophthoralean infections, there is often no marked change in host behavior until directly prior to death. Homoptera, grasshoppers, ants, thrips, and flies infected by entomophthoralean fungi have all been reported to move to elevated positions immediately before host death (Evans, 1989). This behavior has been augmented in the carrot fly/*E. muscae* system; cadavers of carrot flies infected with *E. muscae* are not found at the tops of carrot plants but at the tops of trees in nearby hedgerows. Infected female carrot flies also alter their egg-laying behavior so as to decrease egg survival (Eilenberg, 1986). The physiological basis of climbing behavior just prior to death is unknown. However, it has been hypothesized that negative geotaxis at the time of death enhances subsequent conidial dispersal, e.g., at elevated locations, cadavers could more effectively shower spores over living insects (Wilding, 1981). In fact, growth of fungal rhizoids to attach cadavers in elevated positions such as shrubs or trees is also known from hosts that do not display pre-death negative geotaxis (Evans, 1989). In contrast, late instar gypsy moth larvae infected by the entomophthoralean pathogen *E. maimaiga* descend trees just prior to death, and cadavers containing resting spores are predominantly found adhering to trunks at < 2.5 m (Takamura and Sato, 1973). These cadavers eventually fall from tree trunks and resting spores are subsequently leached into the soil over which potential hosts will walk the next spring. In this case, larval descension instead of climbing might deposit resting spores in a location that will aid in disease transmission the next spring.

In addition to flying to an elevated location, just prior to death house flies exhibit a highly stereotyped sequence of behaviors, including attaching the proboscis to the substrate and raising the wings (Krasnoff *et al.*, 1995). Death of hosts from which an entomophthoralean will produce conidia is often associated with inclination of the abdomen at an angle from the substrate and/or positioning of wings to expose the abdomen, from which conidia are released in abundance (Fig. 8). This positioning of cadavers is thought to maximize exposure of the surface area of the cadaver from which spores will be discharged.

Six aphid/entomophthoralean associations display peak mortality late in photophase, with death in two of these systems occurring daily over time spans as narrow as 4 hour (Milner *et al.*, 1984). House flies infected with *E. muscae* die 0–5 hours before the onset of darkness (Krasnoff *et al.*, 1995). The diurnal pattern of mortality is thought to synchronize the fungus so that cadavers will be ready to sporulate when abiotic conditions are usually more favorable for sporulation, i.e., during scotophase when moisture levels are higher. A circadian rhythm in mortality was observed for house flies exposed to 12L:12D for 3 days and then held in continuous darkness; it is thought that this biological clock is determined by the fungus (Krasnoff *et al.*, 1995).

After hosts die from entomophthoralean infections, the fungus commonly requires a period of growth and/or maturation before conidial discharge begins. For example, entomophthoralean conidial discharge can begin 5 hours after host death for *E. muscae* infecting house flies (Mullens and Rodriguez, 1985), 6–9 hours for *C. obscurus* infecting pea aphids (Wilding, 1971), or 14 hours for *E. maimaiga* infecting gypsy moth larvae (Hajek *et al.*, 1990a). After maturation, under the correct conditions the fungus grows out through the host cuticle. Conidiophores bear conidia, and, among entomophthoralean species, conidia are actively ejected. Conidial production and discharge are sensitive to abiotic conditions and vary widely by fungal species. For three species of Entomophthorales

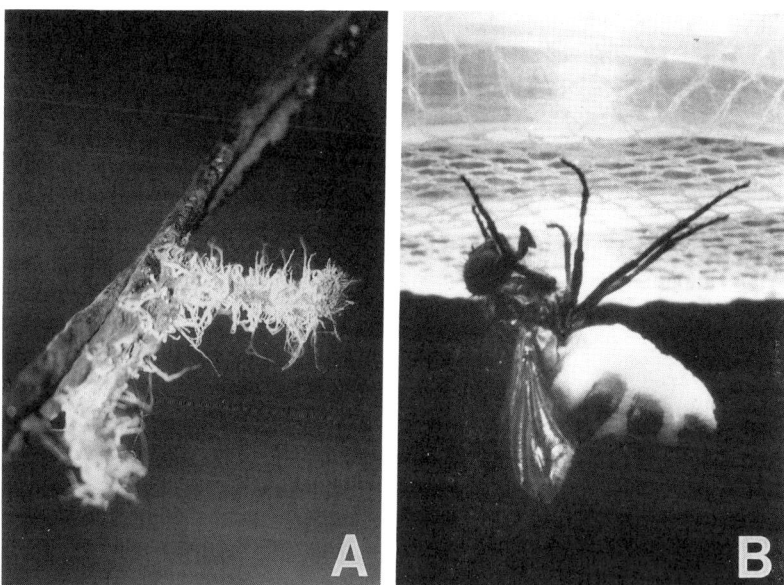

Figure 8. (A) Cadaver of a forest tent caterpillar killed by *Furia crustosa*. Both ends of the body extend away from the substrate to which the cadaver is attached, thereby increasing exposure of the sporulating fungus, (B) cadaver of a housefly killed by *Entomophthora muscae* showing flexion of wings away from dorsum of body and production of conidiophores through intersegmental membranes (Part B courtesy of S. Krasnoff).

infecting the blue-green aphid, numbers of conidia discharged ranged from 6–10 $\times$ 10^3 to 3–4 $\times$ 10^5 conidia/aphid, with sporulation by *C. obscurus* at the same levels across 5–25°C, greatest sporulation by *Entomophthora* sp. nr. *exitialis* only between 15–25°C, and sporulation by *E. planchoniana* only between 10–20°C (Milner, 1981). Many species of Entomophthorales require humidities near 100% to produce and discharge conidia (e.g., Holdom, 1984; Glare *et al.*, 1986; Hajek *et al.*, 1990a; McDonald and Nolan, 1995). In contrast, studies with the fly pathogen *E. muscae* demonstrate the greatest conidial discharge at 50% RH compared with 20 and 80% RH (Mullens and Rodriguez, 1985). This result suggests that *E. muscae* may be better adapted to lower humidity levels than many other species of Entomophthorales.

Discharge of conidia from cadavers by entomophthoralean species can extend over various lengths of time. By 12 hours, *E. muscae* in house fly cadavers held at 20, 50, and 80% RH had released the majority of its conidia (Mullens and Rodriguez, 1985). *Zoophthora phytonomi* in cadavers of alfalfa weevil larvae exposed to still, saturated air released conidia for up to 50 hours (Millstein *et al.*, 1983). In the field, *Z. phytonomi* showered conidia when relative humidities in the alfalfa canopy exceeded 91% for approximately 3 hours (Millstein et al., 1982). When humidity levels were fluctuated in the laboratory, conidial discharge rate and duration decreased, although Z. phytonomi had the ability to resume sporulation if high levels of moisture were restored within < 15 hours of drying (Millstein et al., 1983). Under field conditions, E. grylli in grasshoppers (A. Sawyer, personal communication) and E. maimaiga in gypsy moth (Hajek and Soper, 1992) also can resume sporulation after drying and then rehydrating.

No diurnal periodicities in mortality have been reported from deuteromycete entomopathogens. However, as with Entomophthorales, sporulation by Deuteromycetes is generally sensitive to temperature and humidity levels. B. bassiana only produced spores at > 75% RH with 400 times more spores produced at 100% RH than at 75% RH (Ramoska, 1984). Sporulation by Verticillium lecanii from green peach aphid cadavers was delayed and inhibited below 100% RH (Milner and Lutton, 1986).

4.5. Fungal Dispersal

Studies of dispersal of conidia of entomopathogenic fungi are few compared with this field of study in plant pathology. Most emphasis has been on aerial dispersal, but information is largely empirical. The only experimental investigation of aerial spore dispersal evaluated the settling velocity of conidia of three entomophthoralean species as a precursor to modeling particulate transport for prediction of spore dispersal in the field (Sawyer et al., 1994).

With an active mechanism for conidial ejection, the Entomophthorales are well suited for dispersal. Without the aid of wind, actively ejected entomophthoralean conidia disperse up to several centimeters from cadavers. However, conidia can be extremely abundant in the air spora; it is thought that conidia discharged from sporulating cadavers in exposed locations become airborne (Wilding, 1970). Studies with the cotton aphid pathogen N. fresenii over cotton fields (Steinkraus et al., 1996) the soybean looper pathogen P. gammae over soybean fields (Harper et al., 1984) have documented large quantities of conidia in the air. Both studies documented a diel periodicity in abundance of airborne conidia with peak conidial densities at 0200–0500 for N. fresenii (Steinkraus et al., 1996) and 0000–0600 for P. gammae (Harper et al., 1984). For P. gammae, detection was confined to periods of 100% RH (Harper et al., 1984). Although conidia can be detectable in the air, this does not ensure that these conidia are alive and infective. Densities of airborne conidia of E. maimaiga were positively associated with levels of infection among gypsy moth larvae caged in the forest canopy. However, a closer association was found between infection of caged larvae and temperature and relative humidity, suggesting that although conidia may be airborne, climatic conditions are critical to whether or not conidia are alive and infection occurs (A.E.H., unpublished data).

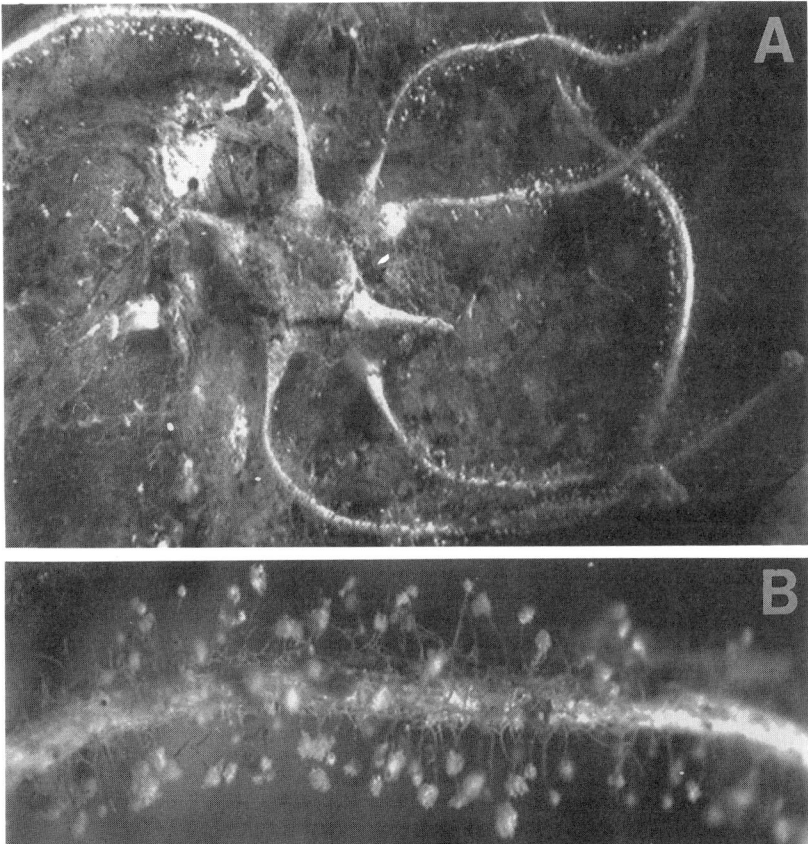

Figure 9. An example of synnemata; (A) synnemata of *Gibellula pulchra* growing from the cadaver of a spider, (B) close-up of synnemata with dry spore masses on apices of erect conidiophores growing at right angles from the synnemata (Courtesy of R. Humber).

Ascospores (sexual spores) of ascomycetes are frequently actively discharged when asci dehisce. Many ascomycetes have pathogenic anamorphic stages (referred to as deuteromycetes) and the teleomorphic stages are uncommon. *Cordyceps* species (teleomorphic ascomycetes) frequently infect soil-dwelling insects and produce long stromata extending up to many centimeters so that ascospores are discharged from perithecia above the soil surface (Roberts and Humber, 1981). Long, fasciculate fungal outgrowths bearing spores (synnemata) are produced from cadavers of insects killed by numerous deuteromycetes (Fig. 9); these structures serve to ensure that conidia are produced in free air where they can attach to passing insects or be dispersed by air currents

(Roberts and Humber, 1981). For *Ascosphaera aggregata* infecting leafcutting bees, adult females acquire spores on emergence through cells containing cadavers of siblings. Large numbers of *A. aggregata* spores can be found in the air around nesting areas where infections by this fungus are prevalent, presumably from spores dislodged from females' bodies (Vandenberg, 1996).

While conidia of Entomopthorales are almost always actively ejected by the fungus, the conidia of entomopathogenic Deuteromycetes are passively released from sporulating cadavers. Inoculation of insects with conidia is usually caused by direct contact with spores on cadavers or substrates. For both *B. bassiana* and *M. anisopliae,* decreasing RH from saturation to 50% was associated with only minimal conidial release (Gottwald and Tedders, 1982); further lowering the RH to < 50% stimulated conidial release. No conidial release occurred when leaf surfaces were wet. Probably the single most important factor stimulating conidial release was vibration, which had a greater effect in darkness and at lower RH for both *B. bassiana* and *M. anisopliae* (Gottwald and Tedders, 1982). The small, dry, singly-produced conidia of *B. bassiana* are thought to be carried on wind currents while the conidia of *M. anisopliae* are produced in large aggregates that remain on or in the soil and are rarely windborne (Roberts and Humber, 1981). Hydrophobic conidia of *N. rileyi* are passively disseminated; no conidia were airborne when the air was still but even low wind velocities dislodged conidia from cadavers (Garcia and Ignoffo, 1977). *N. rileyi* conidia can be abundant in the air over soybean fields, with airborne densities related to infection levels in the velvetbean caterpillar population, daily hours of foliage wetting and drying, and air movement (Kish and Allen, 1978). Almost without exception, rain events were followed by decreased densities of airborne *N. rileyi* conidia (Kish and Allen, 1978).

Rain splash is another important method for dispersal of fungal spores. Fungal spores dispersed by rain splash often are surrounded by mucus, either individually or in groups (Gregory *et al.*, 1959). Mucus coats are considered to be important for adherence to hosts (see above) but mucus is also dissolved by water so that mucus-covered spores become suspended within rain splash droplets. Splash dispersal has not been studied in entomophthoralean species. Among the Deuteromycete entomopathogens, conidia of some species are released in slime drops (e.g., *Verticillium, Acremonium, Synnematium*), covered with slime coats (e.g., *Hirsutella*) (Roberts and Humber, 1981), or produced in aggregations of slime drops (e.g., *Aschersonia*) (Fransen, 1987). Simulated rain falling 11 m from a rain tower onto sporulating cultures or spore suspensions of *Hirsutella cryptosclerotium* yielded dispersal of up to 40 cm, with 50% of droplets carrying spores landing within 10 cm (Fernandez-Garcia and Fitt, 1993). These authors suggest that spore-carrying droplets of 0–100 μm could be windborne, especially because the mealybug host of this pathogen is found in trees. While entomophthoralean conidia are generally mucus-covered, ballistic discharge is considered of primary importance to dispersal and splash dispersal has not been investigated for this group.

Entomopathogenic fungi can disperse under the power of infected insects or as external contaminants on mobile organisms. Some species of entomopathogenic fungi produce spores before host death. The entomophthoralean genus *Massospora* only grows within the abdomens of cicadas, producing resting spores there; as cicadas disperse to new locations, resting spores are deposited in newly invaded habitats (Lloyd *et al.*, 1982). The entomophthoralean fly pathogen *Strongwellsea* grows throughout the abdominal and thoracic hemocoel as well as the nervous system without significantly affecting the life span and behavior of hosts. Conidia are formed within the body and are forcibly discharged through an opening on the ventral surface of the fly abdomen, allowing fungal dispersal from flying hosts (Humber, 1976). *V. lecanii* sporulates on living aphids, and the restless behavior of one host, the green peach aphid, has been hypothesized as promoting conidial spread among aphid populations within glasshouses (Hall and Burges, 1979).

Additional methods for fungal dispersal are myriad, including transfer of conidia from inoculated living males during mating (Watson and Petersen, 1993), the attraction of healthy flies to cadavers of flies that had died from *E. muscae* infections and their attempts to mate with them (Møller, 1993), and potential ingestion of diseased insects or resulting cadavers by predators or scavengers with defecation of spores at new locations later (e.g., Gugnani and Okafor, 1980). Soil from shoes of researchers that had walked through areas with abundant soil-borne resting spores of *E. maimaiga* yielded approximately 250 resting spores/shoe (A.E.H., unpublished data), suggesting yet another method for dispersal.

While long range aerial dispersal of spores has been investigated for numerous species of plant pathogens, this has been proposed in only two instances for entomopathogenic fungi: *Z. phytonomi* was first detected in North America from introduced clover leaf weevil in 1888 but was only found infecting introduced alfalfa weevil populations in 1973 in Ontario (Hajek *et al.*, 1996c). From 1973 through 1981, this pathogen was found progressively further south in North America. Conflicting hypotheses stated that either a new strain of the fungus could have been introduced to North America and spread south from Ontario, or knowledge of how to recognize this pathogen spread southward. Existence of several genotypes of *Z. phytonomi* in North American populations of both weevils (Hajek *et al.*, 1996c), suggests that the hypothesized spread of a newly introduced strain is not impossible although this is far from proven.

In a more closely followed example, *E. maimaiga* was first detected in introduced gypsy moth populations in seven states of northeastern North America in 1989 (Andreadis and Weseloh, 1990; Hajek *et al.*, 1990b). This pathogen was well known in native gypsy moth populations in Japan. During 1990, this

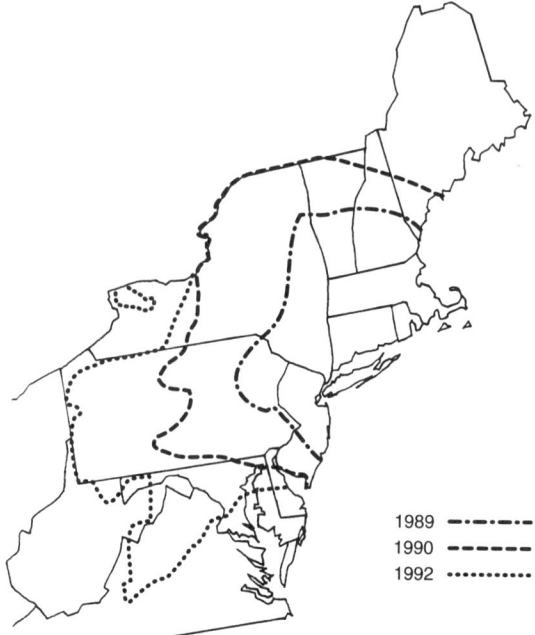

Figure 10. Spread by the gypsy moth pathogen *Entomophaga maimaiga* in the northeastern United States from 1989–1992 (Redrawn by Frances L. Fawcett from Hajek *et al.*, 1995b).

fungus was recovered in many areas bordering the 1989 distribution, apparently due to spread (Elkinton *et al.*, 1991). By 1992, *E. maimaiga* had spread throughout much of the contiguous northeastern distribution of gypsy moth in North America (Hajek *et al.*, 1995b) (Fig. 10). In 1992, this fungus seemed to simultaneously appear across a large area where it had previously not been found. Studies confirming that *E. maimaiga* conidia can be abundant in aerial samples suggest that long range aerial dispersal could have occurred (Hajek *et al.*, 1995b).

4.6. Conidial Survival and Transmission after Dispersal

Studies have frequently been conducted on survival of conidia of entomopathogenic fungi due to interest in utilization of these fungi for insect control; the length of time that conidia remain viable and infective after application can directly affect the number of hosts that will be infected. Studies of spore survival are conducted for various durations, e.g., days to weeks for exposure to sunlight and weeks to years for temperature and moisture relations.

Conidial mortality is thought to occur either through germination without locating a host or through exposure to lethal abiotic conditions. Major factors influencing conidial survival include sunlight, temperature, moisture (rain, dew, humidity), or lack of moisture, and substrate.

Optimal and lethal temperatures for conidia vary by fungal species but, in general, spore viability is lost more slowly at low than at high temperatures (Roberts and Campbell, 1977). Above 40°C, most fungal cells stop growing and soon die (McCoy *et al.*, 1988). Conidia of the entomophthoralean species *C. obscurus* on the surface of unsterilized soil at 20°C were infective for 12 days, at 10°C for > 32 days, while at 5 °C, 23% of aphid hosts were still infected after 90 days (Latteur, 1980). In contrast, conidia of the deuteromycete *M. anisopliae* survived for three months at 37°C and > 24 months at 26°C when stored in plastic boxes in the laboratory (Daoust and Roberts, 1983). Since many deuteromycetes are stored as conidia at sub-zero temperatures, it is no surprise that *P. farinosus, H. thompsonii, M. anisopliae, B. bassiana,* and *N. rileyi* are known to survive temperatures between 8 and -20°C for extended periods, with survival of 2–3 years documented for *N. rileyi* and *B. bassiana* (McCoy *et al.*, 1988). While information documenting survival for > 1 year at low temperatures has little direct relevance to activity in the natural environment where entomopathogenic fungi are active, these data certainly demonstrate that deuteromycete conidia survive through cold winters. Entomophthoralean conidia generally are not thought to survive near or sub-zero temperatures as well as deuteromycetes, e.g., conidia of *E. maimaiga* survived < 4 days at −24°C (A.E.H., unpublished data) and are not known to overwinter in the field.

Solar radiation interacts with temperature and moisture to influence conidial mortality (Zimmermann, 1982). However, because light intensities in the field increase to potentially harmful levels more quickly than temperature or moisture, light is suspected as being the dominant environmental factor affecting conidial survival when conidia land in exposed locations (Carruthers and Haynes, 1986; Galaini, 1984). While ultraviolet radiation has been suggested as detrimental due to photodeactivation of nucleic acids (Ignoffo *et al.*, 1977; Roberts and Campbell, 1977), visible light also can have damaging effects through photooxidation and/or overheating (Leach, 1971). Light is generally detrimental to conidia of *B. bassiana, M. anisopliae, N. rileyi,* and *P. farinosus* (Roberts and Campbell, 1977; Gardner *et al.*, 1977). Far UV light (200–300 nm) is more detrimental than near UV (300–380 nm) to *B. bassiana* conidia, while *H. thompsonii* conidia are reactivated with exposure to far UV (see McCoy *et al.*, 1988). However, since very little far UV reaches the Earth's surface, the influences of near UV and visible light (380–750 nm) are much more relevant (Dickinson, 1986). The half-life of *N. rileyi* conidia was only 2.4 hours when exposed to simulated UV, including both far and near UV (Ignoffo *et al.*, 1977). Screens blocking UV-A (330–400 nm) radiation or wavelengths > 400 nm increased conidial half-life of *N. rileyi* by four times (Fargues *et al.*, 1988). Under field conditions on a sunny day, the half-life of *N. rileyi* conidia was 3.6 hours, but when screens were used to block direct sunlight, the half-life increased to ≥40 hours, suggesting increased survival under protected or cloudy conditions (Fargues *et al.*, 1988). During field studies, blocking sunlight also increased conidial viability from a half-life of <1–2 days in full sunlight to ≥3 weeks in some cases for *B. bassiana* and *M. anisopliae* (Daoust and Pereira, 1986). Using a solar simulator with an irradiance pattern similar to sunlight, duration and intensity of exposure of *E. grylli* conidia interacted with no decrease in viability for conidia exposed to low intensities (≈3 Langleys) but 100% mortality after exposure to higher intensities (20 Langleys) (Carruthers *et al.*, 1988). Field tests with *E. grylli* (Carruthers *et al.*, 1988) and *B. bassiana* (Inglis *et al.*, 1993) corroborate high conidial mortality in the most exposed locations with greatly prolonged survival for conidia in protected locations (e.g., lower canopy foliage).

Rainfall dislodges *B. bassiana* conidia from leaves, supposedly decreasing chances for conidia to contact hosts feeding on leaves (Inglis *et al.*, 1995). Moisture can be critical to the duration of deuteromycete conidial survival, although relationships are not consistent across species. While individual species exhibit optimal survival at either high or low relative humidities, intermediate relative humidities frequently are less favorable. *B. bassiana* survived optimally at low relative humidity while *P. farinosus* survived best at high humidity, although humidities between 75–80% were fatal for both species (see Daoust and Roberts, 1983). *V. lecanii* conidia in distilled water survived up to 160 days at 2°C, while dried conidia at 58% RH died in < 24 hours (Hall, 1981). At 19 and 26°C, *M. anisopliae* conidia survived optimally at 96–97% RH, but survival at 0% RH was greater than survival at 12 through 76% RH (Daoust and Roberts, 1983).

Among the Entomophthorales, high percentages of conidia of several species survived at unsaturated humidities for up to 5 weeks, with greater survival of the specialized capilliconidia of *N. fresenii* compared with primary conidia (Uziel and Kenneth, 1991). At 20°C in the laboratory, 14 days after showering *P. neoaphidis* conidia onto bean leaves or coverslips, more infection was caused by conidia at 40–50% RH than at 70% RH (Brobyn *et al.*, 1987). Under field conditions, conidia of the entomophthoralean *P. neoaphidis* retain infectivity for at least 14 days, with greater survival on abaxial leaves near the bases of bean plants where they were more protected from sunlight and conditions were presumably more humid (Brobyn *et al.*, 1985). When primary conidia of *P. neoaphidis* were showered onto soil and maintained at 3°C in the dark, this fungus survived for at least 9 months due to production of series of replicative conidia (Morgan *et al.*, 1992). Primary conidia of *E. maimaiga* showered onto saturated soil and maintained at 15°C and 14L:10D infected 85.2% of larvae after one week, but infectivity dropped to 1.3% by three weeks (A.E.H., unpublished data).

Non-host substrates (i.e., soil, water, foliage, etc.) upon which conidia are deposited can influence spore survival. Activity of conidia deposited on soil can be impacted by the resident community of soil microorganims as well as other soil attributes (see section 3.2). The few studies describing activity of conidia in standing water generally treat only aquatic fungi and hosts and, therefore, are beyond the scope of this review. Conidia also land on foliage but to date no studies have identified an effect of surface contact with foliage on conidial survival.

4.6.1. Influence of Hosts on Disease Transmission

In general, deposition of conidia onto host cuticle is considered a fairly random process, and it is probably due to the low chance of success in landing on a host that fungi produce so many propagules. However, in a few instances, hosts actively alter this random aspect of disease transmission. Scarab grubs actively avoided soil containing high concentrations of *M. anisopliae* mycelium for up to 20 days (Villani *et al.*, 1994). In contrast, the incorporation of *M. anisopliae* mycelium into soil was associated with increased oviposition by scarab adults.

Social insects are known to alter disease transmission by removing cadavers of insects killed by fungi. For example, honeybees from resistant colonies remove cadavers of larvae killed by *Ascosphaera apis* at a higher rate than susceptible colonies (Gilliam *et al.*, 1983). Interestingly, when colonies resistant or susceptible to *A. apis* were fed this pathogen in pollen patties, *A. apis* was recovered more frequently from stored pollen (for feeding brood) from susceptible than from resistant colonies (Gilliam *et al.*, 1988). Microorganisms producing fungistatic compounds were found in the stored pollen from resistant colonies; these microbes may have been introduced by the bees.

5. Epizootiology

Epizootic is a relative term, defined as "an unusually large number of cases of disease" (Fuxa and Tanada, 1987). While this definition could include a low level of infection in a host population where infection is rarely seen, it usually refers to high levels of infection. The study of epizootiology is the study of causes and forms of disease outbreaks. Historically, investigations of disease patterns were purely descriptive, frequently associating disease prevalence with host density and especially with ambient climatic conditions. More recently, researchers have concentrated on both theoretical and simulation models to better understand epizootic dynamics (see Onstad and Carruthers, 1990).

Fungal pathogens are influenced by their abiotic and biotic environment as well as the host population. While host/pathogen interactions occurring on individual and population levels have been reviewed, community level interactions critical to epizootic development will be discussed below.

5.1. Epizootic Determinants

To conduct studies of the dynamics of entomopathogenic fungi in the field, particular sensitivities of entomopathogenic fungi must be understood. Collections of infected insects for later diagnosis and identification must be maintained at appropriate temperatures so as not to kill fungi. Cadavers of insects that may have been killed by entomophthoralean fungi must be evaluated fairly quickly if conidial stages are transient, as is the case with many Entomophthorales. Fungal mummies can also be overgrown by saprophytic fungi. Direct observations and counts are most frequently used to quantify infection levels although immunoassays have also been developed (e.g., Hajek et al., 1991a). While sampling strategies to achieve known levels of accuracy have been developed for many insect pests, few sampling strategies for entomopathogenic fungi have been developed (but see Feng and Nowierski, 1992; Hollingsworth et al., 1995). In fact, quantification of infection levels is frequently undertaken, while quantification of pathogen propagules is rarely attempted due to the relative difficulty in quantifying fungi in soil, the air, on plant leaves, etc.

Of particular importance to evaluation of epizootics is quantification of host populations. Theories of biological control rely on density dependence of natural enemies. However, *E. maimaiga* causes epizootics in both low- and high-density gypsy moth populations. It has been suggested that rather than being dependent on density, dynamics of fungal pathogens are primarily driven by abiotic conditions (Ullyet and Schonken, 1940). Following this hypothesis, epizootic occurrence would be difficult to predict, especially under highly variable or marginal abiotic conditions. The interaction of pathogen density with host density and microclimate is rarely considered due to the relative difficulty in quantifying the pathogen in the environment. This hypothesis should be explored further considering densities of the pathogen as well as host along with abiotic conditions.

To study epizootic development, it is critical to study the habitat in which fungal interactions take place. Of particular importance to fungi, ambient humidity levels can drastically differ from microclimatic humidity. Small insects such as aphids can live largely within the 0.5 mm boundary layer of leaves where they experience a low level of convection and, due to plant respiration, higher humidity than ambient levels (Ferro and Southwick, 1984). It has been hypothesized that the environment of cotton aphids is close to saturation much of the time. In addition, temperatures of leaf surfaces can vary significantly from ambient levels (Ferro et al., 1979). While microclimate is much more difficult to measure than ambient conditions, it is important to remember that it is microenvironmental rather than ambient conditions that influence disease dynamics.

Spatial aspects of epizootic development have rarely been addressed. It is frequently thought that epizootics begin with foci of infection from which pathogens spread, e.g., fungal growth spreading from a mummy in the soil (Keller and Zimmermann, 1989). Early in the development of an epizootic, cotton aphids infected with *N. fresenii* were clumped in distribution although the within field distribution of infected aphids became much more homogeneous as the epizootic progressed (Hollingsworth *et al.*, 1995). Similarly, the distribution of *B. brongniartii* is initially clumped within an outbreak site, reflecting locations where female scarabs had previously died, but this fungus spreads within a few months to achieve a more homogeneous distribution (Keller and Zimmermann, 1989). Similar spread of *B. bassiana* from foci has been documented in Russian wheat aphids in laboratory microcosms (Knudsen *et al.*, 1994); in this study increased movement of aphids on non-preferred plants is thought to allow increased fungal spread compared with sessile aphid behavior on preferred plants.

Theories of temporal dynamics of epizootics state that infection levels increase and peak after the highest host density and that epizootics thus are evident as host populations decline (Keller and Zimmermann, 1989). Data from *B. brongniartii* in populations of soil-dwelling scarabs fit this pattern well (see Keller and Zimmermann, 1989).

5.1.1. Case Study: Descriptive Epizootiology of Entomophaga maimaiga Infecting Gypsy Moth

E. maimaiga overwinters as azygospores that are already germinating once the primary hosts, gypsy moth larvae, begin hatching during early spring. Infection of univoltine gypsy moth larvae is thought to be closely tied to larval behavior. First instars climb trees and balloon to find acceptable trees; individual larvae can balloon several times before remaining on a tree and, between ascents, larvae walk on the soil over high populations of resting spores at bases of trees. However, the extent to which first instars are infected is questionable because infected first instars have rarely been collected. Second and third instars generally remain in tree canopies feeding, and the level of infection of these instars is generally low. Cadavers of any early instars infected by *E. maimaiga* produce conidia that are actively discharged to readily infect further larvae. Thus, production of only conidia by early instars maximizes secondary transmission early in the season.

Fourth and later instars climb down trees every day to rest in cryptic locations from which they ascend again at dusk. There is also an undocumented amount of wandering by later instars, especially in dense populations that often do not follow the diurnal movement pattern. Much of larval movement is accomplished by following silk trails, so that larvae repeatedly aggregate in the same locations. Several aspects of this highly specialized behavior are tied to disease transmission. The long daily hiding periods provide a time for germinating resting spores to infect. Larval mortality at aggregation locations with subsequent conidial discharge onto neighboring larvae would also aid disease transmission. In these ways, the disease is repeatedly transmitted within a field season to allow epizootic development. Our modeling efforts have estimated that 4–11 cycles from infection to death can occur within one field season (Hajek *et al.*, 1993). The titer of resting spores in the soil has a strong impact on initiation of disease cycling throughout the 2 month field season, but modeling efforts suggest that the logarithmic increase in infection levels within the same season is accomplished by conidial infections. Therefore, conidia are produced from cadavers of later instars as well as resting spores to not only maximize fungal activity during the same season but also to ensure survival and persistence.

Weather conditions are critical to activity of *E. maimaiga*; spore production by cadavers is associated with rainfall on the same day (Weseloh and Andreadis, 1992) or leaf wetness for the previous 3 hours (Hajek and Soper, 1992), while resting spore activity has been associated with rainfall during the previous 1–2 days (Weseloh and Andreadis, 1992). However, rainfall during June was not associated with infection across many sites during 1991 and 1992 (Hajek *et al.*, 1996b). During late instars when explosive increase in infection levels is possible we would expect a positive association between moisture and infection. These results suggest that use of microclimatic moisture by *E. maimaiga* may be critical to transmission when late instars are present.

5.2. Community-Level Interactions

5.2.1. Influence of Host Food

The type of food as well as the amount have been shown to influence fungal infections. Studies of tritrophic interactions have principally focused on systems with phytophagous hosts. Numerous studies have been conducted comparing fungal activity when insects are eating different plants (Table III). While five studies comparing host plant age or species documented host plant effects on fungal entomopathogens, two studies showed no effect. However, the variables measured in different studies do not always allow direct comparison. Mechanisms hypothesized for interactions include an effect of host growth rate and general health or effects of secondary plant compounds on fungal pathogens. For the obligate pathogen *E. maimaiga,* when hosts grew more slowly development of *E. maimaiga* was slower and subsequent fungal sporulation declined (Hajek *et al.*, 1995c). For the more facultatively pathogenic *B. bassiana,* faster growing hosts displayed decreased mortality (Boucias *et al.*, 1984; Hare and Andreadis, 1983), possibly due to enhanced defensive responses to infection. One study hypothesized that allelochemicals in two annual plant species were the cause of

Table III. Tritrophic Level Interactions between Phytophagous Insects, Fungal Entomopathogens, and Food Plants

Insect host	Pathogen	Influence on fungus	Reference
Plant age			
Anticarsia gemmatalis	Nomuraea rileyi	susceptibility: less than with older soybean foliage	Boucias et al., 1984
Leptinotarsa decemlineata	Beauveria bassiana	susceptibility: less than with younger tomato foliage no effect with four Solanum spp.	Hare and Andreadis, 1983
Plant species			
L. decemlineata	B. bassiana	susceptibility: varied among three Solanum spp. and tomato	Hare and Andreadis, 1983
L. decemlineata	B. bassiana	susceptibility: no effect (six Solanum spp.)	Costa and Gaugler, 1989
L. decemlineata	B. bassiana	mortality and sporulation: decreased with α-tomatine in artificial diet	Gallardo et al., 1990
Blissus leucopterus	B. bassiana	mortality and sporulation: barley, wheat > corn, sorghum	Ramoska and Todd, 1985
Helicoverpa armigera	N. rileyi	infection: pigeonpea > tomato, field beans	Gopalakrishnan and Narayanan, 1989
Spodoptera littoralis	N. rileyi	susceptibility: no effect (four plant genera)	Fargues and Maniania, 1992
Lymantria dispar	Entomophaga maimaiga	time to death: oak, pine, larch > maple sporulation: larch > maple	Hajek et al., 1995c

inhibition of *B. bassiana* both *in vivo* and *in vitro* (Ramoska and Todd, 1985). Research with honeybee larvae provided with different types of food demonstrated that susceptibility to *A. apis* is not correlated with bee health or growth rate (Vandenberg, 1994).

Outbreaks of insect populations can result in utilization of all available food with resulting starvation in insect populations. Among gypsy moth larvae inoculated with *E. maimaiga*, limited food intake after infection resulted in earlier larval death and production of fewer resting spores (Hajek, 1989). Among starved aphids, Milner and Soper (1981) found increased mortality due to *Z. radicans* early during infection but a lower total mortality level compared with fully fed aphids. In addition, starved aphids that had been inoculated rarely produced spores after death. In a study of the effect of starvation on predacious insects infected by fungi, mortality and larval development after exposure to *B. bassiana* was evaluated; larval chrysopids were more susceptible to *B. bassiana* if starved for 48 hours before exposure (Donegan and Lighthart, 1989). Larvae fed a suboptimal diet before fungal inoculation developed poorly and were also more susceptible to *B. bassiana* (Donegan and Lighthart, 1989).

5.2.2. Host Species and Specificity

Among other factors, entomopathogenic fungal activity is dependent upon levels of inoculum in the field. The amount of overwintering inoculum will depend on densities of surviving long-lived fungal stages, especially for obligately pathogenic fungal species in temperate climates. For secondary infection within a season and subsequent production of long-lived inoculum during that season, host populations in which the pathogen could reproduce must be present. However, insect populations are frequently ephemeral, and the densities of any one species can vary wildly from year to year. If pathogenic fungal species have broad host ranges or possibly are capable of saprophytic growth in the environment, they should be affected less by the variability from year to year in the presence of certain host species.

How species specific are insect pathogenic fungi and does utilization of multiple hosts increase the pathogen reservoir? Deuteromyctes such as *B. bassiana* and *M. anisopliae* have been isolated from a great diversity of hosts (Humber, 1992) but these species are rarely found infecting several different host species in one location. Field applications of *Beauveria brongniartii* to control a scarab followed by general sampling of all invertebrates yielded only 1.1% infection in the non-targets (Baltensweiler and Cerutti, 1986).

In the Entomophthorales, *Z. radicans* is known to infect species in many insect families in at least seven orders (Papierok *et al.*, 1984). Yet, individual strains of this fungus are host specific; bioassays with an isolate from the potato leafhopper could not infect Lepidoptera and caused only low levels of infection in aphids (McGuire *et al.*, 1987). In this case, *Z. radicans* is at least a very diverse species and probably a species complex (Hodge *et al.*, 1995). Studies testing specificity of individual strains of *Z. radicans* among more closely related hosts have not been conducted. Testing species within the same order, laboratory bioassays challenged 78 species of Lepidoptera with the Lepidoptera-specific *E. maimaiga*, known only from gypsy moth larvae in the field (Hajek *et al.*, 1995a). Under laboratory conditions, *E. maimaiga* infected 35.6% of the species tested, although generally at rather low levels within each susceptible species tested. Species susceptibility was generally not associated with phylogenetic relatedness although high levels of infection were found among all lymantriids (the family including gypsy moth). Because host species can be infected in the laboratory that are never found infected in the field, laboratory bioassays were followed by sampling alternate hosts during *E. maimaiga* epizootics in gypsy moth populations. Field specificity was extremely restricted, with infections in only one individual of each of two species out of a total of 52 species sampled (Hajek *et al.*, 1996a). Of the 20 species assayed in the laboratory and collected in the field, while 8 became infected in the laboratory only one of these (forest tent caterpillar) was infected in the field. Extremely few forest tent caterpillar larvae were infected in the field (0.3%) while 61.0% of individuals challenged became infected in the laboratory (Hajek *et al.*, 1996a). In support of these findings, numerous studies have reported that during epizootics, pathogen species are extremely host specific, often with only one single host species being affected (see Wilding, 1981). Due to the haploid nature of many fungal entomopathogens, host specific lineages could be readily selected in nature. In addition, host range in the field is probably determined by a multitude of factors, e.g., spatial, temporal, as well as the host-specific virulence of fungal isolates.

In one system, infection of multiple hosts by the same fungus has been suggested as crucial for development of epizootics. The pea aphid is very susceptible to the entomophthoralean *P. neoaphidis,* while its congener, the blue-green aphid, is much less susceptible (Pickering and Gutierrez, 1991). Low levels of infection in blue-green aphid populations are suggested as increasing the levels of conidial production sufficiently to cause epizootics in pea aphid, even though populations of pea aphid are not abundant.

While fungal strains vary in virulence and specificity, host strains can also vary in susceptibility. With leaf-cutting bees and pea aphids, coexisting host strains varying in susceptibility to fungal pathogens have been documented (Papierok and Wilding, 1979; Milner, 1982, 1985; Stephen and Fichter, 1990). Occurrence of such variability could stabilize host and fungal populations and would lead to longer-term persistence of both host and pathogen in an area.

5.2.3. Interactions with Predators, Parasites, or Other Pathogens

In comparison with studies of the effect of pathogens on pests, many of which occur in lower trophic levels, relatively few studies have been conducted on the effect of fungal pathogens on higher trophic levels. Predators, parasites, and other pathogens (natural enemies) in the community can have an indirect effect on fungal entomopathogens by decreasing host density. However, these natural enemies can also have a direct relationship with fungal pathogens. Fungal infections of natural enemies could decrease natural enemy populations, but this also would result in increased production of fungal spores. Laboratory studies have generally been conducted testing the effects of pathogens on natural enemies; however, one important factor in evaluating such studies is the host species from which fungi used for testing were isolated, due to the host specificity of fungal isolates. For example, *B. bassiana* isolates from house flies were more virulent against house flies than hymenopterous parasitoids of house flies (Geden *et al.*, 1995).

Not surprisingly, entomopathogenic fungi are known to infect numerous types of insect predators (Goettel et al., 1990; Steenberg et al., 1995), although their impact on predation levels has not been investigated. High levels of overwintering mortality due to *B. bassiana* have been reported for ladybird beetles (Mills, 1981) but fungal infection levels in overwintering ground and rove beetles were low (Steenberg et al., 1995). Laboratory bioassays have demonstrated that *B. bassiana* strains from aphids or beetles can cause high levels of infection in ladybirds (95% and 75%, respectively) (James and Lighthart, 1994). Ladybirds were also susceptible to *M. anisopliae* (isolated from a scarab) and *P. fumosoroseus* (active against Homoptera, thrips, and mites) but not *N. rileyi* (isolated from Lepidoptera) (James and Lighthart, 1994) or *Z. radicans* isolated from leafhoppers (Magalhaes et al., 1988). Interestingly, ladybirds were infected by direct sprays of *B. bassiana* in the laboratory, but when this fungus was applied to leaves one ladybird species was more resistant and a second species did not become infected (Magalhaes et al., 1988).

Studies of fungal/parasitoid interactions have generally focused on the potential impact of fungi on parasitoids and have not emphasized adverse effects of parasitism on fungal infections (Brooks, 1993). In some systems, fungus-infected hosts are avoided by parasitoids (Fransen and van Lenteren, 1993; Velasco, 1983), which thereby circumvent potential competition for nutrients within hosts. In other systems, parasitoids cannot complete development due to fungal infection and, in these cases, co-occurrence of both parasitoid and pathogen must decrease the amount of nutrients available to fungi, unless the fungus also kills and colonizes the parasitoid. Miscellaneous records have documented pupal and adult parasitoids directly infected by fungi (see Brooks, 1993). Parasitized hosts are occasionally more susceptible to fungal infection than non-parasitized hosts, although this may be effective only for a specific period after parasitization (King and Bell, 1978; Führer and El-Sufty, 1979; Powell et al., 1986).

Results of interactions between hymenopterous and dipterous parasitoids and fungal entomopathogens are highly dependent on the sequence of infection and parasitization. Fungal infections generally kill hosts more quickly than parasitoids so that only when insects are parasitized first and then infected does parasitization impact fungal development. Greenhouse whitefly larvae parasitized by *Encarsia formosa* were highly susceptible to *Aschersonia aleyrodis* for up to 4 days after parasitization, after which time susceptibility to the fungus decreased markedly; the change from susceptibility to increased resistance at 4 days coincides with parasitoid egg hatch within hosts, suggesting that physiological changes in the parasitized host confer some degree of resistance to fungal infection (Fransen and van Lenteren, 1994). Application of *B. bassiana* to parasitized Egyptian cotton leafworm larvae did not result in host mortality greater than unparasitized larvae (El-Maghraby et al., 1988). One possible mechanism to explain these anomalous results might be the inhibition of fungal activity. For hosts parasitized by *Cotesia glomerata* (Führer and El-Sufty, 1979), fungistatic factors produced by parasitoid teratocytes can inhibit development of *B. bassiana* after fungal penetration. A similar antimycotic substance is also secreted from the proctodaeum of larval *Pimpla turionellae* (Willers *et al.*, 1982). Surprisingly, one side-effect of parasitism is that larvae confined in the presence of diamondback moth larval cadavers bearing *Z. radicans* conidia were more active if parasitoids were also present, and, therefore, became infected more readily than did the controls (Furlong and Pell, 1996).

Whether predators and parasitoids influence dispersal of fungal entomopathogens has rarely been investigated. The only evidence for transmission of a fungus during parasitoid oviposition was obtained for a few females of *E. formosa* transmitting *A. aleyrodis* to whiteflies after 24 hours of exposure of parasitoids to infected hosts (Fransen, 1987).

Steinernematid and heterorhabditid nematodes parasitize insects and symbiotic bacteria vectored by these nematodes kill insect hosts within several days of penetration. When nematodes and *B. bassiana* were both introduced to waxmoth larvae, the time to death decreased although usually only one of these natural enemies reproduced within a cadaver (Barbercheck and Kaya, 1990). *B. bassiana* did not grow in hosts if nematodes were applied within 24 hours of *B. bassiana* inoculation, while nematodes did not develop well in hosts exposed to *B. bassiana* > 48 hours before nematode exposure. These natural enemies appear to interact antagonistically, resulting in decreased reproductive potential by each.

Even fewer studies have investigated co-occurrence of different fungal entomopathogens or fungal entomopathogens with other pathogens, i.e., bacteria, viruses, and protozoa (last reviewed by Krieg, 1971). It is not uncommon to find different pathogens active in a host population or even occurring within the same host (e.g., Aoki, 1974; Hajek and Roberts, 1992). For pathogens co-occurring within the same insect, the critical issue is whether fungal reproduction is affected. Since some pathogens can kill hosts in 1–2 days (e.g., bacteria) while others cause chronic infections, the nature of pathogenicity is important in determining the outcome. As with fungal/parasitoid interactions, the sequence of infection by two pathogens also influences which pathogen reproduces. When two entomopathogenic fungi coinfected the Colorado potato beetle, overall mortality was not altered but, in general, *P. farinosus* grew faster than *B. bassiana* and spread more quickly within hosts when it infected first or up to 24 hours after *B. bassiana* (Bajan and Kmitowa, 1972).

The gypsy moth nuclear polyhedrosis virus (LdMNPV) and *E. maimaiga* commonly co-occur in the field. *E. maimaiga* develops much faster than LdMNPV at 20–25°C and, generally, when these pathogens coinfect, *E. maimaiga* kills larvae and produces spores and LdMNPV has little chance to reproduce (R. Malakar, personal communication). Gypsy moth is an introduced, outbreak insect with the capacity to increase dramatically in numbers. Since LdMNPV was accidentally introduced to North America at the turn of the century (Hajek et al., 1995b), LdMNPV epizootics have very frequently caused dramatic crashes in extremely dense gypsy moth populations. Since *E. maimaiga* was first found in North America in 1989, this faster-developing pathogen has been causing epizootics in low- to high-density gypsy moth populations, while naturally occurring LdMNPV epizootics have become relatively uncommon. The long-term outcome of this naturally occurring experiment in population level interactions between two pathogens remains to be seen.

6. Relevance to Biological Control

Due to the numerous drawbacks in use of synthetic chemicals for pest control, there is abundant interest in developing entomopathogenic fungi for insect control. There are four major strategies for using fungi for control: classical biological control (inoculative introduction of an exotic fungus, often to control an introduced pest, for long-term control), inoculative release, inundative release (or use as a mycoinsecticide), and conservation (enhancement of naturally occurring microbial populations, e.g., by irrigation, or optimal timing of harvest to allow development of epizootics). All of these strategies are types of biological control in which natural enemies are used to regulate or eliminate pest populations. Of these strategies, classical biological control has been used infrequently, partially due to lack of knowledge of communities of pathogenic fungi in areas from which pests are introduced. Conservation has also rarely been explored because this strategy relies heavily on a detailed understanding of the ecology of both fungal entomopathogen and host, as well as of management practices to understand how systems can be manipulated to favor fungal transmission. Inundative release is generally the strategy employed for control of pests on crops in situations where immediate control is required. Of course, control with entomopathogenic fungi is not as immediate as that with contact insecticides, so use of these pathogens for control requires systems where at least a limited amount of damage or level of pest population is both expected and acceptable. Inoculative release is less seldom emphasized today, but has recently been successfully utilized to release small populations of *E. maimaiga* resting spores into areas at the edge of spreading gypsy moth populations where this fungus did not yet occur (Hajek and Roberts, 1991; Smitley et al., 1995; Hajek et al., 1996b). Presently, many of the fungi being developed for control are Deuteromycetes aimed toward inundative release and further cycling by the pathogen in the insect population is not expected.

When using chemical insecticides, the aim is to blanket the environment so that each insect ingests or is contacted by a lethal dose. Use of entomopathogenic fungi for control of pestiferous insect populations is inherently different from use of synthetic chemical pesticides. Classical biological control, inoculative release, and conservation of entomopathogenic fungi all depend on the death of infected hosts and subsequent transmission of the fungus within the pest population for multiple cycles of infection. Although inundative releases are generally conducted in instances where control is needed immediately, if naturally occurring disease transmission follows fungal release, fewer applications may be necessary for control.

For control strategies using entomopathogenic fungi that rely on repeated cycles of disease transmission after application, studies of the ecology of fungal entomopathogens have a direct impact on use of these fungi for control. It is through understanding the interactions of these fungi with hosts and the environment that infection levels can be improved. While fungi remain sensitive to environmental moisture interacting with temperature, sunlight, etc., this information can be used to apply fungal propagules in formulations retaining moisture or offering protection from sunlight, or to apply fungal propagules to protected locations where they will still contact hosts (e.g., the bottoms of leaves where direct exposure to sunlight is avoided).

Only knowledge of fungal host specificity will help with prediction of the impact of fungal application on non-target species. In recent years, non-target effects of applications of biological control agents, including entomopathogenic fungi, have been of concern. As discussed earlier (section 5.2.2), while much more information is needed, studies to date demonstrate that individual fungal strains can be quite host specific and their activity in the field does not always mirror the levels of infection documented during laboratory bioassays.

Diversity among fungal isolates can have a decided impact on biological control projects using entomopathogenic fungi, e.g., isolates of Z. *radicans* were compared and, based on temperature optima, an Israeli strain of Z. *radicans* was successfully introduced to Australia to control spotted alfalfa aphid (Milner *et al.*, 1982). Differences among fungal strains can therefore be exploited to identify the most virulent isolates for inoculative or inundative release.

At present, mass production and application of the deuteromycetes *A. aleyrodis, B. bassiana, H. thompsonii, M. anisopliae,* and *V. lecanii* has been investigated extensively and these species have been or are being used operationally (Roberts and Hajek, 1992). Most deuteromycetes can be grown *in vitro* much more easily and less expensively than entomophthoralean species, which partially explains why no entomophthoralean fungi have been mass produced, except on an experimental basis.

ACKNOWLEDGMENTS. R. A. Humber, J. D. Vandenberg, and D. W. Watson provided helpful comments on an earlier draft of this manuscript. J. K. Liebherr and R. A. Humber were extremely helpful with figures. This research was supported in part by USDA NRICGP 92–37302–8758.

References

Alexopoulos, C. J., Mims, C. W., and Blackwell, M., 1996, *Introductory Mycology, Fourth Edition*, Wiley, New York.

Allee, L. L., Goettel, M. S., Gol'berg, A., Whitney, H. S., and Roberts, D. W., 1990, Infection by *Beauveria bassiana* of *Leptinotarsa decemlineata* larvae as a consequence of fecal contamination of the integument following *per os* inoculation, *Mycopathologia* **111**:17–24.

Anderson, S. O., 1979, Biochemistry of insect cuticle, *Annu. Rev. Entomol.* **24**:29–61.

Andreadis, T. G., and Weseloh, R. M., 1990, Discovery of *Entomophaga maimaiga* in North American gypsy moth, *Lymantria dispar*, *Proc. Natl. Acad. Sci. U.S.A.* **87**:2461–2465.

Aoki, J., 1974, Mixed infection of the gypsy moth, *Lymantria dispar japonica* Motschulsky (Lepidoptera: Lymantriidae), in a larch forest by *Entomophaga aulicae* (Reich.) Sorok. and *Paecilomyces canadensis* (Vuill.) Brown et Smith, *Appl. Ent. Zool.* **9**:185–190.

Bajan, C., and Kmitowa, K., 1972, Successive infection, *Ekol. Pol.* **20**:433–440.

Baltensweiler, W., and Cerutti, F., 1986, Bericht über die nebenwirkungen einer bekämpfung des maikäfers (*Melolontha melolontha* L.) mit dem pilz *Beauveria brongniartii* (Sacc.) Petch auf die arthropodenfauna des waldrandes, *Mitt. Schweiz. Entomol. Gesell.* **59**:267–274.

Barbercheck, M. E., and Kaya, H. K., 1990, Interactions between *Beauveria bassiana* and the entomogenous nematodes, *Steinernema feltiae* and *Heterorhabditis heliothidis*, *J. Invertebr. Pathol.* **55**:225–234.

Beauvais, A., Latgé, J.-P., Vey, A., and Prevost, M.-C., 1989, The role of surface components of the entomopathogenic fungus *Entomophaga aulicae* in the cellular immune response of *Galleria mellonella* (Lepidoptera), *J. Gen. Microbiol.* **135**:489–498.

Bellini, R., Mullens, B. A., and Jespersen, J. B., 1992, Infectivity of two members of the *Entomophthora muscae* complex (Zygomycetes: Entomophthorales) for *Musca domestica* (Diptera: Muscidae), *Entomophaga* **37**:11–19.

Ben-Ze'ev, I. S., Bitton, S., and Kenneth, R. G., 1990, Induction and inhibition of germination of *Neozygites fresenii* (Entomophthorales: Neozygitaceae) zygospores by various time-temperature stimuli, *J. Invertebr. Pathol.* **55**:1–10.

Berisford, Y. C., and Tsao, C. H., 1975, Appressoria formation by *Aspergillus parasiticus* on bagworm cuticle, *Ann. Entomol. Soc. Amer.* **68**:1111–1112.

Bidochka, M. J., and Hajek, A. E., 1996a, Protoplast plasma membrane glycoproteins in two species of entomophthoralean fungi, *Mycol. Res.* **100**:1094–1098.

Bidochka, M. J., and Hajek, A. E., 1996b, A non-permissive entomophthoralean fungal infection increases activation of insect prophenoloxidase, (submitted).

Bidochka, M. J., and Khachatourians, G. G., 1987, Hemocytic defense response to the entomopathogenic fungus *Beauveria bassiana* in the migratory grasshopper *Melanoplus sanguinipes*, *Entomol. Exp. Appl.* **45**:151–156.

Bidochka, M. J., and Khachatourians, G. G., 1992, Growth of the entomopathogenic fungus *Beauveria bassiana* on cuticular components from the migratory grasshopper, *Melanoplus sanguinipes*, *J. Invertebr. Pathol.* **59**:165–173.

Boucias, D. G., and Latgé, J.-P., 1988, Nonspecific induction of germination of *Conidiobolus obscurus* and *Nomuraea rileyi* with host and non-host cuticle extracts, *J. Invertebr. Pathol.* **51**:168–171.

Boucias, D. G., and Pendland, J. C., 1984, Nutritional requirements for conidial germination of several host range pathotypes of the entomopathogenic fungus *Nomuraea rileyi*, *J. Invertebr. Pathol.* **43**:288–292.

Boucias, D. G., and Pendland, J. C., 1987, Detection of protease inhibitors in the hemolymph of resistant *Anticarsia gemmatalis* which are inhibitory to the entomopathogenic fugnus *Nomuraea rileyi*, *Experientia* **43**:336–339.

Boucias, D. G., and Pendland, J. C., 1991, Attachment of mycopathogens to cuticle: The initial event of mycoses in arthropod hosts, in: *The Fungal Spore and Disease Initiation in Plants and Animals* (G. T. Cole and H. C. Hoch, eds.), Plenum Press, New York, pp. 101–128.

Boucias, D. G., Bradford, D. L., and Barfield, C. S., 1984, Susceptibility of the velvetbean caterpillar and soybean looper (Lepidoptera: Noctuidae) to *Nomuraea rileyi*: Effects of pathotype, dosage, temperature, and host age, *J. Econ. Entomol.* **77:**247–253.

Boucias, D. G., Pendland, J. C., and Latgé, J.-P., 1988, Nonspecific factors involved in the attachment of entomopathogenic deuteromycetes to host insect cuticle, *Appl. Entomol. Microbiol.* **54:**1795–1805.

Brobyn, P. J., Wilding, N., and Clark, S. J., 1985, The persistence of infectivity of conidia of the aphid pathogen *Erynia neoaphidis* on leaves in the field, *Ann. Appl. Biol.* **107:**365–376.

Brobyn, P. J., Wilding, N., and Clark, S. J., 1987, Laboratory observations on the effect of humidity on the persistence of infectivity of conidia of the aphid pathogen *Erynia neoaphidis*, *Ann. Appl. Biol.* **110:**579–584.

Brooks, W. M., 1993, Host–parasitoid–pathogen interactions, in: *Parasites and Pathogens of Insects, Vol. 2: Pathogens* (N. E. Beckage, S. N. Thompson, and B. A. Federici, eds.), Academic Press, San Diego, pp. 231–272.

Butt, T. M., and Humber, R. A., 1989, Response of gypsy moth hemocytes to natural fungal protoplasts of three *Entomophaga* species (Zygomycetes: Entomophthorales), *J. Invertebr. Pathol.* **53:**121–123.

Butt, T. M., Wraight, S. P., Galaini-Wraight, S., Humber, R. A., Roberts, D. W., and Soper, R. S., 1988, Humoral encapsulation of the fungus *Erynia radicans* (Entomophthorales) by the potato leafhopper, *Empoasca fabae* (Homoptera: Cicadellidae), *J. Invertebr. Pathol.* **52:**59–56.

Butt, T. M., Ibrahim, L., Clark, S. J., and Beckett, A., 1995, The germination behaviour of *Metarhizium anisopliae* on the surface of aphid and flea beetle cuticles, *Mycol. Res.* **99:**945–950.

Carroll, G. C., 1991, Fungal associates of woody plants as insect antagonists in leaves and stems, in: *Microbial Mediation of Plant–Herbivore Interactions* (P. Barbosa, V. A. Krischik, and C. G. Jones, eds.), Wiley, New York, pp. 253–271.

Carruthers, R. I., and Haynes, D. L., 1986, Temperature, moisture, and habitat effects on *Entomophthora muscae* conidial germination and survival in the onion agroecosystem, *Environ. Entomol.* **15:**1154–1160.

Carruthers, R. I., and Soper, R. S., 1987, Fungal diseases, in: *Epizootiology of Insect Diseases* (J. R. Fuxa and Y. Tanada, eds.), Wiley, New York, pp. 357–416.

Carruthers, R. I., Feng, Z., Ramos, M. E., and Soper, R. S., 1988, The effect of solar radiation on the survival of *Entomophaga grylli* (Entomophthorales: Entomophthoraceae) conidia, *J. Invertebr. Pathol.* **52:**154–162.

Carruthers, R. I., Larkin, T. S., Firstencel, H., and Feng, Z., 1992, Influence of thermal ecology on the mycosis of a rangeland grasshopper, *Ecology* **73:**190–204.

Charnley, A. K., and St. Leger, R. J., 1991, The role of cuticle-degrading enzymes in fungal pathogenesis in insects, in: *The Fungal Spore and Disease Initiation in Plants and Animals* (G. T. Cole and H. C. Hoch, eds.), Plenum Press, New York, pp. 267–286.

Clarkson, J. M., and Charnley, A. K., 1995, New insights into the mechanisms of fungal pathogenesis in insects, *Tr. Microbiol.* **4:**197–203.

Costa, S. D., and Gaugler, R., 1989, Influence of *Solanum* host plants on Colorado potato beetle (Coleoptera: Chrysomelidae) susceptibility to the entomopathogen *Beauveria bassiana*. *Environ. Entomol.* **18:**531–536.

Daoust, R. A., and Pereira, R. M., 1986, Stability of entomopathogenic fungi *Beauveria bassiana* and *Metarhizium anisopliae* on beetle-attracting tubers and cowpea foliage in Brazil, *Environ. Entomol.* **15:**1237–1243.

Daoust, R. A., and Roberts, D. W., 1983, Studies on the prolonged storage of *Metarhizium anisopliae* conidia: Effect of temperature and relative humidity on conidial viability and virulence against mosquitoes, *J. Invertebr. Pathol.* **41:**143-150.

Delalibera, I., Sosa Gomes, D. R., de Moraes, G. J., de Alencar, J. A., and Farias Araujo, W., 1992, Infection of *Mononychellus tanajoa* (Acari: Tetranychidae) by the fungus *Neozygites* sp. (Entomophthorales) in northeastern Brazil, *Fla. Entomol.* **75:**145-147.

Dickinson, C. H., 1986, Adaptations of micro-organisms to climatic conditions affecting aerial plant surfaces, in: *Microbiology of the Phyllosphere* (N. J. Fokkema and J. van den Heuvel, eds.), Cambridge University Press, Cambridge, pp. 77-100.

Dillon, R. J., and Charnley, A. K., 1991, The fate of fungal spores in the insect gut, in: *The Fungal Spore and Disease Initiation in Plants and Animals* (G. T. Cole and H. C. Hoch, eds.), Plenum Press, New York, pp. 129-156.

Doberski, J. W., and Tribe, H. T., 1980, Isolation of entomogenous fungi from elm bark and soil with reference to ecology of *Beauveria bassiana* and *Metarhizium anisopliae*, *Trans. Br. Mycol. Soc.* **74:**95-100.

Donegan, K., and Lighthart, B., 1989, Effect of several stress factors on the susceptibility of the predatory insect, *Chrysoperla carnea* (Neuroptera: Chrysopidae), to the fungal pathogen *Beauveria bassiana*, *J. Invertebr. Pathol.* **54:**79-84.

Eilenberg, J., 1986, Effect of *Entomophthora muscae* (C.) Fres. on egg-laying behaviour of female carrot-flies (*Psila rosae* F.), in: *Fundamental and Applied Aspects of Invertebrate Pathology* (R. A. Samson, J. M. Vlak, and D. Peters, eds.), Foundations of the Fourth International Colloquium Invertebrate Pathology, Wageningen, Netherlands, p. 235.

Elkinton, J. S., Hajek, A. E., Boettner, G. H., and Simons, E. E., 1991, Distribution and apparent spread of *Entomophaga maimaiga* (Zygomycetes: Entomophthorales) in gypsy moth (Lepidoptera: Lymantriidae) population in North America, *Environ. Entomol.* **20:**1601-1605.

El-Maghraby, M. M. A., Hebag, A., and Yousif-Khalil, S. I., 1988, Interactions between *Bacillus thuringiensis* Berl., *Beauveria bassiana* (Bals.) Vuill. and the host/parasitoid system *Spodoptera littoralis* (Boisd.)/*Microplitis rufiventris* Kok., *J. Appl. Ent.* **106:**417-421.

Emmett, R. W., and Parbery, D. G., 1975, Appressoria, *Annu. Rev. Phytopathol.* **13:**147-167.

Evans, H. C., 1989, Mycopathogens of insects of epigeal and aerial habitats, in: *Insect-Fungus Interactions* (N. Wilding, N. M. Collins, P. M. Hammond, and J. F. Webber, eds.), Academic Press, London, pp. 205-238.

Fargues, J., 1984, Adhesion of the fungal spore to the insect cuticle in relation to pathogenicity, in: *Infection Processes of Fungi* (D. W. Roberts and J. R. Aist, eds.), Rockefeller Foundation Conference Report, pp. 90-110.

Fargues, J., and Maniania, J. K., 1992, Variabilité de la sensibilité de *Spodoptera littoralis* [Lep.: Noctuidae] a l'hyphomycete entomopathogène *Nomuraea rileyi*. *Entomophaga* **37:**545-554.

Fargues, J., Rougier, M., Goujet, R., and Itier, B., 1988, Effet du rayonnement solaire sur la persistance des conidiospores de l'hyphomycète entomopathogène, *Nomuraea rileyi*, a la surface d'un couvert végétal, *Entomophaga* **33:**357-370.

Feng, M. G., and Nowierski, R. M., 1992, Spatial patterns and sampling plans for cereal aphids [Hom: Aphididae] killed by entomophthoralean fungi and hymenopterous parasitoids in spring wheat, *Entomophaga* **37:**265-272.

Feng, M. G., Poprawski, T. J., and Khachatourians, G. G., 1994, Production, formulation and application of the entomopathogenic fungus *Beauveria bassiana* for insect control: current status, *Biocontrol Sci. Technol.* **4:**3-34.

Fernandez-Garcia, E., and Fitt, B. D. L., 1993, Dispersal of the entomopathogen *Hirsutella cryptosclerotium* by simulated rain, *J. Invertebr. Pathol.* **61:**39-43.

Ferro, D. N., and Southwick, E. E., 1984, Microclimates of small arthropods: estimating humidity within the leaf boundary layer, *Environ. Entomol.* **13**:926–929.
Ferro, D. N., Chapman, R. B., and Penman, D. R., 1979, Observations on insect microclimate and insect pest management, *Environ. Entomol.* **8**:1000–1003.
Ferron, P., Fargues, J., and Riba, G., 1991, Fungi as microbial insecticides, in: *Handbook of Applied Mycology. Vol. 2: Humans, Animals, and Insects* (D. K. Arora, L. Ajello, and K. G. Mukerji, eds.), Marcel Dekker, New York, pp. 665–706.
Fransen, J. J., 1987, *Aschersonia aleyrodis as a Microbial Control Agent of Greenhouse Whitefly*, Ph.D. dissertation, Agricultural University of Wageningen, Netherlands.
Fransen, J. J., and van Lenteren, J. C., 1993, Host selection and survival of the parasitoid *Encarsia formosa* on greenhouse whitefly, *Trialeurodes vaporariorum*, in the presence of hosts infected with the fungus *Aschersonia aleyrodis*, *Entomol. Exp. Appl.* **69**:239–249.
Fransen, J. J., and van Lenteren, J. C., 1994, Survival of the parasitoid *Encarsia formosa* after treatment of parasitized greenhouse whitefly larvae with fungal spores of *Aschersonia aleyrodis*, *Entomol. Exp. Appl.* **71**:235–243.
Führer, E., and El-Sufty, R., 1979, Production of fungistatic metabolites by teratocytes of *Apanteles glomeratus* L. (Hym., Braconidae), *Z. Parasitenkd.* **59**:21–25.
Funk, C. J., Ramoska, W. A., and Bechtel, D. B., 1993, Histopathology of *Entomophaga grylli* pathotype 2 infections in *Melanoplus differentialis*, *J. Invertebr. Pathol.* **61**:196–202.
Furlong, M. J., and Pell, J. K., 1996, Interactions between the fungal entomopathogen *Zoophthora radicans* Brefeld (Entomophthorales) and two hymenopteran parasitoids attacking the diamondback moth, *Plutella xylostella* L., *J. Invertebr. Pathol.* **68**:15–21.
Fuxa, J. R., and Tanada, Y., 1987, Epidemiological concepts applied to insect epizootiology, in: *Epizootiology of Insect Diseases*, John Wiley, New York, pp. 3–21.
Galaini, S., 1984, *The Efficacy of Foliar Applications of Beauveria bassiana Conidia against Leptinotarsa decemlineata*, M.S. dissertation, Cornell University, Ithaca, NY.
Gallardo, F., Boethel, D. J., Fuxa, J. R., and Richter, A., 1990, Susceptibility of *Heliothis zea* (Boddie) larvae to *Nomuraea rileyi* (Farlow) Samson: Effects of α-tomatine at the third trophic level. *J. Chem. Ecol.* **16**:1751–1759.
Garcia, C., and Ignoffo, C. M., 1977, Dislodgement of conidia of *Nomuraea rileyi* from cadavers of cabbage looper, *Trichoplusia ni*, *J. Invertebr. Pathol.* **30**:114–116.
Gardner, W. A., Sutton, R. M., and Noblet, R., 1977, Persistence of *Beauveria bassiana*, *Nomuraea rileyi*, and *Nosema necatrix* on soybean foliage, *Environ. Entomol.* **5**:616–618.
Gaugler, R., Costa, S. D., and Lashomb, J., 1989, Stability and efficacy of *Beauveria bassiana* soil inoculations, *Environ. Entomol.* **18**:412–417.
Geden, C. J., Rutz, D. A., and Steinkraus, D. C., 1995, Virulence of different isolates and formulations of *Beauveria bassiana* for house flies and the parasitoid *Muscidifurax raptor*, *Biol. Control* **5**:615–621.
Gilliam, M., Taber III, S., and Richardson, G. V., 1983, Hygienic behavior of honey bees in relation to chalkbrood disease, *Apidologie* **14**:29–39.
Gilliam, M., Taber III, S., Lorenz, B. J., and Prest, D. B., 1988, Factors affecting development of chalkbrood disease in colonies of honey bees, *Apis mellifera*, fed pollen contaminated with *Ascosphaera apis*, *J. Invertebr. Pathol.* **52**:314–325.
Glare, T. R., and Milner, R. J., 1991, Ecology of entomopathogenic fungi, in: *Handbook of Applied Mycology. Vol. 2: Humans, Animals, and Insects* (D. K. Arora, L. Ajello, and K. G. Mukerji, eds.), Marcel Dekker, New York, pp. 547–612.
Glare, T. R., Milner, R. J., and Chilvers, G. A., 1986, The effect of environmental factors on the production, discharge, and germination of primary conidia of *Zoophthora phalloides* Batko, *J. Invertebr. Pathol.* **48**:275–283.

Glare, T. R., Milner, R. J., and Chilvers, G. A., 1989, Factors affecting the production of resting spores by *Zoophthora radicans* in the spotted alfalfa aphid, *Therioaphis trifolii* f. *maculata*, *Can. J. Bot.* **67**:848–855.

Goettel, M. W., and Inglis, G. D., 1997, Fungi: Hyphomycetes, in: *Manual of Techniques in Insect Pathology* (L. A. Lacey, ed.), Academic Press, San Diego, pp. 213–249.

Goettel, M. S., St. Leger, R. J., Rizzo, N. W., Staples, R. C., and Roberts, D. W., 1989, Ultrastructural localization of a cuticle-degrading protease produced by the entomopathogenic fungus *Metarhizium anisopliae* during penetration of host (*Manduca sexta*) cuticle, *J. Gen. Microbiol.* **135**:2233–2239.

Goettel, M. S., Poprawski, T. J., Vandenberg, J. D., Li, Z., and Roberts, D. W., 1990, Safety to nontarget invertebrates of fungal biocontrol agents, in: *Safety of Microbial Insecticides* (M. Laird, L. A. Lacey, and E. W. Davidson, eds.), CRC Press, Boca Raton, Florida, pp. 209–231.

Gopalakrishnan, C., and Narayanan, K., 1989, Epizootiology of *Nomuraea rileyi* (Farlow) Samson in field populations of *Helicoverpa* (= *Heliothis*) *armigera* (Hübner) in relation to three host plants, *J. Biol. Contr.* **3**:50–52.

Gottwald, T. R., and Tedders, W. L., 1982, Studies on condia release by the entomogenous fungi *Beauveria bassiana* and *Metarhizium anisopliae* (Deuteromycotina: Hyphomycetes) from adult pecan weevil (Coleoptera: Curculionidae) cadavers, *Environ. Entomol.* **11**:1274–1279.

Gottwald, T. R., and Tedders, W. L., 1984, Colonization, transmission, and longevity of *Beauveria bassiana* and *Metarhizium anisopliae* (Deuteromycotina: Hyphomycetes) on pecan weevil larvae (Coleoptera: Curculionidae) in the soil, *Environ. Entomol.* **13**:557–560.

Gregory, P. H., Guthrie, E. J., and Bunce, M. E., 1959, Experiments on splash dispersal of fungus spores, *J. Gen. Microbiol.* **20**:328–354.

Groden, E., and Lockwood, J. L., 1991, Effects of soil fungistasis on *Beauveria bassiana* and its relationship to disease incidence in the Colorado potato beetle, *Leptinotarsa decemlineata*, in Michigan and Rhode Island soils, *J. Invertebr. Pathol.* **57**:7–16.

Gugnani, H. C., and J. I. Okafor, 1980, Mycotic flora of the intestine and other internal organs of certain reptiles and amphibians with special reference to characterization of *Basidiobolus* isolates, *Mykosen* **23**:260–268.

Hajek, A. E., 1989, Food consumption by *Lymantria dispar* (Lepidoptera: Lymantriidae) larvae injected with *Entomophaga maimaiga* (Zygomycetes: Entomophthorales), *Environ. Entomol.* **18**:723–727.

Hajek, A. E., 1993, New options for insect control using fungi, in: *Pest Management: Biologically Based Technologies* (R. D. Lumsden and J. L. Vaughn, eds.), American Chemical Society, pp. 54–62.

Hajek, A. E., 1995, Persistence of *Entomophaga maimaiga* in the environment, *Proc. U.S. Dept. Agric. Interagency Gypsy Moth Res. Forum*, U.S.D.A. Forest Service, Northeastern Forest Experiment Station, General Technical Report NE-213, p. 73.

Hajek, A. E., 1997, *Entomophaga maimaiga* reproductive output determined by spore type initiating infection, *Mycol. Res.* (in press).

Hajek, A. E., and Roberts, D. W., 1991, Pathogen reservoirs as a biological control resource: Introduction of *Entomophaga maimaiga* to North American gypsy moth, *Lymantria dispar*, populations, *Biol. Contr.* **1**:29–34.

Hajek, A. E., and Roberts, D. W., 1992, Field diagnosis of gypsy moth (Lepidoptera: Lymantriidae) larval mortality caused by *Entomophaga maimaiga* and the gypsy moth nuclear polyhedrosis virus, *Environ. Entomol.* **21**:706–713.

Hajek, A. E., and St. Leger, R. J., 1994, Interactions between fungal pathogens and insect hosts, *Annu. Rev. Entomol.* **39**:293–322.

Hajek, A.E., and Shimazu, M., 1996, Types of spores produced by *Entomophaga maimaiga* infecting the gypsy moth *Lymantria dispar*, *Can. J. Bot.* **74**:708–715.
Hajek, A. E., and Soper, R. S., 1992, Temporal dynamics of *Entomophaga maimaiga* after death of gypsy moth (Lepidoptera: Lymantriidae) larval hosts, *Environ. Entomol.* **21**:129–135.
Hajek, A. E., and Wheeler, M. M., 1994, Application of techniques for quantification of soil-borne entomophthoralean resting spores, *J. Invertebr. Pathol.* **64**:71–73.
Hajek, A. E., Carruthers, R. I., and Soper, R. S., 1990a, Temperature and moisture relations of sporulation and germination by *Entomophaga maimaiga* (Zygomycetes: Entomophthoraceae), a fungal pathogen of *Lymantria dispar* (Lepidoptera: Lymantriidae), *Environ. Entomol.* **19**:85–90.
Hajek, A. E., Humber, R. A., Elkinton, J. S., May, B., Walsh, S. R. A., and Silver, J. C., 1990b, Allozyme and RFLP analyses confirm *Entomophaga maimaiga* responsible for 1989 epizootics in North American gypsy moth populations, *Proc. Natl. Acad. Sci. U.S.A.* **87**:6979–6982.
Hajek, A. E., Butt, T. M., Strelow, L. I., and Gray, S. M., 1991a, Detection of *Entomophaga maimaiga* (Zygomycetes: Entomophthorales) using enzyme-linked immunosorbent assay (ELISA), *J. Invertebr. Pathol.* **58**:1–9.
Hajek, A. E., Humber, R. A., Walsh, S. R. A., and Silver, J. C., 1991b, Sympatric occurrence of two *Entomophaga aulicae* (Zygomycetes: Entomophthorales) complex species attacking forest Lepidoptera, *J. Invertebr. Pathol.* **58**:373–380.
Hajek, A. E., Larkin, T. S., Carruthers, R. I., and Soper, R. S., 1993, Modeling the dynamics of *Entomophaga maimaiga* (Zygomycetes: Entomophthorales) epizootics in gypsy moth (Lepidoptera: Lymantriidae) populations, *Environ. Entomol.* **22**:1172–1187.
Hajek, A. E., Butler, L., and Wheeler, M. M., 1995a, Laboratory bioassays testing the host range of the gypsy moth fungal pathogen *Entomophaga maimaiga*, *Biol. Contr.* **5**:530–544.
Hajek, A. E., Humber, R. A., and Elkinton, J. S., 1995b, Mysterious origin of *Entomophaga maimaiga* in North America, *Amer. Entomol.* **41**:31–42.
Hajek, A. E., Renwick, J. A. A., and Roberts, D. W., 1995c, Effects of larval host plant on the gypsy moth (Lepidoptera: Lymantriidae) fungal pathogen, *Entomophaga maimaiga* (Zygomycetes: Entomophthorales), *Environ. Entomol.* **24**:1307–1314.
Hajek, A. E., Butler, L., Walsh, S. R. A., Silver, J. C., Hain, F. P., Hastings, F. L., O'Dell, T. M., And Smitley, D. R., 1996a, Host range of the gypsy moth (Lepidoptera: Lymantriidae) pathogen *Entomophaga maimaiga* (Zygomycetes: Entomophthorales) in the field versus laboratory, *Environ. Entomol.* **25**:709–721.
Hajek, A. E., Elkinton, J. S., and Witcosky, J. J., 1996b, Introduction and spread of the fungal pathogen *Entomophaga maimaiga* (Zygomycetes: Entomophthorales) along the leading edge of gypsy moth (Lepidoptera: Lymantriidae) spread, *Environ. Entomol.* **25**:1235–1247.
Hajek, A. E., Hodge, K. T., Liebherr, J. K., Day, W. H., and Vandenberg, J. D., 1996c, Use of RAPD analysis to trace the origin of the weevil pathogen *Zoophthora phytonomi* in North America, *Mycol. Res.* **100**:349–355.
Hall, R. A., 1981, The fungus *Verticillium lecanii* as a microbial insecticide against aphids and scales, in: *Microbial Control of Pests and Plant Diseases 1970–1980* (H. D. Burges, ed.), Academic Press, New York, pp. 483–498.
Hall, R. A., and Burges, H. D., 1979, Control of aphids in glasshouses with the fungus, *Verticillium lecanii*, *Ann. Appl. Biol.* **93**:235–246.
Hall, R. A., and Papierok, B., 1982, Fungi as biological control agents of arthropods of agricultural and medical importance, *Parasitology* **84**:205–240.
Hare, J. D., and Andreadis, T. G., 1983, Variation in the susceptibility of *Leptinotarsa decemlineata* (Coleoptera: Chrysomelidae) when reared on different host plants to the fungal pathogen, *Beauveria bassiana* in the field and laboratory, *Environ. Entomol.* **12**:1892–1897.
Harper, J. D., Herbert, D. A., and Moore, R. E., 1984, Trapping patterns of *Entomophthora*

gammae (Weiser) (Entomophthorales: Entomophthoraceae) conidia in a soybean field infested with the soybean looper, *Pseudoplusia includens* (Walker) (Lepidoptera: Noctuidae), *Environ. Entomol.* **13:**1186–1190.

Hodge, K. T., Sawyer, A. J., and Humber, R. A., 1995, RAPD-PCR for identification of *Zoophthora radicans* isolates in biological control of the potato leafhopper, *J. Invertebr. Pathol.* **65:**1–9.

Holdom, D. G., 1984, *Studies on the Biology, Nutrition and Physiology of Entomophthora planchoniana Cornu (Zygomycetes: Entomophthorales), a Pathogen of the Bluegreen aphid, Acyrthosiphon kondoi Shinji (Homoptera: Aphididae)*, Ph.D. dissertation, University of Queensland, Australia.

Hollingsworth, R. G., Steinkraus, D. C., and McNew, R. W., 1995, Sampling to predict fungal epizootics in cotton aphids (Homoptera: Aphididae), *Environ. Entomol.* **24:**1414–1421.

Humber, R. A., 1976, The systematics of the genus *Strongwellsea* (Zygomycetes: Entomophthorales), *Mycologia* **68:**1042–1060.

Humber, R. A., 1992, *Collection of Entomopathogenic Fungal Cultures: Catalog of Strains, 1992*, U.S. Department of Agriculture, Agricultural Research Service. ARS-110.

Humber, R. A., 1996, Fungal pathogens of the Chrysomelidae and prospects for their use in biological control, in: *Chrysomelidae Biology. Vol. 2: Ecological Studies* (P. H. A. Jolivet & M. L. Cox, eds.), Academic Press, Netherlands, pp. 93–115.

Hung, S.-Y., and Boucias, D. G., 1992, Influence of *Beauveria bassiana* on the cellular defense response of the beet armyworm, *Spodoptera exigua*, *J. Invertebr. Pathol.* **60:**152–158.

Hung, S.-Y., Boucias, D. G., and Vey, A., 1993, Effect of *Beauveria bassiana* and *Candida albicans* on the cellular defense response of the beet armyworm, *Spodoptera exigua*, *J. Invertebr. Pathol.* **61:**179–187.

Ignoffo, C. M., Hostetter, D. L., Sikorowski, P. P., Sutter, G., and Brooks, W. M., 1977, Inactivation of representative species of entomopathogenic viruses, a bacterium, fungus, and protozoan by an ultraviolet light source, *Environ. Entomol.* **6:**411–415.

Ignoffo, C. M., Garcia, C., Hostetter, D. L., and Pinnell, R. E., 1978, Stability of conidia of an entomopathogenic fungus, *Nomuraea rileyi*, in and on soil, *Environ. Entomol.* **7:**724–727.

Iijima, R., Kurata, S., and Natori, S., 1993, Purification, characterization, and cDNA cloning of an antifungal protein from the hemolymph of *Sarcophaga peregrina* (flesh fly) larvae, *J. Biol. Chem.* **268:**12055–12061.

Inglis, G. D., Goettel, M. S., and Johnson, D. L., 1993, Persistence of the entomopathogenic fungus, *Beauveria bassiana*, on phylloplanes of crested wheatgrass and alfalfa, *Biol. Contr.* **3:**258–270.

Inglis, G. D., Johnson, D. L., and Goettel, M. S., 1995, Effects of simulated rain on the persistence of *Beauveria bassiana* conidia on leaves of alfalfa and wheat, *Biocontrol Sci. Technol.* **5:**365–369.

James, R. R., and Lighthart, B., 1994, Susceptibility of the convergent lady beetle (Coleoptera: Coccinellidae) to four entomogenous fungi, *Environ. Entomol.* **23:** 190–192.

Keller, S., 1983, Die mikrobiologische Bekämpfung des Maikäfers (*Melolontha melolontha* L.) mit dem Pilz *Beauveria brongniartii*, *Mitt. Schweiz. Landwirtsch.* **31:**61–64.

Keller, S., 1991, Arthropod pathogenic Entomophthorales of Switzerland. II. *Erynia, Eryniopsis, Zoophthora*, and *Tarichium*, *Sydowia Ann. Mycol.* **43:**39–122.

Keller, S., and Zimmermann, G. 1989. Mycopathogens of soil insects, in: *Insect–Fungus Interactions* (N. Wilding, N. M. Collins, P. M. Hammond, and J. F. Webber, eds.), Academic Press, London, pp. 240–270.

Kenneth, R., Wallis, G., Gerson, U., and Plaut, H. N., 1972, Observations and experiments on

Triplosporium floridanum (Entomophthorales) attacking spider mites in Israel, *J. Invertebr. Pathol.* **19**:366–369.

Kerwin, J. L., 1984, Fatty acid regulation of the germination of *Erynia variabilis* conidia on adults and puparia of the lesser housefly, *Fannia canicularis*, *Can. J. Microbiol.* **30**:158–161.

Khachatourians, G. G., 1991, Physiology and genetics of entomopathogenic fungi, in: *Handbook of Applied Mycology. Vol. 2: Humans, Animals, and Insects* (D. K. Arora, L. Ajello, and K. G. Mukerji, eds.), Marcel Dekker, New York, pp. 613–663.

King, D. S., and Humber, R. A., 1981, Identification of the Entomophthorales, in: *Microbial Control of Pests and Plant Diseases 1970–1980* (H. D. Burges, ed.), Academic Press, London, pp. 107–127.

King, E. G., and Bell, J. V., 1978, Interactions between a braconid, *Microplitis croceipes*, and a fungus, *Nomuraea rileyi*, in laboratory-reared bollworm larvae, *J. Invertebr. Pathol.* **31**:337–340.

Kish, L. P, and Allen, G. E., 1978, The biology and ecology of *Nomuraea rileyi* and a program for predicting its incidence on *Anticarsia gemmatalis* in soybean, *Florida Agric. Exp. Sta. Tech. Bull.* **795**.

Knudsen, G. R., Schotzko, D. J., and Krag, C. R., 1994, Fungal entomopathogen effect on numbers and spatial patterns of the Russian wheat aphid (Homoptera: Aphididae) on preferred and nonpreferred host plants, *Environ. Entomol.* **23**:1558–1567.

Kramer, J. P., 1980, The housefly mycosis caused by *Entomophthora muscae*: Influences of relative humidity on infectivity and conidial germination, *J. N. Y. Ent. Soc.* **88**:236–240.

Krasnoff, S. B., Watson, D. W., Gibson, D. M., and Kwan, E. C., 1995, Behavioral effects of the entomopathogenic fungus *Entomophthora muscae* on its host *Musca domestica*: postural changes in dying hosts and gated pattern of mortality, *J. Ins. Physiol.* **41**:895–903.

Krejzova, R., 1971, Resistance and germinability of resting spores of some species of the genus *Entomophthora*, *Ceska Mykol.* **25**:231–238.

Krieg, A., 1971, Interactions between pathogens, in: *Microbial Control of Insects and Mites* (H. D. Burges, and N. W. Hussey, eds.), Academic Press, London, pp. 459–468.

Krueger, S. R., Nechols, J. R., and Ramoska, W. A., 1991a, Infection of chinch bug, *Blissus leucopterus leucopterus* (Hemiptera: Lygaeidae), adults from *Beauveria bassiana* (Deuteromycotina: Hyphomycetes) conidia in soil under controlled temperature and moisture conditions, *J. Invertebr. Pathol.* **58**:19–26.

Krueger, S. R., Villani, M. G., Nyrop, J. P., and Roberts, D. W., 1991b, Effect of soil environment on the efficacy of fungal pathogens against scarab grubs in laboratory bioassays, *Biol. Control* **1**:203–209.

Latgé, J.-P., Sampedro, L., Brey, P., and Diaquin, M., 1987, Aggressiveness of *Conidiobolus obscurus* against the pea aphid: Influence of cuticular extracts on ballistospore germination of aggressive and non-aggressive strains, *J. Gen. Microbiol.* **133**:1987–1997.

Latteur, G., 1980, The persistence of infectivity of conidia of *Entomophthora obscura* at different temperatures on the surface of unsterilized soil, *Acta Oecol., Oecol. Applic.* **1**:29–34.

Latteur, G., Destain, J., Oger, R., and Godefroid, J., 1982, Ètude de la production de conidies par les spores durables de *Conidiobolus obscurus* (Hall et Dunn) Remaud. et Kell., une entomophthorale pathogène de pucerons, *Parasitica* **38**:139–161.

Leach, C. M., 1971, A practical guide to the effects of visible and ultraviolet light on fungi, in: *Methods in Microbiology, Vol. 4* (C. Booth, ed.), Academic Press, New York, pp. 609–664.

Li, Z., Soper, R. S., and Hajek, A. E., 1988, A method for recovering resting spores of Entomophthorales (Zygomycetes) from soil, *J. Invertebr. Pathol.* **52**:18–26.

Li, Z., Butt, T. M., Beckett, A., and Wilding, N., 1993, The structure of dry mycelia of the entomophthoralean fungi *Zoophthora radicans* and *Erynia neoaphidis* following different preparatory treatments, *Mycol. Res.* **97**:1315-1323.

Lloyd, M., White, J., and Stanton, N., 1982, Dispersal of fungus-infected periodical cicadas to new habitat, *Environ. Entomol.* **11**:852-858.

MacDonald, R. M., and Spokes, J. R., 1981, *Conidiobolus obscurus* in arable soil: a method for extracting and counting azygospores, *Soil Biol. Biochem.* **13**:551-553.

Magalhaes, B. P., Lord, J. C., Wraight, S. P., Daoust, R. A., and Roberts, D. W., 1988, Pathogenicity of *Beauveria bassiana* and *Zoophthora radicans* to the coccinellid predators *Coleomegilla maculata* and *Eriopis connexa*, *J. Invertebr. Pathol.* **52**:471-473.

Magalhaes, B. P., Humber, R. A., Shields, E. J., and Roberts, D. W., 1991a, Effects of environment and nutrition on conidium germination and appressorium formation by *Zoophthora radicans* (Zygomycetes: Entomophthorales): a pathogen of the potato leafhopper (Homoptera: Cicadellidae), *Environ. Entomol.* **20**:1460-1468.

Magalhaes, B. P., Wayne, R., Humber, R. A., Shields, E. J., and Roberts, D. W., 1991b, Calcium-regulated appressorium formation of the entomopathogenic fungus *Zoophthora radicans*, *Protoplasma* **160**:77-88.

Mazet, I., Hung, S.-Y., and Boucias, D. G., 1994, Detection of toxic metabolites in the hemolymph of *Beauveria bassiana*-infected *Spodoptera exigua* larvae, *Experientia* **50**:142-147.

McCoy, C. W., Samson, R. A., and Boucias, D. G. 1988. Entomogenous fungi, in: *Handbook of Natural Pesticides, Vol. 5, Microbial Insecticides, Part A* (C. M. Ignoffo, ed.), CRC Press, Boca Raton, FL, pp. 151-236.

McDonald, D. M., and Nolan, R. A., 1995, Effect of relative humidity and temperature on *Entomophaga aulicae* conidium discharged from infected eastern hemlock looper larvae and subsequent conidium development, *J. Invertebr. Pathol.* **65**:83-90.

McGuire, M. R., 1985, *Erynia radicans*: Studies on its Distribution, Pathogenicity, and Host Range in relation to Potato Leafhopper, *Empoasca fabae*, Ph.D. dissertation, University of Illinois, Urbana-Champaign.

McGuire, M. R., Maddox, J. V., and Armbrust, E. J., 1987, Host range studies of an *Erynia radicans* strain (Zygomycetes: Entomophthoraceae) isolated from *Empoasca fabae* (Homoptera: Cicadellidae), *J. Invertebr. Pathol.* **50**:75-77.

Mills, N. J., 1981, The mortality and fat content of *Adalia bipunctata* during hibernation, *Entomol. Exp. Appl.* **30**:265-268.

Millstein, J. A., Brown, G. C., and Nordin, G. L., 1982, Microclimatic humidity influence on conidial discharge in *Erynia* sp. (Entomophthorales: Entomophthoraceae), an entomopathogenic fungus of the alfalfa weevil (Coleoptera: Curculionidae), *Environ. Entomol.* **11**:1166-1169.

Millstein, J. A., Brown, G. C., and Nordin, G. L., 1983, Microclimatic moisture and conidial production in *Erynia* sp. (Entomophthorales: Entomophthoraceae): In vivo production rate and duration under constant and fluctuating moisture regimes, *Environ. Entomol.* **12**:1344-1349.

Milner, R. J., 1981, Patterns of primary spore discharge of *Entomophthora* spp. from the blue green aphid, *Acyrthosiphon kondoi*, *J. Invertebr. Pathol.* **38**:419-425.

Milner, R. J., 1982, On the occurrence of pea aphids, *Acyrthosiphon pisum*, resistant to isolates of the fungal pathogen *Erynia neoaphidis*, *Entomol. Exp. Appl.* **32**:23-27.

Milner, R. J., 1985, Distribution in time and space of resistance to the pathogenic fungus *Erynia neoaphidis* in the pea aphid *Acyrthosiphon pisum*, *Entomol. Exp. Appl.* **37**:235-240.

Milner, R. J., and Lutton, G. G., 1986, Dependence of *Verticillium lecanii* (Fungi: Hyphomycetes) on high humidities for infection and sporulation using *Myzus persicae* (Homoptera: Aphididae) as host, *Environ. Entomol.* **15**:380-382.

Milner, R. J., and Soper, R. S., 1981, Bioassay of *Entomophthora* against the spotted alfalfa aphid *Therioaphis trifolii* f. *maculata*, *J. Invertebr. Pathol.* **37**:168-173.

Milner, R. J., Soper, R. S., and Lutton, G. G., 1982, Field release of an Israeli strain of the fungus *Zoophthora radicans* (Brefeld) Batko for biological control of *Therioaphis trifolii* (Monell) f. *maculata, J. Aust. Ent. Soc.* **21:**113–118.

Milner, R. J., Holdom, D. G., and Glare, T. R., 1984, Diurnal patterns of mortality in aphids infected by entomophthoran fungi, *Entomol. Exp. Appl.* **36:**37–42.

Møller, A. P., 1993, A fungus infecting domestic flies manipulates sexual behaviour of its host, *Behav. Ecol. Sociobiol.* **33:**403–407.

Morgan, L. W., Boddy, L., and Wilding, N., 1992, The survival and germination of *Erynia neoaphidis* conidia, *Proc. 25th Ann. Mtg. Soc. Invertebr. Pathol., Heidelberg, Germany,* p. 220.

Morgan, L. W., Boddy, L., Clark, S. J., and Wilding, N., 1995, Influence of temperature on germination of primary and secondary conidia of *Erynia neoaphidis* (Zygomycetes: Entomophthorales), *J. Invertebr. Pathol.* **65:**132–138.

Mullens, B. A., and Rodriguez, J. L., 1985, Dynamics of *Entomophthora muscae* (Entomophthorales: Entomophthoraceae) conidial discharge from *Musca domestica* (Diptera: Muscidae) cadavers, *Environ. Entomol.* **14:**317–322.

Nadeau, M. P., Dunphy, G. B., and Boisvert, J. L., 1996, Development of *Erynia conica* (Zygomycetes: Entomophthorales) on the cuticle of the adult black flies *Simulium rostratum* and *Simulium decorum* (Diptera: Simuliidae), *J. Invertebr. Pathol.* **68:**50–58.

Newman, G. G., and Carner, G. R., 1975, An *Entomophthora* infection of the adult cluster fly *Pollenia rudis, J. Ga. Ent. Soc.* **10:**315–326.

Nolan, R. A., Dunphy, G. B., and MacLeod, D. M., 1976, In vitro germination of *Entomophthora egressa* resting spores, *Can. J. Bot.* **54:**1131–1134.

Onstad, D. W., and Carruthers, R. I., 1990, Epizootiological models of insect diseases, *Annu. Rev. Entomol.* **35:**399–419.

Oduor, G. I., 1995, *Abiotic Factors and the Epizootiology of Neozygites cf. floridana, a Fungus Pathogenic to the Cassava Green Mite,* Ph.D. dissertation, University of Amsterdam, Holland.

Papierok, B., and Hajek, A. E., 1997, Fungi: Entomophthorales, in: *Manual of Techniques in Insect Pathology* (L. A. Lacey, ed.), Academic Press, San Diego, pp. 187–212.

Papierok, B., and Wilding, N., 1979, Mise en evidence d'une différence de sensibilité entre 2 clones du puceron du pois *Acyrthosiphon pisum* Harr. (Homoptères: Aphidiidae), exposé à 2 souches du champignon Phycomycète *Entomophthora obscura* Hall & Dunn, *C. R. Acad. Sci. Ser. D* **288:**93–95.

Papierok, B., Torres, B. V. L., and Arnault, M., 1984, Contribution a l'étude de la spécificité parasitaire du champignon entomopathogène *Zoophthora radicans* (Zygomycètes, Entomophthorales), *Entomophaga* **29:**109–119.

Pell, J. K., and Wilding, N., 1992, The survival of *Zoophthora radicans* (Zygomycetes: Entomophthorales) isolates as hyphal bodies in mummified larvae of *Plutella xylostella* (Lep.: Yponomeutidae), *Entomophaga* **37:**649–654.

Pendland, J. C., 1982, Resistant structures in the entomogenous hyphomycete, *Nomuraea rileyi*: an ultrastructural study, *Can. J. Bot.* **60:**1569–1576.

Pendland, J. C., Heath, M. A., and Boucias, D. G., 1988, Function of a galactose-binding lectin from *Spodoptera exigua* larval hemolymph: Opsonization of blastospores from entomogenous Hyphomycetes, *J. Ins. Physiol.* **34:**533–540.

Pendland, J. C., Hung, S.-Y., and Boucias, D. G., 1993, Evasion of host defense by in vivo-produced protoplast-like cells of the mycopathogen *Beauveria bassiana, J. Bacteriol.* **175:**5962–5969.

Pendland, J. C., Hung, S.-Y., and Boucias, D. G., 1995, In vivo development of the entomogenous hyphomycete *Paecilomyces farinosus* in host *Spodoptera exigua* (beet armyworm) larvae, *Mycopathologia* **130:**151–158.

Perry, D. F., 1988, Germination of *Pandora bullata* resting spores, *J. Invertebr. Pathol.* **51:**161–162.
Perry, D. F., and Fleming, R. A., 1989a, The timing of *Erynia radicans* resting spore germination in relation to mycosis of *Choristoneura fumiferana, Can. J. Bot.* **67:**1657–1663.
Perry, D. F., and Fleming, R. A., 1989b, *Erynia crustosa* zygospore germination, *Mycologia* **81:**154–158.
Perry, D. F., and Latgé, J.-P., 1982, Dormancy and germination of *Conidiobolus obscurus* azygospores, *Trans. Br. Mycol. Soc.* **78:**221–225.
Pickering, J., and Gutierrez, A. P., 1991, Differential impact of the pathogen *Pandora neoaphidis* (R. & H.) Humber (Zygomycetes: Entomophthorales) on the species composition of *Acyrthosiphon* in alfalfa, *Can. Entomol.* **123:**315–310.
Powell, W., Wilding, N., Brobyn, P. J., and Clark, S. J., 1986, Interference between parasitoids (Hym.: Aphidiidae) and fungi (Entomophthorales) attacking cereal aphids. *Entomophaga* **31:**293–302.
Ramoska, W. A., 1984, The influence of relative humidity on *Beauveria bassiana* infectivity and replication in the chinch bug, *Blissus leucopterus, J. Invertebr. Pathol.* **43:**389–394.
Ramoska, W. A., and Todd, T., 1985, Variation in efficacy and viability of *Beauveria bassiana* in the chinch bug (Hemiptera: Lygaeidae) as a result of feeding activity on selected host plants. *Environ. Entomol.* **14:**146–148.
Rath, A. C., Worledge, D., Koen, T. B., and Rowe, B. A., 1995a, Long-term field efficacy of the entomogenous fungus *Metarhizium anisopliae* against the subterranean scarab, *Adoryphorus couloni, Biocontrol Sci. Technol.* **5:**439–451.
Rath, A. C., Anderson, G. C., Worledge, D., and Koen, T. B., 1995b, The effect of low temperatures on the virulence of *Metarhizium anisopliae* (DAT F-001) to the subterranean scarab, *Adoryphorus couloni, J. Invertebr. Pathol.* **65:**186–192.
Roberts, D. W., and Campbell, A. S., 1977, Stability of entomopathogenic fungi, *Misc. Publ. Entomol. Soc. Amer.* **10:**1–80.
Roberts, D. W., and Hajek, A. E., 1992, Entomopathogenic fungi as bioinsecticides, in: *Frontiers in Industrial Mycology* (G. F. Leatham, ed.), Chapman & Hall, New York, pp. 144–159.
Roberts, D. W., and Humber, R. A., 1981, Entomogenous fungi, in: *Biology of Conidial Fungi, Vol. 2* (G. T. Cole and B. Kendrick, eds.), Academic Press, New York, pp. 201–236.
Roberts, D. W., Fuxa, J. R., Gaugler, R., Goettel, M., Jaques, R., and Maddox, J., 1991, Use of pathogens in insect control, in: *CRC Handbook of Pest Management in Agriculture, 2nd Ed., Vol. II* (D. Pimentel, ed.), CRC Press, Boca Raton, FL, pp. 243–278.
St. Leger, R. J., 1991, Integument as a barrier to microbial infections, in: *Physiology of the Insect Epidermis* (K. Binnington and A. Retnakaran, eds.), CSIRO, Australia, pp. 284–306.
St. Leger, R. J., 1993, Biology and mechanisms of insect-cuticle invasion by Deuteromycete fungal pathogens, in: *Parasites and Pathogens of Insects. Vol. 2: Pathogens* (N. E. Beckage, S. N. Thompson, and B. A. Federici, eds.), Academic Press, San Diego, pp. 211–229.
St. Leger, R. J., 1995, The role of cuticle-degrading proteases in fungal pathogenesis of insects, *Can. J. Bot., Suppl. 1* **73:**1119–1125.
St. Leger, R. J., Cooper, R. M., and Charnley, A. K., 1987, Distribution of chymoelastases and trypsin-like enzymes in five species of entomopathogenic Deuteromycetes, *Arch. Biochem. Biophys.* **258:**123–131.
St. Leger, R. J., Durrands, P. K., Charnley, A. K., and Cooper, R. M., 1988, The role of extracellular chymo-elastase in the virulence of *Metarhizium anisopliae* for *Manduca sexta, J. Invertebr. Pathol.* **52:**285–293.
St. Leger, R. J., Butt, T. M., Goettel, M. S., Roberts, D. W., and Staples, R. C., 1989, Production in vitro of appressoria by the entomopathogenic fungus *Metarhizium anisopliae, Exp. Mycol.* **13:**274–288.

St. Leger, R. J., Butt, T. M., Staples, R. C., and Roberts, D. W., 1990, Second messenger involvement in differentiation of the entomopathogenic fungus *Metarhizium anisopliae*, *J. Gen. Microbiol.* **136:**1779–1789.

St. Leger, R. J., Goettel, M., Roberts, D. W., and Staples, R. C., 1991, Prepenetration events during infection of host cuticle by *Metarhizium anisopliae*, *J. Invertebr. Pathol.* **58:**168–179.

St. Leger, R. J., May, B., Allee, L. L., Frank, D. C., and Roberts, D. W., 1992, Genetic differences in allozymes and in formation of infection structures among isolates of the entomopathogenic fungus *Metarhizium anisopliae*, *J. Invertebr. Pathol.* **60:**89–101.

Saito, T., and Aoki, J., 1983, Toxicity of free fatty acids on the larval surfaces of two lepidopterous insects towards *Beauveria bassiana* (Bals.) Vuill. and *Paecilomyces fumoso-roseus* (Wize) Brown et Smith (Deuteromycetes: Moniliales), *Appl. Entomol. Zool.* **18:**225–233.

Samson, R. A., Evans, H. C., and Latgé, J.-P., 1988, *Atlas of Entomopathogenic Fungi*, Springer-Verlag, Berlin.

Samuels, R. I., Charnley, A. K., and Reynolds, S. E., 1988a, The role of destruxins in the pathogenicity of three strains of *Metarhizium anisopliae* for the tobacco hornworm *Manduca sexta*, *Mycopathologia* **104:**51–58.

Samuels, R. I., Reynolds, S. E., Charnley, A. K., 1988b, Calcium channel activation of insect muscle by destruxins, insecticidal compounds produced by the entomopathogenic fungus *Metarhizium anisopliae*, *Comp. Biochem. Physiol.* **90C:**403–412.

Sawyer, A. J., Griggs, M. H., and Wayne, R., 1994, Dimensions, density, and settling velocity of entomophthoralean conidia: Implications for aerial dissemination of spores, *J. Invertebr. Pathol.* **63:**43–55.

Shimazu, M., 1987, Effect of rearing humidity of host insects on the spore types of *Entomophaga maimaiga* Humber, Shimazu et Soper (Entomophthorales; Entomophthoraceae), *Appl. Entomol. Zool.* **22:**394–397.

Shimazu, M., Koizumi, C., Kushida, T., and Mitsuhashi, J., 1987, Infectivity of hibernated resting spores of *Entomophaga maimaiga* Humber, Shimazu et Soper (Entomophthorales: Entomophthoraceae), *Appl. Ent. Zool.* **22:**216–221.

Shimazu, M., Mitsuhashi, W., Hashimoto, H., and Ozawa, T., 1993, Persistence of *Metarhizium anisopliae* (Deuteromycotina: Hyphomycetes) as a control agent of *Anomala cuprea* (Coleoptera: Scarabaeidae) in a forestry nursery, *Appl. Entomol. Zool.* **28:**103–105.

Smith, R. J., and Grula, E. A., 1981, Nutritional requirements for conidial germination and hyphal growth of *Beauveria bassiana*, *J. Invertebr. Pathol.* **37:**222–230.

Smith, R. J., and Grula, E. A., 1982, Toxic components on the larval surface of the corn earworm (*Heliothis zea*) and their effects on germination and growth of *Beauveria bassiana*, *J. Invertebr. Pathol.* **39:**15–22.

Smitley, D. R., Bauer, L. S., Hajek, A. E., Sapio, F. J., and Humber, R. A., 1995., Introduction and establishment of *Entomophaga maimaiga*, a fungal pathogen of gypsy moth (Lepidoptera: Lymantriidae) in Michigan, *Environ. Entomol.* **24:**1685–1695.

Soper, R. S., Holbrook, F. R., Majchrowicz, I., and Gordon, C. C., 1975, Production of *Entomophthora* resting spores for biological control of aphids, *Univ. Maine Orono, Life Sci. Agric. Exp. Stn., Tech. Bull.* **76:**15 pp.

Sprenkel, R. K., and Brooks, W. M., 1977, Winter survival of the entomogenous fungus *Nomuraea rileyi* in North Carolina, *J. Invertebr. Pathol.* **29:**262–266.

Steenberg, T., Langer, V., and Esbjerg, P., 1995, Entomopathogenic fungi in predatory beetles (Col.: Carabidae and Staphylinidae) from agricultural fields, *Entomophaga* **40:**77–85.

Steinhaus, E. A., 1949. *Principles of Insect Pathology*, McGraw-Hill, New York.

Steinhaus, E. A., 1956. Microbial control—The emergence of an idea, *Hilgardia* **26:**107–160.

Steinkraus, D. C., and Slaymaker, P. H., 1994, The effect of temperature and humidity on formation, germination, and infectivity of conidia of *Neozygites fresenii* (Zygomycetes: Neozygitaceae) from *Aphis gossypii* (Homoptera: Aphididae), *J. Invertebr. Pathol.* **64**:130–137.

Steinkraus, D. C., Kring, T. J., and Tugwell, N. P., 1991, *Neozygites fresenii* in *Aphis gossypii* on cotton, *Southwest. Entomol.* **16**:118–122.

Steinkraus, D. C., Hollingsworth, R. G., and Boys, G. O., 1996, Aerial spores of *Neozygites fresenii* (Entomophthorales: Neozygitaceae): Density, periodicity, and potential role in cotton aphid (Homoptera: Aphididae) epizootics, *Environ. Entomol.* **25**:48–57.

Stephen, W. P., and Fichter, B. L., 1990, Chalkbrood (*Ascosphaera apis*) resistance in the leafcutting bee. I. Challenge of selected lines, *Apidologie* **21**:209–219.

Storey, G. K., and Gardner, W. A., 1987, Vertical movement of commercially formulated *Beauveria bassiana* conidia through four Georgia soil types, *Environ. Entomol.* **16**:178–181.

Storey, G. K., Gardner, W. A., and Tollner, E. W., 1989, Penetration and persistence of commercially formulated *Beauveria bassiana* conidia in soil of two tillage systems, *Environ. Entomol.* **18**:835–839.

Stoy, W. M., Valovage, W. D., Frye, R. D., and Carlson, R. B., 1988, Germination of resting spores of the grasshopper (Orthoptera: Acrididae) pathogen, *Entomophaga grylli* (Zygomycetes: Entomophthorales) pathotype 2, in selected environments, *Environ. Entomol.* **17**:238–245.

Studdert, J. P., and Kaya, H. K., 1990a, Water potential, temperature, and clay-coating of *Beauveria bassiana* conidia: Effect on *Spodoptera exigua* pupal mortality in two soil types, *J. Invertebr. Pathol.* **56**:327–336.

Studdert, J. P., and Kaya, H. K., 1990b, Water potential, temperature, and soil type on the formation of *Beauveria bassiana* colonies, *J. Invertebr. Pathol.* **56**:380–386.

Studdert, J. P., Kaya, H. K., and Duniway, J. M., 1990, Effect of water potential, temperature, and clay-coating on survival of *Beauveria bassiana* conidia in a loam and peat soil, *J. Invertebr. Pathol.* **55**:417–427.

Takamura, N., and Sato, H., 1973, Observations on the epizootic of an entomophthoralean disease in outbreak populations of the gypsy moth, II, in: *Transactions of the 84th Annual Meeting of the Japanese Forestry Society*, Japanese Forestry Society (in Japanese), pp. 355–357.

Tyrrell, D., 1988, Survival of *Entomophaga aulicae* in dried insect larvae, *J. Invertebr. Pathol.* **52**:187–188.

Ullyet, G. C., and Schonken, D. B., 1940, A fungus disease of *Plutella maculipennis* Curt. in South Africa, with notes on the use of entomogenous fungi in insect control, *For. Sci. Bull., S. Afr. Dep. Agric.* **218**:1–24.

Uziel, A., and Kenneth, R. G., 1991, Survival of primary conidia and capilliconidia at different humidities in *Erynia* (subgen. *Zoophthora*) spp. and in *Neozygites fresenii* (Zygomycotina: Entomophthorales), with special emphasis on *Erynia radicans*, *J. Invertebr. Pathol.* **58**:118–126.

Vandenberg, J. D., 1993, Division on microbial control workshop: recent activities in product development and registration: Fungi, *Society for Invertebrate Pathology Newsletter* **25** (3):31–32.

Vandenberg, J. D., 1994, Chalkbrood susceptibility among larvae of the alfalfa leafcutting bee (Hymenoptera: Megachilidae) reared on different diets, *J. Econ. Entomol.* **87**:350–355.

Vandenberg, J. D., 1996, Spore loads of the chalkbrood fungus *Ascosphaera aggregata* in the air within commercially-managed nest shelters of the alfalfa leafcutting bee, *Megachile rotundata*, *Bee Science* **4**:106–110.

Velasco, L. R. I., 1983, Field parasitism of *Apanteles plutellae* Kurdj. (Braconidae: Hymenoptera) on the diamond-back moth of cabbage, *Philipp. Entomol.* **6**:539–553.

Villani, M. G., Krueger, S. R., Schroeder, P. C., Consolie, F., Consolie, N. H., Preston-Wilsey,

L. M., and Roberts, D. W., 1994, Soil application effects of *Metarhizium anisopliae* on Japanese beetle (Coleoptera: Scarabaeidae) behavior and survival in turfgrass microcosms, *Environ. Entomol.* **23**:502–513.

Wainwright, M., 1992, *An Introduction to Fungal Biotechnology*, Wiley, Chichester, England.

Wallace, D. R., MacLeod, D. M., Sullivan, C. R., Tyrrell, D., and DeLyzer, A. J., 1976, Induction of resting spore germination in *Entomophthora aphidis* by long-day light conditions, *Can. J. Bot.* **54**:1410–1418.

Walstad, J. D., Anderson, R. F., and Stambough, W. J., 1970, Effects of environmental conditions on two species of muscardine fungi (*Beauveria bassiana* and *Metarrhizium anisopliae*), *J. Invertebr. Pathol.* **16**:221–226.

Watson, D. W., and Petersen, J. J., 1993, Sexual activity of male *Musca domestica* (Diptera: Muscidae) infected with *Entomophthora muscae* (Entomophthoraceae: Entomophthorales), *Biol. Control* **3**:22–26.

Watson, D. W., Mullens, B. A., and Petersen, J. J., 1993, Behavioral fever response of *Musca domestica* (Diptera: Muscidae) to infection by *Entomophthora muscae* (Zygomycetes: Entomophthorales). *J. Invertebr. Pathol.* **61**:10–16.

Wellings, M., Zelazny, B., Scherer, R., and Zimmermann, G., 1995, First record of the entomopathogenic fungus *Sorosporella* sp. (Deuteromycotina: Hyphomycetes) in *Locusta migratoria* (Orthoptera: Acrididae) from Madagascar: symptoms of infection, morphology and infectivity, *Biocontrol Sci. Technol.* **5**:465–474.

Weseloh, R. M., and Andreadis, T. G., 1992, Mechanisms of transmission of the gypsy moth (Lepidoptera: Lymantriidae) fungus, *Entomophaga maimaiga* (Entomophthorales: Entomophthoraceae) and effects of site conditions on its prevalence, *Environ. Entomol.* **21**:901–906.

Weseloh, R. M., and Andreadis, T. G., 1997, Persistence of resting spores of *Entomophaga maimaiga*, a fungal pathogen of gypsy moth, *Lymantria dispar*, *J. Invertebr. Pathol.* **69**:195–196.

Wilding, N., 1970, *Entomophthora* conidia in the air-spora, *J. Gen. Microbiol.* **62**:149–157.

Wilding, N., 1971, Discharge of conidia of *Entomophthora thaxteriana* Petch from the pea aphid *Acyrthosiphon pisum* Harris, *J. Gen. Microbiol.* **69**:417–422.

Wilding, N., 1973, The survival of *Entomophthora* spp. in mummified aphids at different temperatures and humidities, *J. Invertebr. Pathol.* **21**:309–311.

Wilding, N., 1981, Pest control by Entomophthorales, in: *Microbial Control of Pests and Plant Diseases 1970–1980* (H. D. Burges, ed.), Academic Press, London, pp. 539–554.

Wilding, N., and Latteur, G., 1987, The Entomophthorales—Problems relative to their mass production and their utilization, *Med. Fac. Landbouww. Rijksuniv. Gent* **52**:159–164.

Willers, D., Lehmann-Danzinger, H., and Führer, E., 1982, Antibacterial and antimycotic effect of a newly discovered secretion from larvae of an endoparasitic insect, *Pimpla turionellae* L. (Hym.), *Arch. Microbiol.* **133**:225–229.

Wraight, S. P., Butt, T. M., Galaini-Wraight, S., Allee, L., Soper, R. S., and Roberts, D. W., 1990, Germination and infection processes of the entomophthoralean fungus *Erynia radicans* on the potato leafhopper, *Empoasca fabae*, *J. Invertebr. Pathol.* **56**:157–174.

Zimmermann, G., 1982, Effect of high temperature and artificial sunlight on the viability of conidia of *Metarhizium anisopliae*, *J. Invertebr. Pathol.* **40**:36–40.

6

Mahoney Lake: A Case Study of the Ecological Significance of Phototrophic Sulfur Bacteria

JÖRG OVERMANN

1. Introduction

Phototrophic sulfur bacteria require light as an energy source and reduced inorganic sulfur compounds as electron-donating substrates for growth. Dense accumulations of these bacteria can develop where light reaches sulfide-containing layers of stratified water bodies and sediments. Frequently, such blooms are visible with the naked eye as purple to pink, peach, brown, or green layers. If they occur in the water column of lakes, such "bacterial plates" can extend over a depth of several meters (Takahashi and Ichimura, 1968; Biebl and Pfennig, 1979; Parkin and Brock, 1980b; Guerrero *et al.*, 1985; Gorlenko, 1988). In sediments, the gradients of light intensity and sulfide concentration are much steeper (Jørgensen and Revsbech, 1983; van Gemerden *et al.*, 1989; Visscher *et al.*, 1990) and, as a result, the layers of phototrophic sulfur bacteria are only millimeters to centimeters thick (Nicholson *et al.*, 1987; van Gemerden *et al.*, 1989).

Although blooms of anoxygenic photosynthetic bacteria are observed in the pelagial of many lakes and reservoirs (see the recent compilation by van Gemerden and Mas, 1995), their ecological significance in these systems remains unclear.

The doubling times of phototrophic bacteria in their natural habitat are in the order of days and weeks (e.g., Eichler and Pfennig, 1990; Overmann *et al.*, 1991). This is because the light intensity reaching anoxic water layers is very low and controls photosynthesis of the bacteria (Parkin and Brock, 1980b; van Gem-

---

JÖRG OVERMANN • Institut für Chemie und Biologie des Meeres, Universität Oldenburg, D-26111 Oldenburg, Germany.
Advances in Microbial Ecology, Volume 15, edited by Jones. Plenum Press, New York, 1997.

erden and Mas, 1995), despite their efficient light-harvesting structures (Fowler et al., 1971; Drews, 1985) and low-maintenance energy requirements (e.g., Overmann et al., 1992a).

In some lakes, however, phototrophic sulfur bacteria contribute significantly to total primary production (up to 85%, Culver and Brunskill, 1969) and theoretically could support the aerobic food web. Zooplankton grazing on phototrophic bacteria has been observed (e.g., Sorokin, 1970; Caumette et al., 1983), but the presence of hydrogen sulfide might restrict the role of aerobic grazers. Indeed, the analysis of stable isotope ratios ($\delta^{13}C$ and $\delta^{34}S$) in one meromictic lake indicated that the biomass of phototrophic sulfur bacteria did not enter the food web (Fry, 1986). So far it remains entirely unclear if phototrophic sulfur bacteria have any impact on the overall trophic structure of their habitats (van Gemerden and Mas, 1995).

Another function of phototrophic sulfur bacteria is the oxidation of the sulfide generated by sulfate- or sulfur-reducing bacteria. A closed sulfur cycle is thus established under anoxic conditions (Pfennig, 1978; Biebl and Pfennig, 1979) and probably represents a significant part of the overall biogeochemical sulfur cycle in many stratified lakes. So far, we have very little information on the turnover of sulfur compounds and the coupling of sulfate reduction and photosynthetic sulfide oxidation in such ecosystems.

In view of these numerous open questions, the curiosity of many scientists was excited when an extraordinary dense population of purple sulfur bacteria was detected in the small Canadian meromictic lake named Mahoney Lake. The tremendous accumulation and the sharp stratification of purple sulfur bacteria so far appears to be unique to Mahoney Lake. In 1989 a comprehensive study of the diversity and physiology of bacteria and of the sulfur and carbon cycles in this lake was initiated. This work has yielded new insights into the ecophysiology and biogeochemical significance of phototrophic sulfur bacteria.

2. The Habitat

Mahoney Lake (Fig. 1) is a small lake (surface area, 11.5 ha; maximum depth, 14.5 m) located 470 m above sea level near Penticton in the Okanagan Valley (British Columbia, Canada). The lake has no outflow and receives only occasional surface inflow via discharge channels at the north shore. Mahoney Lake is saline, which became obvious during the last decade when its water level dropped, leaving behind white salt crusts on the exposed littoral sediments (Fig. 1).

With respect to salinity, Mahoney Lake is no exception on the Southern Interior Plateau. Saline lakes are numerous in this semiarid part of British Columbia (Scudder, 1969), which belongs to the Ponderosa Pine/ Bunchgrass bio-

Figure 1. View of Mahoney Lake from the northern hillside. The white salt crusts are visible on the exposed littoral sediments.

geoclimatic zone and is an extension of the larger climatic region of the North American Great Basin. Due to the rain-shadow effect of the western mountains, annual precipitation is significantly lower than potential evaporation (< 410 mm versus > 810 mm) (Hammer, 1986), which is a prerequisite for the formation of saline lakes.

The limnological study of Mahoney Lake began in 1961. Initially it was part of a survey of several inland lakes in which rainbow trout populations could not be maintained by stocking (Northcote and Halsey, 1969). Similar to many other saline lakes in south-central British Columbia (Scudder, 1969), fish are absent from Mahoney Lake. It was during the first investigations that high-frequency echo soundings revealed the presence of a horizontally continuous scattering

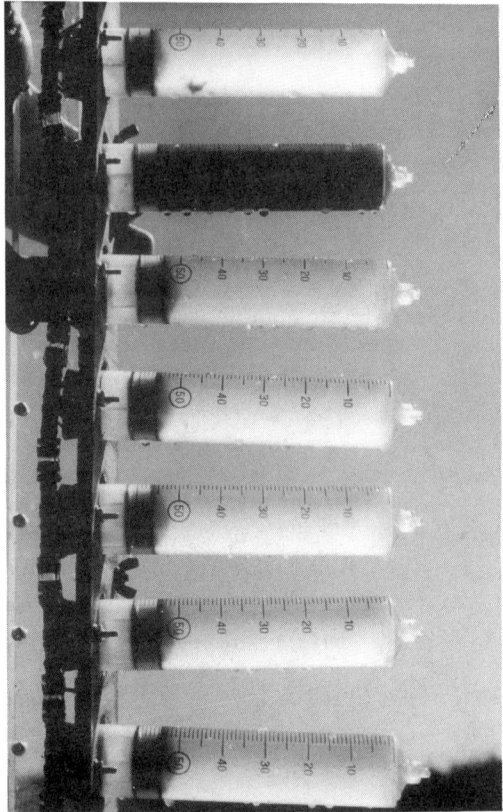

Figure 2. Vertical distribution of purple sulfur bacteria in a syringe sampler. Distance between syringes is 5 cm in this prototype. The sampler is held against the light to demonstrate the enormous biomass density reached in the chemocline and its sharp stratification.

layer in the chemocline (Northcote and Halsey, 1969). Water samples from this layer contained purple-colored bacteria in high concentrations (compare Fig. 2). Seasonal changes in the vertical position of the layer were observed, but the purple bacteria were always found at a depth where dissolved oxygen was below the detection limit. At times this "bacterial plate" had a jellylike consistency because of its extreme biomass density.

It was found that Mahoney Lake is meromictic, with a 6 m deep oxygenated mixolimnion overlying permanently anoxic bottom waters (the monimolimnion). In the monimolimnion, sulfide concentrations reach extraordinarily high concentrations of 60 mM (Overmann *et al.*, 1996a). The cause for the permanent stratification of the lake is the difference in salinity (hence density) between the mixolimnion (7.5–16 ‰) and monimolimnion (up to 39 ‰) water. This salinity gradient has been attributed to the great relative depth of the lake and to the protection from wind by the surrounding hills (Northcote and Hall, 1983). The steepest increase in salinity was found in the chemocline (presently at about 7 m depth), exactly where the dense accumulation of purple bacteria had been detected. Because of the stable stratification, the water temperature in the monimolimnion remains almost constant (9 to 13°C) throughout the year compared to that in the mixolimnion (-1.4 to 23°C).

The salinity of Mahoney Lake is comparable to that of the marine environment. Thus Mahoney Lake is saline but not hypersaline. However, the ionic composition of Mahoney Lake water (Northcote and Halsey, 1969) differs from that of seawater and is typical for lakes in south-central British Columbia (Scudder, 1969, 1983) and for those of the plains region of Saskatchewan (Hammer, 1978). The low chloride concentration is characteristic of an athalassohaline system not associated with the sea or with marine deposits of the geological past. In the case of Mahoney Lake, the dissolved salts originate from the alkali-rich lavas of the Tertiary Marron formation that prevail in the watershed (Northcote and Hall, 1983). Sulfate is the dominant anion in the lake water (sulfate:carbonate ≈ 4; sulfate:chloride ≈ 8) and sulfate concentrations amount to 250 mM in the mixolimnion and exceed 400 mM in monimolimnion water.

Mahoney Lake is not an alkaline environment. The molar ratio of (Ca^{2+} + Mg^{2+}) to carbonate (HCO_3^- + CO_3^{2-}) is about 4 (calculated from Northcote and Halsey, 1969). This great excess of divalent cations over carbonate prevents the formation of a soda brine and pH values remain below 9.0 (Northcote and Hall, 1983). Due to the excess of divalent ions, calcite precipitation occurs and causes occasional whitings (lakewide precipitation of $CaCO_3$) in the mixolimnion, and layers of calcareous laminated sediments at the lake bottom (Lowe *et al.*, 1996).

Aquatic habitats rich in Mg^{2+} typically are inimical to life (Oren, 1983; Javor, 1989) and $SO_4^=$ can be more inhibitory than the Cl^- anion (Javor, 1984). The coincidence of high sulfate and high Mg^{2+} concentrations is the most likely reason for the low species diversity of eukaryotes observed in Mahoney Lake. The fauna comprises mainly two species, the rotifer *Brachionus plicatilis* and the calanoid copepod *Diaptomus connexus* (Northcote and Hall, 1983). In littoral areas only corixids (*Cenocorixa* sp.), chironomids, and very few individuals of the dytiscid beetle genera *Laccophilus* and *Hygrotus* are present. In earlier years, nymphs of the circumpolar bluet damselfly (*Enallagma cyathigerum*) had been observed (Hall and Northcote, 1986). This species was not found during more recent investigations.

In athalassohaline systems the species diversity of algae and aquatic invertebrates is inversely related to salinity (Javor, 1983; Hammer, 1986). Compared to Mahoney Lake, waters of even higher salinity (30 ‰), but with Na^+ and Cl^- as the dominating ions, harbor as much as 64 different animal species (Thienemann,

1925). In Waldsea Lake, which has an ionic composition comparable to Mahoney Lake, the species diversity of zooplankton (5 species), zoobenthos (2 major species), and macrophytes (2 species) is very limited (Hammer, 1986). With respect to the diversity and species composition of its fauna, Mahoney Lake resembles hypersaline (Grant and Tindall, 1986), not saline, waters.

Studies in microbial ecology are often hampered by the fact that a single species or physiological group of interest is present only in low cell concentrations in its natural environment. The population density of purple sulfur bacteria in Mahoney Lake so far had been attained only in pure laboratory cultures and provides the unique opportunity to study their ecological significance under natural conditions. In contrast to microbial mat systems (Jørgensen, 1982), where measurements are complicated by the sediment matrix, the physiological activities in Mahoney Lake can be assessed much easier by *in situ* incubations. In addition, it can be expected that due to the low species diversity of higher organisms the food web in Mahoney Lake is much simpler than in other lakes.

In sulfate-rich ecosystems, sulfur compounds are the main electron carriers employed by microorganisms during anaerobic mineralization. Mahoney Lake provides the opportunity to elucidate the pelagial sulfur cycle in a system dominated by phototrophic bacteria.

Because of the steep gradients in physicochemical parameters and microbial biomass, the sampling strategy had to be adapted. Water samples were collected in a syringe sampler modified after Baker *et al.* (1985) (Fig. 2). Using this device, the chemocline was sampled at vertical intervals of 2.5 cm. A special incubation rack was designed that allowed the placement of incubation tubes at a vertical distance of 5 cm, but avoided shading of the tubes by the overlying ones (Overmann *et al.*, 1991). All sampling, incubation experiments, and underwater light measurements were performed with a calibrated steel cable. With this cable, the measurements were highly reproducible.

3. Species Composition of the Bacterial Plate

Light microscopic examination of water samples from the chemocline indicated that one type of purple sulfur bacteria was dominant. It had a coccal shape, the cells accumulated sulfur globules intracellularly (Fig. 3), and some contained gas vacuoles. Pure cultures were isolated and their absorption spectrum was identical to that of chemocline water samples (Overmann *et al.*, 1991). The isolates contained three different isomers of the carotenoid okenone (Overmann *et al.*, 1993) and were capable of using glucose, acetate, and pyruvate in the presence of sulfide for phototrophic growth. On the basis of these features, the dominating phototrophic species was identified as *Amoebobacter purpureus*, which belongs to the family Chromatiaceae. The species reaches cell concentra-

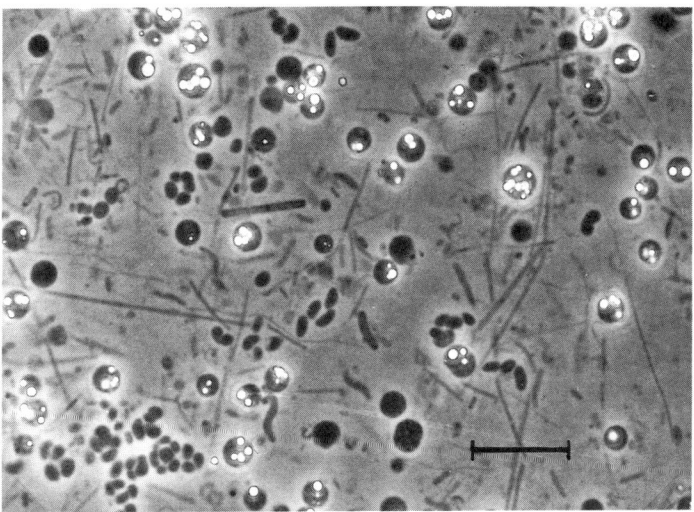

Figure 3. Phase contrast photomicrograph of a sample from the bacterial plate. Chemotrophic bacteria were concentrated by differential centrifugation. (Bar = 10 μm.)

tions of 4×10^8 ml^{-1} during summer. Of these, 27% were viable in standard sulfide-reduced culture media (Overmann et al., 1991).

Of all phototrophic microorganisms in the layer, *Amoebobacter purpureus* comprised 97.9%. Several other species of phototrophic microorganisms were detected in chemocline water samples using dilution series in deep agar media or selective enrichment procedures in liquid media. The purple sulfur bacterium *Thiocapsa roseopersicina* accounted for 2%, the purple nonsulfur bacterium *Rhodobacter capsulatus* for 0.02%, and the green sulfur bacterial species *Chloroherpeton thalassium* and *Prosthecochloris aestuarii* for ≤ 0.002%. Brown-colored species of the green sulfur bacteria could not be detected, even by three different methods (Overmann et al., 1991). *Thiorhodovibrio winogradskyi*, a newly described species and genus of the Chromatiaceae (Overmann et al., 1992b) was found in high numbers in the Mahoney Lake littoral but not in the chemocline of the lake.

The outstanding feature of Mahoney Lake is the pronounced dominance and extreme biomass (see Fig. 4A) of *A. purpureus*. It became evident that the success of the different species of phototrophic sulfur bacteria to colonize the chemocline is related to their specific capacity to absorb underwater irradiance. Due to the progressive narrowing of the spectrum with depth, light of wavelengths between 515 and 625 nm reaches the bacterial plate (Overmann et al., 1991). Within this wavelength range, the absorption spectra of the different species differ significantly (Fig. 5A). If the quanta absorbed between 515 and

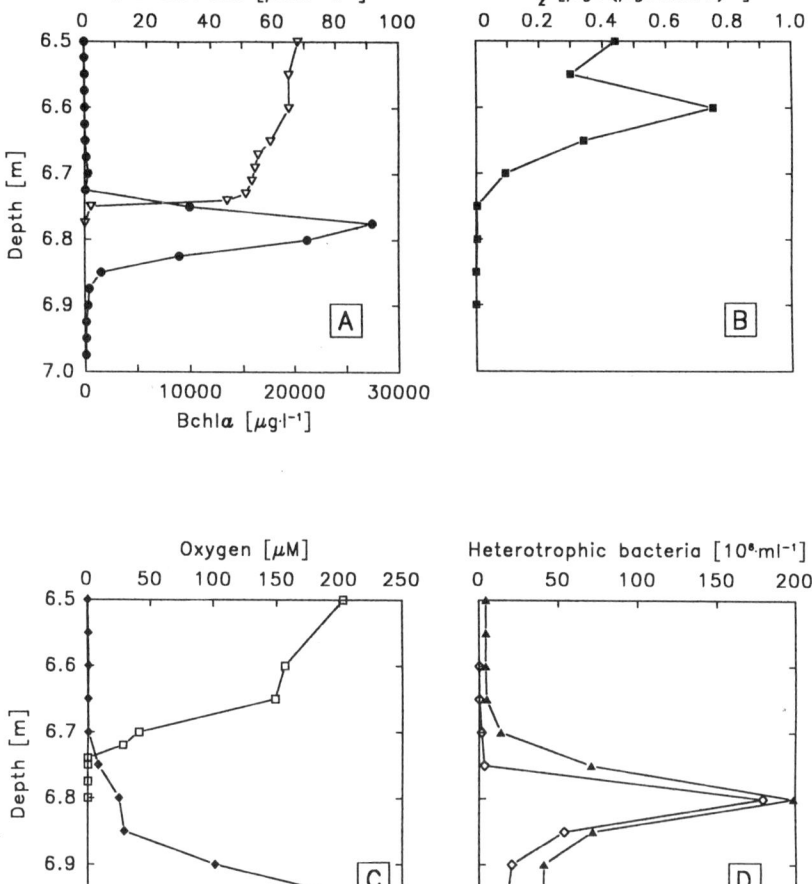

Figure 4. Vertical distribution of (A) underwater light (∇), biomass of *A. purpureus* (●), (B) specific photosynthetic activity (assimilation number P_z after addition of sulfide to a final concentration of 200 μM) (■), (C) oxygen (□) and sulfide (♦) concentrations, and (D) numbers of heterotrophic bacteria (▲) and sulfate reduction rates (◊) (on August 4, 1992).

625 nm are standardized for the quanta absorbed over the whole spectrum ("relative absorption"), *A. purpureus* cells exhibit the highest values. Relative absorption of the other species decreased in parallel with decreasing abundance in the plate (Fig. 5B).

Green light is harvested preferentially by carotenoids. Those Chromatiaceae

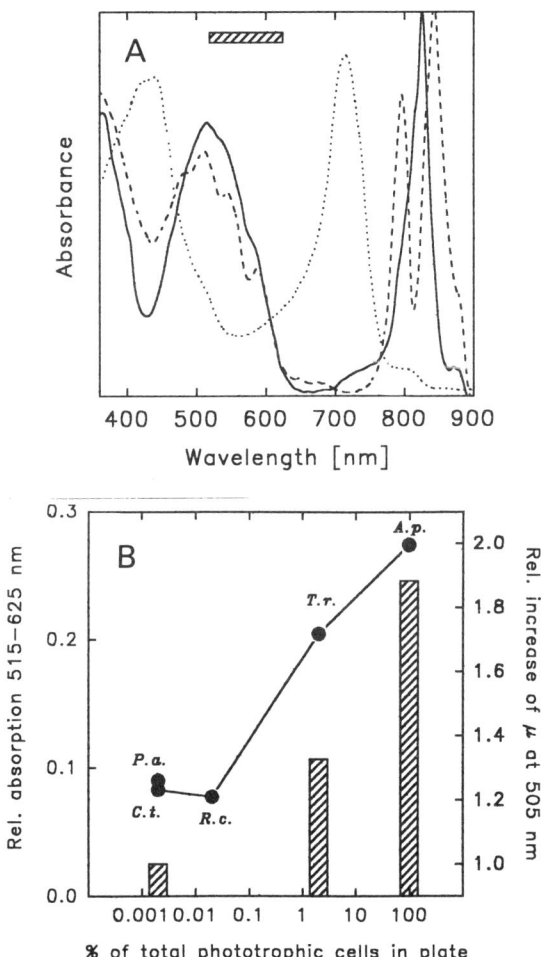

Figure 5. (A) Absorption spectra of whole cells in pure cultures of *Amoebobacter purpureus* (———),
Thiocapsa roseopersicina (---), and *Prosthecochloris aestuarii* (· · · ·) isolated from the chemocline
of Mahoney Lake. (B) Correlation between abundance of *A. purpureus* (A.p.), *T. roseopersicina*
(T.r.), *Rhodobacter capsulatus* (R.c.), *P. aestuarii* (P.a.), and *Chloroherpeton thalassium* (C.t.) and
their relative absorption at 515–625 nm (●). For comparison, the stimulation of light-limited growth
by green light (a filter with transmittance at 515–625 nm not beeing commercially available) versus
daylight is shown (bars).

species that dominate in lakes mostly contain okenone (Guerrero et al., 1987; van Gemerden and Mas, 1995). The competitive advantage of okenone as the light-harvesting carotenoid could be demonstrated in the laboratory using three isolates from Mahoney Lake. In green light, the growth rates of pure cultures of *A. purpureus* and *T. roseopersicina* increased by 190% and 135%, respectively, as compared to artificial daylight (Fig. 5B, hatched bars). Growth rates of *P. aestuarii* remained unchanged (Overmann et al., 1991). In conclusion, the light conditions in Mahoney Lake seem more favorable for *A. purpureus* because this species absorbs and uses light energy in the carotenoid portion of the absorption spectrum most efficiently for growth. As in other environments (Parkin and Brock, 1980a; Montesinos et al., 1983), the shading of the enormous standing biomass of *A. purpureus* in Mahoney Lake then keeps the biomass of green species of the green sulfur bacteria at very low levels and completely prevents growth of the brown-colored species.

Apart from the cells of *A. purpureus*, many different morphotypes of colorless bacteria were present in the chemocline (Fig. 3). The total cell numbers of chemotrophic bacteria reached 2×10^8 ml^{-1} in summer (Overmann et al., 1996a) and were similar to cell numbers of *A. purpureus*. In order to understand possible microbial interactions in the bacterial plate it was essential to identify the major physiological groups of heterotrophs present.

Within the chemocline a high rate of sulfate reduction was measured during summer. The minimum fraction of heterotrophic cells that are active sulfate-reducing bacteria can be estimated from this rate (5.93 µmol $SO_4^{2-} \times l^{-1} \times h^{-1}$, Fig. 4D) and the cell-specific rate of sulfate reduction determined in pure cultures (mean: 3.31×10^{-10} µmol $SO_4^{2-} \times$ cell$^{-1} \times h^{-1}$; Jørgensen, 1978). The calculated cell concentration was 0.18×10^8 ml^{-1}, which indicates that sulfate-reducing bacteria comprise a significant fraction of the non-photosynthetic cells in the chemocline. Most probable numbers of sulfate-reducing bacteria in the layer were much lower, however (maximum 1.4×10^5 ml^{-1}; M. Rodrigo, J. T. Beatty, and K. J. Hall, personal communication).

In the chemocline, high concentrations of polysulfides were detected in spring. The concentrations decreased rapidly over the following summer. Sulfur-reducing bacteria could be enumerated by employing a selective growth medium. Their most probable numbers increased from 2×10^3 ml^{-1} to 4.6×10^4 ml^{-1} in the lower part of the chemocline (Overmann et al., 1996a), parallel to the decrease in polysulfide concentrations. Obviously, a fraction of the chemotrophic bacteria in the chemocline are physiologically active sulfur reducers.

4. Factors Controlling Growth of *Amoebobacter purpureus*

In the chemocline of Mahoney Lake, the vertical profiles of microbial biomass and physiological activity reflect the steep gradients of light, oxygen, and sulfide in the environment (Fig. 4). Before considering the causes for the extraordinary high biomass accumulation of purple sulfur bacteria, the factors limiting anoxygenic photosynthesis of *A. purpureus* will be discussed.

From the steep gradient of underwater light (Fig. 4A) it can be deduced that most of the *A. purpureus* cells thrive in the dark. It was calculated that only 10% of the *Amoebobacter* cells in the chemocline can harvest sufficient light energy to perform anoxygenic photosynthesis (Overmann *et al.*, 1991). The majority of *A. purpureus* cells must persist within the plate in a photosynthetically inactive state due to extreme self-shading.

During most of the year, the photosynthetic activity of *A. purpureus* continuously declined with depth in the bacterial plate, indicating that light is the major factor limiting anoxygenic photosynthesis. This is also supported by the fact that the biomass of purple sulfur bacteria and the average light intensity in the chemocline were significantly correlated (Overmann *et al.*, 1994). For the summer months (July–August), however, light saturation of anoxygenic photosynthesis was observed at the top of the plate (maximum of the assimilation number P_z in Fig. 4B). During this time interval, factors other than light obviously become limiting for the most metabolically active part of the population.

In Mahoney Lake, sulfide as the electron donor of anoxygenic photosynthesis is supplied by upward diffusion from the monimolimnion and by sulfate reduction within the chemocline (Overmann *et al.*, 1994). During summer the sulfide supply by both processes and its demand by anoxygenic photosynthesis are balanced (Overmann *et al.*, 1991, 1994). The supply of sulfide thus controls anoxygenic photosynthesis during the peak of the purple bacterial bloom in summer and limits photosynthetic rates at the top of the plate. Sulfide limitation of anoxygenic photosynthesis is also indicated by the low sulfur content of *A. purpureus* cells (Overmann *et al.*, 1994) and the formation of cell aggregates *in situ* during this time (Fig. 6). In pure cultures such aggregation is observed only after the depletion of sulfide (Overmann and Pfennig, 1992).

When the demand for sulfide was calculated from the photosynthetic carbon fixation rates measured, it was found that supply and demand of sulfide must have already been balanced in the month of May (Overmann *et al.*, 1994). Therefore, anoxygenic photosynthesis should have been limited by sulfide much earlier than observed. Sulfide limitation of bacterial photosynthesis was not observed before July, however. This strongly indicates that besides sulfide, additional electron-donating substrate(s) must be available for anoxygenic photosynthesis during spring. Furthermore, if it is assumed that *A. purpureus* oxidizes sulfide completely to sulfate during all seasons, anoxygenic photosynthesis would account for only a very small fraction of total sulfide oxidation between November and April (Overmann *et al.*, 1994). Yet, the vertical profiles of oxygen and sulfide did not overlap on certain sampling dates (Fig. 7), therefore oxidation of sulfide by molecular oxygen cannot be the reason for the disappearance of sulfide during these times.

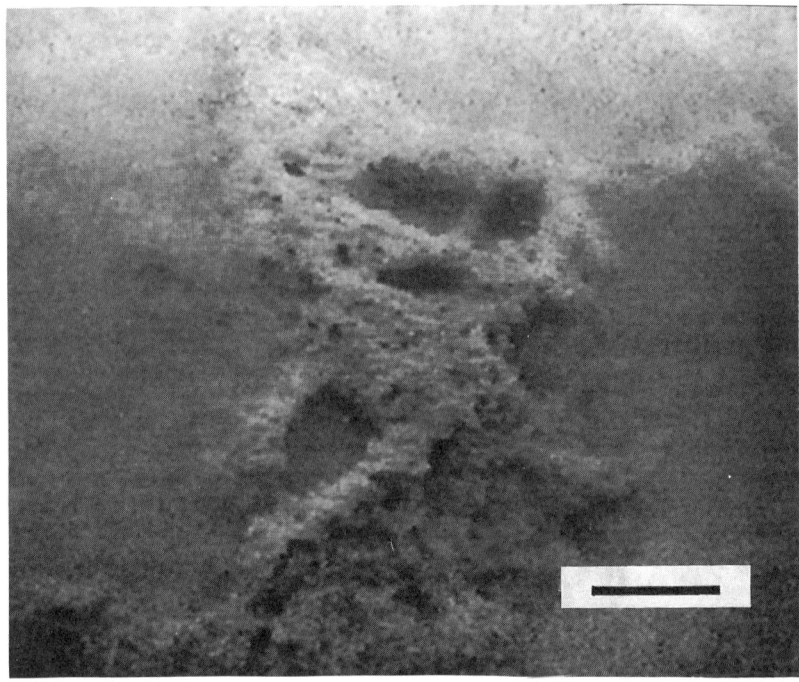

Figure 6. Surface of the bacterial layer at 6.7 m depth viewed at an angle of 45° from above with an underwater video camera in Mahoney Lake on August 19, 1993. (Bar = 10 cm.) After Overmann *et al.* (1994).

This seemingly contradictory evidence can be explained if the sulfur cycle in the chemocline of Mahoney Lake is evaluated by a balance calculation (Overmann *et al.*, 1996a). It was found that the pool of intracellular sulfur plays a major role as a buffer of electron-donating potential in *A. purpureus*.

During winter the photosynthetic oxidation of sulfide was at its minimum (Overmann *et al.*, 1994). The imbalance between supply and consumption of sulfide then led to sulfide concentrations above 100 μM at the top of the bacterial plate. At the same time, the intracellular sulfur content of *A. purpureus* increased to its maximum of 34% of dry cell mass (Fig. 8B), a value that compares well with the maximum sulfur content determined for pure cultures of *Chromatium* (van Gemerden, 1968). With elemental sulfur as the oxidation product, the rate of sulfide oxidation of anoxygenic photosynthesis increases by a factor of 4. This explains why in spring anoxygenic photosynthesis can balance the sulfide supply despite low carbon-fixation rates.

After sulfide was depleted in the upper part of the layer, intracellular sulfur

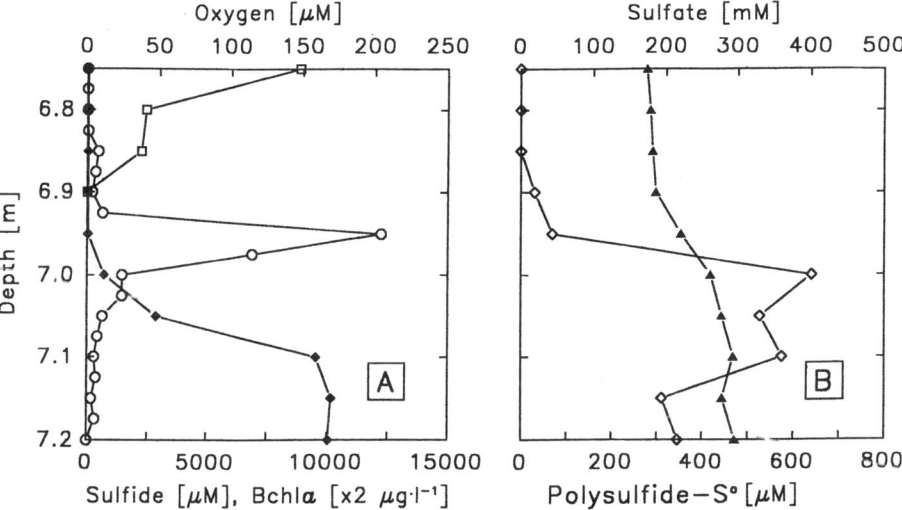

Figure 7. Vertical concentration profiles of (A) oxygen (□), sulfide (♦), biomass of *A. purpureus* (○), and (B) sulfate (▲) and polysulfide-sulfur (◊) in the chemocline on April 21, 1992.

transiently became the main electron donor for photosynthesis (Overmann *et al.*, 1996a). Within two months, the sulfur content of *A. purpureus* cells decreased to 3.5% of dry cell mass (Overmann *et al.*, 1994) (Fig. 8 B).

The number of times a phototrophic sulfur bacterium can double in the absence of sulfide with such an amount of intracellular sulfur (34–3.5 = 30% of dry weight; or 1 mol S per 4 mol carbon) available for carbon fixation can be calculated. During oxidation of elemental sulfur to sulfate, the ratio of sulfur oxidized to CO_2 assimilated is 1:1.5. Therefore, a phototrophic sulfur bacterium can double 0.4 times under the conditions observed in the chemocline of Mahoney Lake. Clearly, the intracellular sulfur can serve as an electron donor only for a limited period of time.

Whereas sulfide limitation of bacterial photosynthesis during summer (Parkin and Brock, 1981b) and light limitation during the rest of the year is a pattern observed also in other lakes (Culver and Brunskill, 1969), the present studies suggest a much more dynamic pattern of electron flow in anoxygenic photosynthesis. The accumulation of intracellular sulfur in purple sulfur bacteria during times of light limitation permits a transient increase in the rates of anoxygenic photosynthesis in subsequent weeks. The formation of intracellular sulfur acts as a buffer against seasonal fluctuations of sulfide supply. While sulfide is completely oxidized to sulfate (Parkin and Brock, 1981b) in some enviroments, photosynthetic oxidation of sulfide—at least during certain times—stops at the

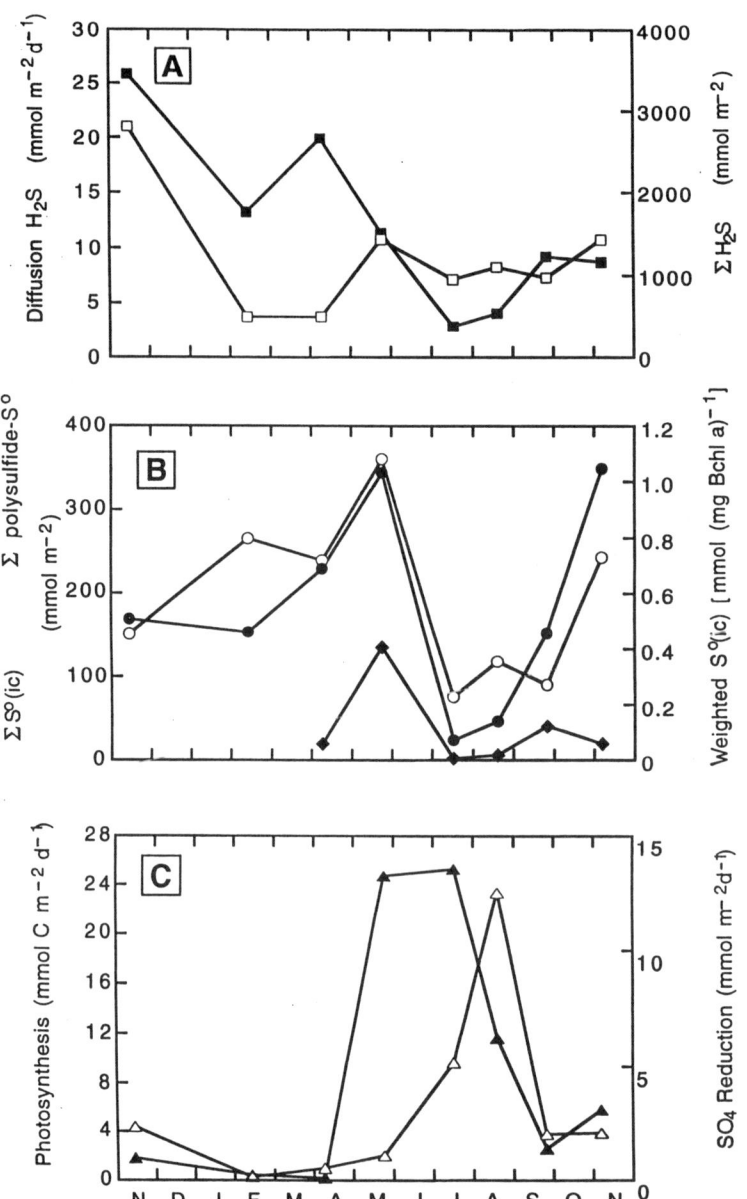

Figure 8. (A) Diffusive flux of sulfide (□) and amount of sulfide present in the zone of photosynthetic activity above the biomass maximum of *A. purpureus* (■). (B) Total amount of intracellular sulfur (○) and polysulfides (♦) integrated over the whole bacterial layer. (●), weighted intracellular sulfur content of *A. purpureus*. (C) Areal rates of anoxygenic photosynthesis (▲) and of sulfate reduction (△) in the layer between November 1992 and 1993. Left and right ordinate in (C) correspond to each other. After Overmann *et al.* (1996a).

level of elemental sulfur in other ecosystems (van Gemerden et al., 1985). A transient shift to intracellular sulfur as the main electon donor supporting photosynthesis has not been described before for a natural population of photosynthetic bacteria, but may be important in other environments as well. This shift also alters the transformations within the sulfur cycle in Mahoney Lake, especially by formation of polysulfides (Section 6).

With a maximum concentration of 27,500 µg Bchl$a$ × l^{-1} (Fig. 4A), the biomass of phototrophic sulfur bacteria in Mahoney Lake exceeds that of any other natural water body of the world by a factor of 7 (compare van Gemerden and Mas, 1995). The maximum integrated amount of 1773 mg Bchl$a$ × m^{-2} is matched only by Lake Cisó in Spain. Even under the thick ice cover in winter, bacteriochlorophyll a concentrations in Mahoney Lake still exceed 10,000 µg Bchl$a$ × l^{-1}. This high biomass of *A. purpureus* could be due to either a high specific growth rate or to low loss rates (lysis, sedimentation, grazing), or both. The growth rates in the bacterial plate will be considered first.

The annual production of *A. purpureus* was 33.5 g C × m^{-2} × yr^{-1} (Overmann et al., 1994). If compared to other lakes (van Gemerden and Mas, 1995) the productivity of *A. purpureus* in Mahoney Lake is rather normal and in sharp contrast to its exceptionally high biomass. Consequently, the growth rate must be comparatively low in Mahoney Lake. Growth rates, µ, and doubling times, t_d, can be calculated from assimilation numbers P_z (the photosynthetic rate divided by the concentration of Bchl$a$), according to the formulas

$$\mu = \frac{1}{T} \times \ln(1 + P_z * F * T) \quad (1)$$

$$t_d = \frac{\ln 2}{\mu} \quad (2)$$

where F is a factor for the conversion of Bchl$a$ into cellular carbon [$F = 0.01991$ mg Bchl$a$ × (mg C)$^{-1}$ in pure cultures of *A. purpureus*] and T the unit of time of the assimilation number (usually 1 hour).

The shortest doubling times determined in this way were 2.8 days for *A. purpureus* cells at the top of the layer and 43.6 days for the whole layer. Thus the physiological activity of *A. purpureus* cells *in situ* is very low. Growth rates of the whole population were inversely correlated with its biomass (Fig. 9). Such an inverse relation between standing crops of phototrophic sulfur bacteria and their productivity was also found in Deadmoose Lake (Hammer, 1986) and Waldsea Lake (Lawrence et al., 1978) and reflects the increasing degree of limitation by self-shading and sulfide oxidation at higher biomass levels.

In fact, the physiological activity of *A. purpureus* in Mahoney Lake is far too low to be the cause for its biomass accumulation as revealed by a simple

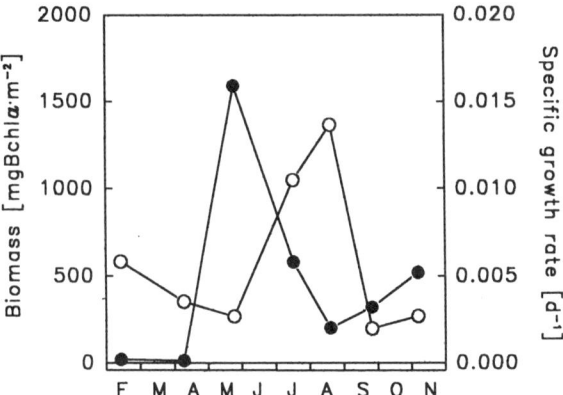

Figure 9. Comparison of population growth rates (●) and biomass (○) of *A. purpureus* in the chemocline in 1993.

calculation (compare van Gemerden and Mas, 1995). A maximum of anoxygenic photosynthesis of A_{max} = 304 mg C × m^{-2} × d^{-1} was measured (summer 1993). Estimates of the loss rate, k, of phototrophic sulfur bacteria in other environments are in the order of about − 0.1 d^{-1}. At such a loss rate and at steady state, the maximum biomass B_{max} supported by photosynthesis A_{max} is (rearranging Eq. 1):

$$B = A \times F \times T / (e^{-k \times T} - 1) \qquad (3)$$

In the case of Mahoney Lake, the theoretical maximum would be 58 mg Bchl$a$ × m^{-2}. The maximum that was actually measured was 1773 mg m^{-2} (Overmann et al., 1994). Furthermore, a comparison of the maximum photosynthetic rates measured in summer with those of pure cultures revealed that even under optimum conditions at the very top of the purple layer, only 14% of the *Amoebobacter* cells were physiologically active (Overmann et al., 1994).

It follows that the losses of *A. purpureus* in the Mahoney Lake ecosystem must be significantly lower than in other environments, or that *A. purpureus* must obtain metabolic energy from processes other than photosynthesis.

5. What Are the Mechanisms of the Accumulation and Disappearance of *A. purpureus*?

Measurements of respiration rates of water samples from the bacterial layer and of pure cultures demonstrated that *A. purpureus* is not capable of obtaining additional energy by chemolithotrophic metabolism in the dark (Overmann and Pfennig, 1992; Overmann et al., 1994). In Mahoney Lake, phototrophic bacteria depend on photosynthesis as the only means for growth. Only if the low growth rates of A. purpureus are accompanied by even lower loss rates can the high biomass persisting in the chemocline be explained.

For an Amoebobacter population in another lake it was demonstrated that sedimentation and washout were the major loss processes (Mas et al., 1990). In Mahoney Lake, losses due to washout are very unlikely because the lake has no outflow. Unlike phototrophic bacterial populations in many other lakes the bacterial layer in Mahoney Lake exhibited a sharp lower boundary with only very small amounts of Bchla below (Fig. 4A). The vertical position of the layer of purple sulfur bacteria corresponded exactly with the depth of the steep vertical density gradient of the lake water. This indicates that the high salt concentration in the monimolimnion prevents the sedimentation of A. purpureus cells from the layer. Only the cells from the very top of the layer had a higher buoyant density than the surrounding water, while most of the cells in the plate showed neutral buoyancy (Overmann et al., 1991). Throughout the year, only very small fractions of $\leq 2.8\%$ of the cells were recovered in sedimentation traps positioned just below the bacterial plate at 8 m depth. Both results indicate that the losses due to sedimentation of A. purpureus cells to the lake bottom are very low.

The sinking velocity of A. purpureus cells is reduced due to a light-dependent buoyancy regulation. Studies with pure cultures revealed that the cells increase their gas vesicle content by a factor of 9 when shifted from saturating light intensities to the dark. In the stationary phase, the cellular carbohydrate content is much lower than during exponential growth (Overmann and Pfennig, 1992). The buoyant density of A. purpureus cells with a high gas vesicle and low carbohydrate content was 1002 kg $\times$ m^{-3}, which is considerably lower than the density of the surrounding saline lake water (1015 kg $\times$ m^{-3}). Therefore, the buoyancy regulation of the A. purpureus cells together with the steep density stratification of the lake water explain the massive accumulation of phototrophic sulfur bacteria in Mahoney Lake.

Washout and sedimentation losses of A. purpureus are very small in comparison to another loss process, the upwelling into upper water layers. In water samples taken from the upper part of the bacterial plate, large, purple cell aggregates were visible. Aggregates with dimensions of up to 10 cm and protruding out of the layer were observed by video camera imaging in a SCUBA dive (Fig. 6; Overmann et al., 1994). Microscopic examination revealed aggregates of cells that remained firmly attached to each other and could not be disintegrated by mechanical shearing. Dense cell aggregates were formed after the depletion of sulfide, but could be disintegrated within less than 1 second by addition of sulfide or various thiol compounds. This novel mechanism of cell-to-cell adhesion in A. purpureus is mainly caused by a hydrophobic effect and includes a specific mechanism possibly mediated by a surface protein (Overmann and Pfennig, 1992).

The major loss of *A. purpureus* biomass occurred between August and September, when the amount of Bchl*a* in the layer declined sharply (Fig. 9, Overmann *et al.*, 1994). At the same time numerous purple bacterial aggregates appeared in the mixolimnion, changing the color of Mahoney Lake from its usual blue-green to purplish. Aggregates floating at the lake surface after this upwelling event had a cellular gas vesicle content very similar to cells sampled in the chemocline, but were less dense than ambient lake water (Overmann *et al.*, 1994). Thus the upwelling of aggregates to the mixolimnion cannot be explained by an unusual increase in cellular gas vesicle content. If cells with a high gas-vesicle content and low intracellular concentrations of ballast compounds rise to the sulfide-depleted upper part of the chemocline during summer, it is expected that they aggregate rapidly before the cellular gas-vesicle content can decrease. The specific photosynthetic activity of *A. purpureus* is very low during this time, therefore ballast compounds such as carbohydrate and intracellular sulfur do not accumulate to a significant degree. Indeed, the purple aggregates observed in summer extended further into the mixolimnion water than the top of the plate had before (Overmann *et al.*, 1994). Because of their bigger diameter, aggregates formed at the top of the plate in August would rise to higher water layers much more rapidly than individual cells.

In addition, isothermal and isohaline conditions from the lake surface down to the bacterial layer were observed in August and September. During this time the stability of the water column, i.e., its resistance to vertical mixing by wind and by convective cooling, was very small (Overmann *et al.*, 1996b). The aggregates must have risen out of the steep vertical density gradient and thereby entered the mixing currents in the mixolimnion. Thus a coincidence of the low buoyant density of *Amoebobacter* aggregates and the low stability of the water layers in late summer causes the upwelling of *A. purpureus* aggregates into oxic water layers. Wind-driven and convective mixing aid in the sudden vertical distribution of *A. pupureus* throughout the mixolimnion but are not the reason for the rise of the cells in upper water layers.

Autumnal upwelling and subsequent degradation in the mixolimnion and littoral sediments represents a major loss process of purple sulfur bacteria in Mahoney Lake. Eighty-six percent of the biomass of *A. purpureus* that had been present in the plate in August were lost by this process within one month, but only 3.5% of the lost biomass were still suspended in the mixolimnion in September. After the upwelling event the littoral sediment of Mahoney Lake was densely covered with a thick layer of purple sulfur bacteria. As there is no indication of markedly increased degradation of *A. purpureus* cells within the bacterial plate, the remainder (> 80% of the biomass) must have been degraded in the mixolimnion or deposited in the littoral. The upwelling of *A. purpureus* biomass between August and September (54.8 gC m^{-2}) approached the total net biomass accumulation of the whole year 1993 (68.4 gC m^{-2} y^{-1}). The massive input of bacterial biomass into upper water layers has far-reaching consequences for the carbon and nutrient cycles in the oligotrophic mixolimnion (Section 8).

However, before discussing the relevance of the bacterial plate to microbial processes in oxic water layers, the significance of anoxygenic phototrophs for the accompanying bacteria in the chemocline will be elucidated. Numbers of chemotrophic bacteria in the chemocline were very high (Section 3) and changed parallel to those of *A. purpureus* (Overmann et al., 1996a). Interactions between purple sulfur bacteria and chemotrophic bacteria are therefore likely.

6. Bacterial Interactions in the Chemocline: Anoxygenic Phototrophs and Sulfate Reducers

The isotopic composition of sulfate (δ^{34}S = + 22.2 ‰ to + 27.5 ‰) and of sulfide (− 29.4 ‰ to − 24.1 ‰) confirm that sulfide is formed via biological sulfate reduction in Mahoney Lake. Sulfide concentrations reached their maximum at 12 m depth (Overmann et al., 1996a). An increase in concentrations further down towards the sediment surface at 14.5 m was never observed. This demonstrates that sulfate reduction mainly occurs in the free water column and probably is very low in the bottom sediments.

In the chemocline, the high biomass accumulation of *A. purpureus*, together with the steep sulfide gradient and the high rates of sulfate reduction, indicates a rapid cycling of sulfur compounds. Within the purple layer, sulfide (concentrations up to 10 mM), elemental sulfur (up to 10 mM) and polysulfide-sulfur (up to 600 μM) were the main species of reduced sulfur. Other sulfur compounds of intermediate redox state are unimportant for the sulfur cycle in Mahoney Lake (Overmann et al., 1996a). Between April and November 1993 the oxidative part of the sulfur cycle was dominated by anoxygenic photosynthesis of *A. purpureus*. Up to 98% (mean 75%) of the sulfide in the chemocline was reoxidized via anoxygenic photosynthesis.

Sulfate reduction within the bacterial plate contributes as much as 60% of the sulfide for anoxygenic photosynthesis during summer. Compared to other aquatic environments the sulfate reduction rates in the chemocline were very high (Table I). By inference, the supply of carbon substrates for sulfate-reducing bacteria also must be very high in the chemocline. Organic carbon substrates theoretically could be provided by phototrophic bacteria, which can excrete substantial amounts of photosynthetically fixed carbon (up to 30%; Czeczuga and Gradzki, 1973).

Such a syntrophic interaction seems to exist between cyanobacteria and sulfate-reducing bacteria in microbial mats (Jørgensen and Cohen, 1977; Sky-

Table I. Sulfate Reduction Rates in the Chemocline of Various Lakes

Site	Sulfate reduction (μmol $SO_4^{2-} \cdot l^{-1} \cdot d^{-1}$)	Reference
Lake Gek Gel	<0.19	Sorokin, 1970
Lake Belovod	0.93	Sorokin, 1970
Solar Lake	1.0	Jørgensen et al., 1979
Lake Veisovo	2.35	Gorlenko et al., 1983
Big Soda Lake	0.7–3.2	Smith and Oremland, 1987
Knaack Lake	8.16	Parkin and Brock, 1981b
Lake Pomyaretskoe	9.71	Gorlenko et al., 1983
Plußsee	13.7	van Gemerden, 1967
Lake Cisó	232.8	van Gemerden et al., 1985
Mahoney Lake	0.96–297.6	Overmann et al., 1991, 1996a

ring and Bauld, 1990; Fründ and Cohen, 1992). Oxygenic photosynthesis and sulfate reduction proceed in close proximity and are quantitatively related, indicating that the DOC excreted by cyanobacteria is rapidly oxidized by sulfate-reducing bacteria.

Although the annual carbon requirement of sulfate-reducing bacteria in the chemocline of Mahoney Lake (22.5 gC $\times$ m^{-2} $\times$ y^{-1}) was indeed met by the photosynthetic carbon fixation of *A. purpureus* (33.5 gC $\times$ m^{-2} $\times$ y^{-1}), there is no evidence for a close syntrophic relation between sulfate reducers and purple sulfur bacteria. Sulfate reduction and anoxygenic photosynthesis proceeded asynchronously (Fig. 8C) and the areal rates of sulfate reduction did not show any correlation with those of anoxygenic photosynthesis ($r = 0.108$). *A. purpureus* cells excreted only between 0 and 16.2% of the photosynthetically fixed carbon (Overmann et al., 1996a), but in August 1993, the carbon demand of sulfate-reducing bacteria temporarily exceeded even total photosynthetic carbon fixation (Fig. 8C).

Obviously, processes other than excretion of photosynthates by *A. purpureus* are important in the chemocline carbon cycle. In fact, *A. purpureus* uses some of the volatile fatty acids released by other processes for its own photoheterotrophic growth rather than excreting these compounds, because a peak of light-dependent incorporation of ^{14}C-acetate was measured at the Bchl*a* maximum (Overmann et al., 1991).

The origin of the carbon substrates for sulfate-reducing bacteria was elucidated by a combination of stimulation and inhibition studies. During spring, the rate of sulfate reduction was substantially increased by the addition of the volatile fatty acids formate, acetate, propionate and butyrate (Fig. 10A). A slight stimulation was also observed with lactate, methanol, and a mixture of yeast extract plus casamino acids. According to the spectrum of substrates utilized, the domi-

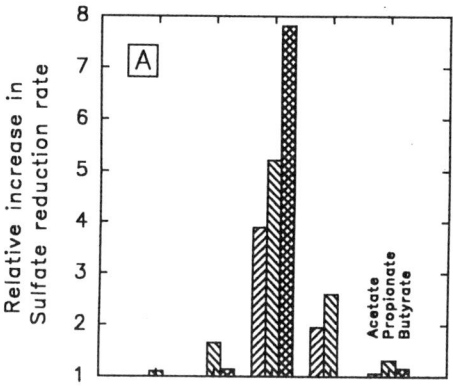

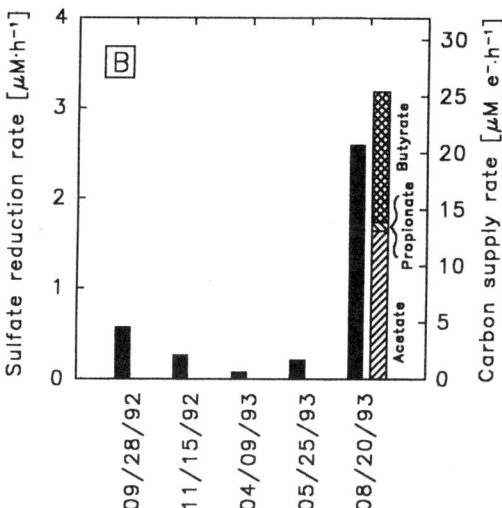

Figure 10. (A) Stimulation of sulfate reduction in the chemocline by acetate, propionate or butyrate (final concentration, 5 mM) at five different dates in 1993. (B) Sulfate reduction rate in the chemocline at the same dates (filled bars) and release of acetate, propionate, and butyrate after inhibition of sulfate reduction in a water sample from the chemocline (addition of 400 mM sodium molybdate). Left and right ordinate in B. correspond to each other.

nating species of sulfate-reducing bacteria belong to the subgroup of complete oxidizers. During summer when the sulfate reduction rate, and thus carbon demand, was at its maximum (Fig. 10B), none of the substrates tested had a stimulating effect, which at first glance seems contradictory.

The addition of sodium molybdate as an inhibitor of sulfate reduction resulted in a significant accumulation of acetate, propionate, and butyrate in chemocline water samples. The accumulation rates of the three volatile fatty acids can be converted to rates of supply of reducing equivalents, assuming a complete oxidation to CO_2. In August, the supply rate of reducing equivalents was in the same order of magnitude as the demand by sulfate reduction (Fig. 10B). In conclusion, the release of carbon substrates within the bacterial plate in late summer is sufficient to explain the sulfate reduction rates measured.

Several lines of evidence indicate that particulate carbon in the bacterial layer, including the biomass of *A. purpureus*, is the source of the carbon compounds for sulfate reduction. First, the rate of sulfate reduction in the purple layer was positively correlated to the biomass of *A. purpureus* ($r = 0.909$, $p < 0.01$), but not to anoxygenic photosynthesis. Secondly, the activities of the biomass-degrading ectoenzymes β-glucosidase, lysozyme, and protease showed maximum activity in samples from the chemocline and also were much higher as compared to eutrophic environments (Overmann *et al.*, 1996a). The release of carbon monomers by these enzymes exceeded the carbon demand of sulfate-reducing bacteria by a factor of three. Thus the degradation potential for microbial biomass in the chemocline of Mahoney Lake is sufficient to provide the precursors of the volatile fatty acid intermediates, and therefore indirectly drives a major part of the anoxygenic photosynthesis.

7. Bacterial Interactions in the Chemocline: Anoxygenic Phototrophs and Sulfur Reducers

Filtered water samples from the lower part of the purple layer and from the monimolimnion exhibited a characteristic yellow color and absorption spectra exhibited a shoulder at 293 nm, which is characteristic for polysulfides. Polarographic measurements showed that the purple layer contained high concentrations of polysulfide-sulfur exceeding 600 μM during spring (Fig 7B). Between 12 and 14 m, a second maximum of polysulfide concentrations was measured, with values similar to those in the chemocline.

Most likely, the polysulfides form by chemical reaction of the intracellular sulfur of *A. purpureus* with dissolved sulfide. This was concluded from the fact that elemental sulfur is found exclusively in cells of *A. purpureus* and that the time course of the intracellular sulfur content and the integrated amount of polysulfides in the chemocline are strikingly similar (Fig. 8B). Furthermore, the $\delta^{34}S$ values for polysulfide-sulfur and the intracellular sulfur of *A. purpureus* were almost identical (-20.5 versus -20.2 ‰, respectively).

Polysulfide-sulfur present in the anoxic waters of Mahoney Lake could serve as an electron acceptor for sulfur-reducing bacteria. Because of isotopic exchange reactions, the physiological activity of sulfur-reducing bacteria cannot be assessed directly. However, the polysulfide concentrations in the lower part of the bacterial plate decreased over the summer (Fig. 8B), and at the same time most probable numbers of sulfur-reducing bacteria increased by a factor of 23. In the monimolimnion both parameters remained constant over the whole year.

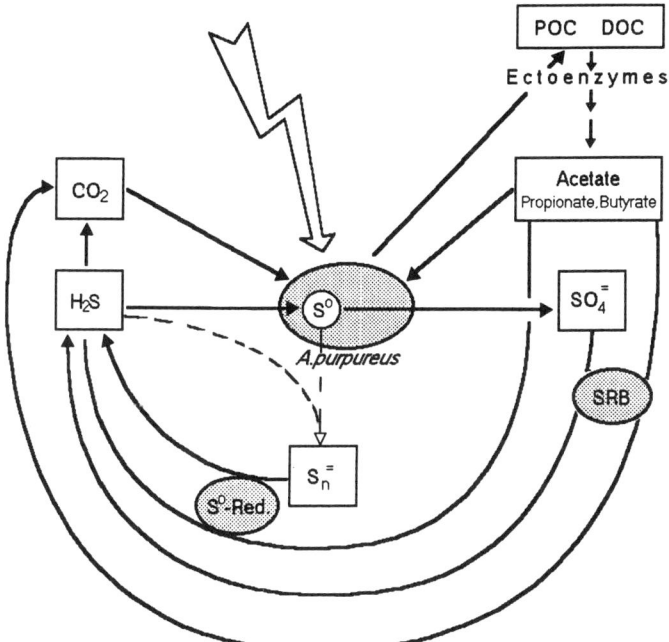

Figure 11. Combined carbon and sulfur cycles in the chemocline of Mahoney Lake. POC, particulate organic carbon; DOC, dissolved organic carbon; SRB, sulfate-reducing bacteria; S^o-Red., sulfur-reducing bacteria.

Obviously, the reduction of polysulfide-sulfur is confined to the purple layer in the chemocline, where polysulfides represent significant intermediates in the sulfur cycle. Below, sulfur-reducing bacteria were probably inhibited by the high sulfide concentrations of up to 60 mM (Pfennig and Biebl, 1976).

The information about the major physiological processes in the chemocline of Mahoney Lake, gathered during the last years of field work, can be combined in a flow diagram for carbon and sulfur depicted in Fig. 11.

Unexpectedly, anoxygenic photosynthesis and sulfate reduction are only indirectly coupled via degradation and autolysis of photosynthetically formed biomass and of other organic carbon materials. Anoxygenic photosynthesis is also only loosely coupled to sulfide formation by sulfate-reducing bacteria in the plate because the diffusion from below supplies a major fraction of sulfide, and reducing power is stored transiently by *A. purpureus* in the form of intracellular sulfur.

When intracellular sulfur is accumulated by *A. purpureus* cells, polysulfides form abiotically, and in turn are used for energy conservation by sulfur-reducing bacteria. Thus, the formation and utilization of polysulfide are part of the sulfur cycle in the ecosystem Mahoney Lake. These processes very likely are important also in other anoxic aquatic environments.

8. The Coupling between Phototrophic and Sulfate-Reducing Bacteria: General Implications

The carbon dioxide fixation of phototrophic sulfur bacteria depends on the availability of reduced sulfur compounds, mostly sulfide, as the electron-donating substrates. With the exception of hydrothermal environments, this sulfide originates from sulfate or sulfur reduction during the terminal degradation of organic matter.

In the course of sulfate or sulfur reduction, about 9% of the organic carbon substrates utilized are assimilated (Pfennig and Biebl, 1976; Widdel, 1988) and electrons from the remaining 91% of the substrate end up in H_2S. Because of the electrons constantly diverted towards bacterial biomass, the cycling of electrons between sulfate reducers and phototrophs would finally come to an end, even if all the carbon fixed in anoxygenic photosynthesis would again serve as substrate for sulfate or sulfur reducers and if the degradation and assimilation of organic carbon by fermenting bacteria are disregarded. Consequently, an input of organic carbon is required to drive the carbon and sulfur cycles in the anoxic part of such ecosystems (Fig. 12). Ultimately, anoxygenic photosynthesis is fueled indirectly by carbon that already has been fixed by oxygenic phototrophs within or outside

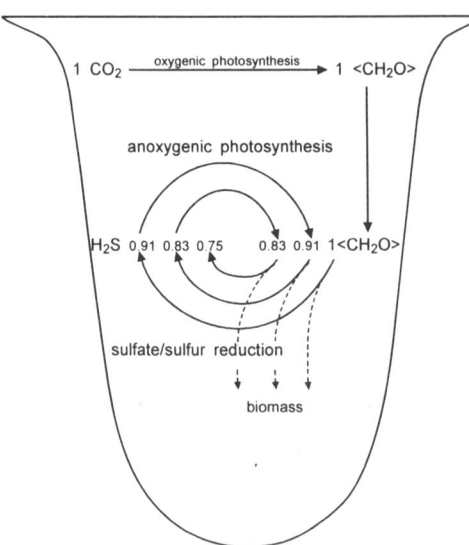

Figure 12. Dependence of anoxygenic CO_2-fixation on oxygenic photosynthesis. Numbers refer to the fraction of electrons originally present in carbon fixed by oxygenic phototrophs. These electrons are transferred to sulfate (yielding H_2S) and back to CO_2 (during anoxygenic photosynthesis). Dotted arrows indicate the fraction of electrons constantly diverted towards biomass of sulfate-reducing bacteria. At equilibrium the theoretical maximum (s_n) of anoxygenic photosynthesis is 10.11 relative to oxygenic photosynthesis.

$$s_n = \sum_1^n 0.91^n = \frac{1 - 0.91^{n+1}}{1 - 0.91} - 1$$

$$\lim_{n \to \infty} s_n = \frac{1}{1 - 0.91} - 1 = 10.11$$

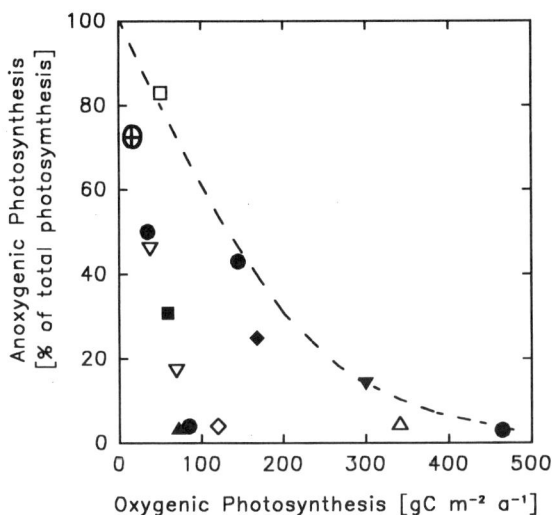

Figure 13. Relation between annual integrals of oxygenic photosynthesis and anoxygenic photosynthesis (as percentage of total primary production) in lakes. Data obtained or calculated from Biebl and Pfennig (1979) (●), Parker *et al.* (1983) (∇), Cloern *et al.* (1983) (▼), Culver and Brunskill (1969) (□), Cohen *et al.* (1977) (■), Parkin and Brock (1981a) (△), Steenbergen (1982) (▲), Overmann and Tilzer (1989) (◇), and van Gemerden and Mas (1995) (♦)

the ecosystem. Anoxygenic photosynthesis thus represents secondary, and not new, primary production (Pfennig, 1978).

The calculation in Fig. 12 demonstrates that the rate of anoxygenic photosynthesis can exceed that of oxygenic photosynthesis if the cycling of electrons is very efficient. If 100% of the photosynthates of anoxygenic phototrophs were recycled, the ratio of anoxygenic to oxygenic photosynthesis would be 10.1, i.e., 91% of the total photosynthesis in the ecosystem would be anoxygenic. Such a high proportion indeed has been observed (in the case of Fayetteville Green Lake, New York, see open rectangle in Fig. 13). If the fraction of anoxygenically fixed carbon which is reutilized is only 20%, the ratio of anoxygenic to oxygenic photosynthesis would be 1.1, i.e., 52% of total photosynthesis would be anoxygenic.

Anoxygenic photosynthesis exceeds that of the phytoplankton in many oligotrophic lakes (Fig. 13). This raises the question of whether phytoplankton carbon is indeed recycled as efficiently in these systems as indicated in Fig. 12. Usually not all of the phytoplankton carbon is available for degradation in the chemocline of natural water bodies. About 50% of the primary production of phytoplankton can be lost from the pelagial by sedimentation and be permanently buried at the lake bottom (e.g., in Lawrence Lake; Wetzel, 1983). In the meromictic Black Sea <10% of the particulate organic carbon sinking down from the photic zone reaches the chemocline (Karl and Knauer, 1991).

Consequently, the high ratios of anoxygenic to oxic photosynthesis observed in many lakes may also be caused by the input of additional (allochthonous) organic matter. It became clear that this seems to be the major reason for the high rate of anoxygenic photosynthesis observed in Mahoney Lake. To the knowledge of the author this aspect of the carbon cycle has not been investigated in any other ecosystem as yet.

In Fig. 14 the fluxes of organic carbon in Mahoney Lake as calculated from the rates of photosynthesis, sulfate reduction, and sedimentation are combined in an overall scheme of carbon flow. Because upwelling is the major loss process for *A. purpureus* (80% of its production leaves the carbon cycle of anoxic water layers), anoxygenic photosynthesis can provide only a small fraction of the carbon substrate of sulfate-reducing bacteria (7.9 of about 47 g C $\times$ m^{-2} $\times$ yr^{-1}). Of the total secondary primary production, only 4.3 g C $\times$ m^{-2} $\times$ yr^{-1} remains within the chemocline. Therefore only a third of the sulfate reduction within the bacterial plate (carbon demand of 13.9 g C $\times$ m^{-2} $\times$ yr^{-1}) can be explained by recycling of *A. purpureus* carbon.

A major fraction of the sulfide used in anoxygenic photosynthesis is supplied by sulfate reduction in the monimolimnion. Because sedimentation of phytoplankton carbon into the anoxic bottom layers is also very small, the input of allochthonous organic carbon into Mahoney Lake must be substantial. As expected, the rate of sedimentation of POC into the bacterial plate was high (Fig. 14). The δ^{13}C-values of total POC from the bacterial plate indicate that most of it is of allochthonous origin (Overmann *et al.*, 1996b). Additional evidence comes from the observation of high amounts of plant debris in the mixolimnion (Northcote and Hall, 1990), and from the extremely high concentrations of DOC (up to 90 mg C $\times$ l^{-1}) in the lake water (Overmann *et al.*, 1996b). Similar DOC concentrations have been detected in other salt lakes (Waiser and Robarts, 1995). δ^{13}C determinations indicate that the DOC in Mahoney Lake is of terrestrial origin and is derived mainly from the litter of conifers that grow in great numbers in the watershed of Mahoney Lake (Fig. 1). Similar to Mahoney Lake, C-mineralization in Big Soda Lake also exceeds sedimentation of autotrophically formed biomass (Smith and Oremland, 1987). It was suggested that additional carbon sources like the abundant dissolved organic carbon or the benthic production were the source of organic carbon.

Figure 14 suggests that the function of purple sulfur bacteria in the Mahoney Lake ecosystem is the formation of biomass by recycling electrons derived from the degradation of allochthonous matter. In this way, refractory organic carbon is converted into easily degradable bacterial biomass, which is then returned to the oxic part of the system. Phototrophic bacteria use light as energy source and all sulfide is utilized in the formation of biomass. If light would not reach the

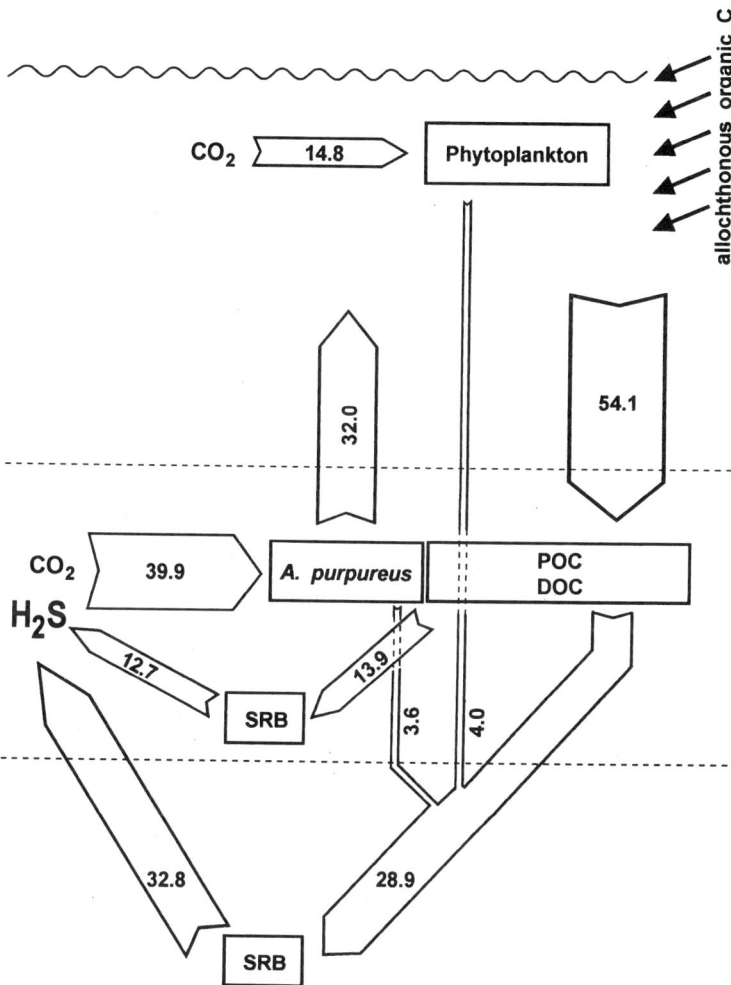

Figure 14. Scheme of the carbon cycle in the anoxic part of Mahoney Lake based on the rates of photosynthesis, sulfate reduction, sedimentation, or upwelling of *A. purpureus*. Numbers are in g C $\times$ m^{-2} $\times$ yr^{-1} and all calculated per m² lake surface. Horizontal arrows, photosynthesis; oblique arrows, carbon demand of sulfate reduction; vertical arrows, sedimentation of POC or upwelling of the POC of *A. purpureus*. Dotted lines indicate upper and lower boundary of the bacterial plate. Data are for the year 1993.

chemocline of Mahoney Lake, and sulfide would be reoxidized by chemolithotrophic bacteria instead of phototrophs; a much smaller fraction of bacterial biomass would be formed from the amount of sulfide available. The specific physicochemical conditions in Mahoney Lake ultimately are the reason for the efficient conversion of allochthonous carbon.

It can be expected that phototrophic sulfur bacteria are significant for the overall carbon cycle in all those systems where a considerable amount of light (well above 1% surface intensity) reaches the chemocline, anoxygenic photosynthesis contributes a high proportion of total primary production, and losses of phototrophic biomass by sedimentation and washout are small.

9. Significance of Purple Sulfur Bacteria for Oxic Water Layers

The upwelling of *A. purpureus* biomass fundamentally alters the microbial processes in the mixolimnion.

In many aquatic habitats the growth of aerobic heterotrophic bacterioplankton appears to be controlled by the photosynthetic activity of the accompanying phytoplankton (Cole *et al.*, 1988; Cho and Azam, 1990). Other authors have argued that growth of both algae and heterotrophic bacteria is limited by the supply of phosphorus or by temperature (e.g., Morris and Lewis, 1992; Shiah and Ducklow, 1994).

In the mixolimnion of Mahoney Lake, the biomass of heterotrophic bacteria exceeds that of the phytoplankton more than 10 times. Similarly, heterotrophic bacterial production was 7 times higher than oxygenic photosynthesis (Overmann *et al.*, 1996b). In contrast to other ecosystems, the seasonal changes in phytoplankton biomass (measured as chlorophyll *a*) were not accompanied by changes in bacterial cell numbers and both parameters were not correlated. Also, oxygen or water temperature do not seem to influence bacterial growth rates. Thus other factors must control the growth and biomass of heterotrophic bacteria in oxic water layers. Mahoney Lake is not the only meromictic lake where primary and bacterial production are not balanced. A similar situation was reported for saline Big Soda Lake, Nevada (Cloern *et al.*, 1987; Zehr *et al.*, 1987). An analysis of the factors that control growth of aerobic heterotrophic bacteria in Mahoney Lake would aid in better understanding of other lakes systems with anoxic bottom waters.

Until summer, bacterial growth rates (measured as thymidine incorporation rates) in the mixolimnion of Mahoney Lake increased significantly after addition of HPO_4^{2-} or NH_4^+. Evidently, heterotrophic bacteria were limited by inorganic nutrients during this time. Throughout the whole year, the addition of acetate plus glucose never caused any effect.

Total phosphorus concentrations in the mixolimnion of Mahoney Lake are mostly below the detection limit (<0.1 µM, Overmann et al., 1996b; Northcote and Hall, 1983) and concentrations of nitrate plus nitrite are below 0.7 µM (Northcote and Hall, 1983). The extremely low levels of total phosphorus in the mixolimnion of Mahoney Lake are characteristic for hardwater brines and have been attributed to adsorption of phosphate to calcite and subsequent coprecipitation (Northcote and Hall, 1983). Similarly to the case with bacterioplankton, the growth of phytoplankton was limited by phosphorus (Hall and Northcote, 1990), which prevents the accumulation of a high standing stock of phytoplankton biomass in Mahoney Lake.

The heterotrophic bacterial production in the mixolimnion of 104.7 g C × m^{-2} × yr^{-1} appears much too high for such an oligotrophic lake. Similar to the situation in the chemocline (Fig. 14), the high amounts in POC and DOC present in Mahoney Lake are likely to be major substrates for heterotrophic bacterial growth in the lake. The DOC in Mahoney Lake consists mainly of compounds related to humic matter and therefore is very poor in phosphorus (C:P > 10000:1, by molarity) (Overmann et al., 1996b).

Heterotrophic bacterial production and cell numbers, as well as mean bacterial cell volumes increased to very high values in September and November, concomitant with the upwelling of *A. purpureus* (Overmann et al., 1996b). After the upwelling, bacterial growth in the mixolimnion could be stimulated only by addition of yeast extract plus casamino acids. Obviously, the supply of inorganic nutrients during this period was sufficient for heterotrophic bacterial growth. At the same time, total soluble phosphorus concentrations still remained below the detection limit. At first sight, the extremely low concentrations would contradict the results of the stimulation studies and would indicate severe nutrient limitation of planktonic microorganisms in the mixolimnion.

Another way of assessing the limitation of bacterial growth in aquatic environments is the measurement of the activity of relevant ectoenzymes. The activity of the ectoenzyme alkaline phosphatase has been suggested as a measure for detecting phosphorus deficiency of algae. Alkaline phosphatase is derepressed only after depletion of extracellularly available inorganic phosphate and of intracellularly stored phosphorus. The enzyme is induced by phosphoester substrates (Chróst, 1991). The activity of alkaline phosphatase in the mixolimnion of Mahoney Lake always exceeded that of protease, lysozyme, α- and β- glucosidase, and glucuronidase by three orders of magnitude, and reached values more than 50 times higher than in other nonacidic environments (Overmann et al., 1996b). Phosphoesters must therefore represent a major source of phosphate for microbial growth. In autumn, during the time of maximum heterotrophic bacterial production, the specific alkaline phosphatase activity (activity per unit microbial biomass) decreased 10-fold, indicating a significant improvement in the supply of phosphorus.

In fact, only 16% of the phosphorus that was organically bound in the upwelling biomass of *A. purpureus* would be required to explain the autumnal increase in the biomass of heterotrophic bacteria in the mixolimnion. This, the extremely low concentration of inorganic phosphorus, and the high phosphatase activity, all indicate a rapid turnover of the soluble phosphorus pool and a rapid incorporation of the phosphorus released from *A. purpureus* into heterotrophic bacterial biomass.

The upwelling of purple bacterial biomass minimizes the losses of organically bound nutrients and represents a recycling process, especially for phosphorus. The processes elucidated in Mahoney Lake might be generally significant for bacterioplankton in those lakes where phototrophic bacterial biomass frequently enters the oxic water layers during mixing events (e.g., Mas *et al.*, 1990).

In addition, purple sulfur bacteria may represent a significant food source for higher organisms in Mahoney Lake. Peak numbers of the rotifer *Brachionus plicatilis* and adults of the copepod *Diaptomus connexus* were found shortly above the purple sulfur bacterial layer, whereas the nauplii and copepodites of the latter species reached maximum numbers at or close to the lake surface (Northcote and Halsey, 1969; Northcote and Hall, 1983). At the onset of the limnological studies of Mahoney lake, Northcote and Halsey (1969) noted that individuals of *D. connexus*, as well as of *B. plicatilis*, had yellow-stained mouth parts and anus during all seasons. They concluded that purple sulfur bacteria might be a significant food source for the zooplankton. Several lines of evidence indicate that copepods feed on *A. purpureus* cells as a food supply (Overmann *et al.*, in preparation). This is yet another aspect of the general role of phototrophic sulfur bacteria in aquatic habitats, which can be interpreted as a means to recycle inorganic nutrients and dissolved organic matter, and to return them to the food chain—even of the oxic part of the ecosystem.

10. Paleomicrobiology of Mahoney Lake

In the foregoing chapters, the transformations in the carbon and sulfur cycles as they presently occur in Mahoney Lake (Fig. 11, Fig. 14) have been analyzed. Major microbial processes depend either directly or indirectly on *A. purpureus*. Mahoney Lake was formed after the retreat of the Wisconsin ice sheet about 13,000 years ago and probably occupies a kettle basin. The question that remains is: at which point in the lake's history, did purple sulfur bacteria become significant for the biogeochemistry of the lake.

The past changes in its limnology and microbiology should have left records in the lake sediment. Mahoney Lake is well suited for a paleomicrobiological analysis of the ontogeny of a meromictic lake system over a time interval of more than 10,000 years. As in other meromictic lakes, the stable and deoxygenated

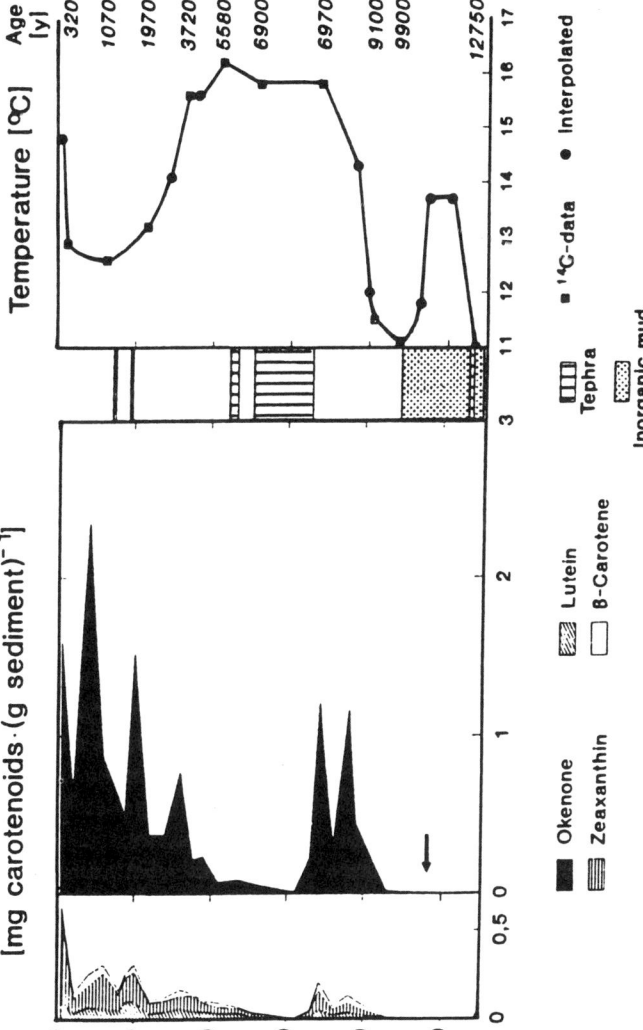

Figure 15. Vertical distribution of okenone, zeaxanthin, lutein and β-carotene in a sediment core from Mahoney Lake. Concentrations given per dry mass of sediment. Arrow indicates the oldest sediment layer containing detectable concentrations of okenone and β-carotene. Ash layers in the sediment were attributed to the following eruptions (from top of sediment core): 70 cm, 95 cm, 230 cm—Mount St. Helens; 260–330 cm—Mazama Mountain; 530 cm—Glacier Peak. The lowermost meter of sediment contains mainly inorganic mud. Paleotemperatures (mean summer values) for Masset (British Columbia, Canada) and Humptulips (Washington). ■, values for ^{14}C data; ●, temperature values determined after linear interpolation of age between adjacent ^{14}C data. After Overmann et al. (1993).

bottom waters ensure the deposition of sediments that are undisturbed by currents or bioturbation so that organic material is well preserved in discrete layers. A 6-m-long sediment core, obtained with a piston corer from the deepest part in the center of the lake, served as a source of material for all analyses.

The nature and timing of sediment deposition in the lake were inferred from stratigraphic analysis of the core (Lowe et al., 1996). At a depth of 260 to 330 cm in the core a conspicuously thick layer of volcanic ash was found. This layer formed 6900 years ago after the eruption of Mazama Mountain (now Oregon) (Fig. 15). Within 70 years, a 80-cm-thick ash layer was deposited in the lake. This layer, albeit of less thickness, was also detected in many soils of southern British Columbia. Two additional tephras (Mount St. Helens and Bridge River) could be identified in more recent sediment layers of the core.

According to the stratigraphic analysis, the lake became saline early in its development and meromixis developed probably 8300–9150 years ago (Lowe et al., 1996). A transition from freshwater to saline conditions is also supported by analysis of fossil remains of chironomids. Head capsules that are characteristic of a freshwater community were deposited during the early postglacial, but were replaced by those indicative of saline environments (*Cricotopus/Orthocladius, Tanypus*) in the early Holocene (Heinrichs, 1995). Chironomid-inferred salinity values reflect the shift from freshwater to saline water and indicate that Mahoney Lake reached peak salinities during the mid-Holocene, 6400–5700 years before present.

As concluded from the layers of calcareous laminated sediments deposited, the lake underwent periodical meromixis with relatively high water levels (> ca. 12 m depth) during the past 9000 years. The laminated sediments consist of non-glacial varves with alternating light and dark layers. The dark layers may have resulted from sedimentary accumulation of dead algae or purple sulfur bacteria or from allochthonous organic deposits, while the light layers may have formed by precipitation of $CaCO_3$ (Lowe et al., 1996). At other times the lake even may have approached dryness.

The role of anoxygenic phototrophic bacteria in past lake metabolism was investigated by an analysis of fossil carotenoids isolated from sediment samples (Overmann et al., 1993). In paleolimnological studies, carotenoids of phototrophic bacteria have served as a sensitive measure of past biomass changes (Brown et al., 1984; Züllig, 1985). Carotenoids are preserved well in the absence of light and oxygen (Leavitt and Carpenter, 1990).

Carotenoids were detected down to a depth of 4.9 m in the sediment core (Fig. 15). Six different carotenoids and their isomers could be separated and quantified. In all samples okenone was the dominating carotenoid and was present in three different isomers (all-*trans*- and *cis*-okenone and demethylated okenone). In addition, lutein, zeaxanthin, α-carotene and β-carotene were identified. *Amoebobacter purpureus* ML1 that was isolated from the chemocline of the lake contains the same three okenone isomers.

The oldest sediment layers containing okenone were deposited 11,000 years ago during the Alleröd. Okenone is found exclusively in 10 Chromatiaceae species and it was concluded that purple sulfur bacteria colonized Mahoney Lake shortly after its formation.

In contrast to the high okenone concentrations, no carotenoids of purple nonsulfur bacteria or green sulfur bacteria were detected. Similar to the present situation, okenone-containing Chromatiaceae species obviously dominated the phototrophic bacterial population in the past. Chromatiaceae have a selective advantage over purple nonsulfur bacteria in the presence of sulfide as an electron-donating substrate. Therefore it appears most likely that Mahoney Lake had anoxic, sulfide-containing bottom water during most of its history.

A comparison with paleo-temperatures prevailing during the deposition of different sediment layers (Fig. 15, right panel) indicates that ambient temperatures did not influence the amount of okenone deposited. However, the deposition of carotenoids seems to be influenced by the input of volcanic tephras in the lake sediment. Within the ash layers only low carotenoid concentrations were measured. Obviously, the markedly increased sedimentation of Mazama ash simply diluted the concurrently deposited carotenoids.

In all sediment zones okenone exceeded the concentrations of those carotenoids that are characteristic of oxygenic phototrophs. Mahoney Lake probably also has been oligotrophic during the past. The environmental conditions for growth of phototrophic sulfur bacteria during the past 10,000 years must have resembled those of present day and the lake must have harbored a well-developed population of okenone-bearing purple sulfur bacteria.

The maximum okenone content of Mahoney Lake sediment was about 2.4 mg $\times$ (g dry sediment)$^{-1}$, which is 2.3 times more than the highest concentration reported so far (in Lago di Cadagno; Züllig, 1985). Conversely, the sedimentation rates in Lago di Cadagno were 2.6 times higher than those in Mahoney Lake. Therefore the absolute amount of okenone deposited per time interval is very similar in both lakes. However, the biomass of purple sulfur bacteria found in the chemocline of Lago di Cadagno is 340 times smaller than that in Mahoney Lake. Consequently, the losses due to sedimentation to the lake bottom must have been extremely reduced throughout the whole history of Mahoney Lake. As today, sedimentation in the past may have been restricted by a steep density gradient and regulation of the buoyant density of purple bacterial cells.

Presently, 10 Chromatiaceae species are known that contain okenone. One or several of these species most likely have lived in Mahoney Lake in the past 10,000 years. Carotenoid analysis does not permit a more precise taxonomic identification, however. Therefore, the species composition of bacteria buried in bottom sediments recently has been investigated by molecular methods (Overmann et al., unpublished). DNA could be isolated from various sediment layers of an age between 700 and 9900 years. 16S rDNA sequences were analyzed by polymerase chain reaction (PCR) and subsequent denaturing gradient gel electrophoresis (DGGE). In fact, the first results indicate that the majority of purple sulfur bacteria buried in the past belong to the species *A. purpureus*.

Epilogue

The study of Mahoney Lake not only revealed that purple sulfur bacteria are of major significance for the flux of carbon and the transformations of sulfur compounds, but also demonstrated that a small group of Chromatiacea, probably even one single species, has persisted in this rather extraordinary ecosystem for about 10,000 years.

Because of its unusual features and its potential for scientific investigations it was proposed to convert Mahoney Lake into an ecological reserve (Proposal No. 369 submitted by T. G. Northcote to British Columbia Ministry of Lands, Parks and Housing; Parks and Outdoor Recreation Division, December, 1983). Meanwhile the ecological reserve has been established. There are still many open questions with regard to the functioning of the ecosystem Mahoney Lake. The lake will therefore remain a fascinating study site in the future.

ACKNOWLEDGMENTS. The work of the author at Mahoney Lake was supported by a grant of the Deutsche Forschungsgemeinschaft. I am indebted to Prof. N. Pfennig who introduced me to the world of anoxygenic phototrophic bacteria. I would like to thank Profs. Tom Beatty, Ken Hall, and Tom Northcote for the excellent cooperative work and for maintaining good spirits in the field, as well as in the lab.

References

Baker, A. L., Baker, K. K., and Tyler, P. A., 1985, A family of pneumatically-operated thin layer samplers for replicate sampling of heterogenous water columns, *Hydrobiologia* **22**:107–211.
Biebl, H., and Pfennig, N., 1979, Anaerobic CO_2 uptake by phototrophic bacteria. A review, *Arch. Hydrobiol. Beih. Ergeb. Limnol.* **12**:48–58.
Brown, S. R., McIntosh, H. J., and Smol, J. P., 1984, Recent paleolimnology of a meromictic lake: Fossil pigments of photosynthetic bacteria, *Verh. Int. Ver. Limnol.* **22**:1357–1360.
Caumette, P., Pagano, M., and Saint-Jean, L., 1983, Répartition verticale du phytoplancton, des bactéries et du zooplancton dans un milieu stratifié en Baie de Biétri (Langune Ebrié, Cote d'Ivoire). Relations trophiques, *Hydrobiologia* **106**:135–148.
Cho, B. C., Azam, F., 1990, Biogeochemical significance of bacterial biomass in the ocean's euphotic zone, *Mar. Ecol. Prog. Ser.* **63**:253–259.
Chróst, R. J., 1991, Environmental control of the synthesis and activity of aquatic microbial ectoenzymes, in: *Microbial Enzymes in Aquatic Environments* (R. J. Chróst, ed.), Springer, New York, pp. 29–59.
Cloern, J. E., Cole, B. E., and Oremland, R. S., 1983, Autotrophic processes in meromictic Big Soda Lake, Nevada, *Limnol. Oceanogr.* **28**:1049–1061.
Cloern, J. E., Cole, B. E., and Wienke, S. M., 1987, Big Soda Lake (Nevada). 4. Vertical fluxes of

particulate matter: seasonality and variations across the chemocline, *Limnol. Oceanogr.* **32:**815–824.
Cohen, Y., Krumbein, W. E., and Shilo, M., 1977, Solar Lake (Sinai) 2. Distribution of photosynthetic microorganisms and primary production, *Limnol. Oceanogr.* **22:**609–620.
Cole, J. J., Findlay, S., and Pace, M. L., 1988, Bacterial production in fresh and saltwater ecosystems: a cross-system overview, *Mar. Ecol. Prog. Ser.* **43:**1–10.
Culver, D. A., and Brunskill, G. J., 1969, Fayetteville Green Lake, New York. V. Studies of primary production and zooplankton in a meromictic marl lake, *Limnol. Oceanogr.* **14:**862–873.
Czeczuga, B., and Gradzki, F., 1973, Relation between extracellular and cellular production in the sulfuric green bacterium *Chlorobium limicola* Nads. as compared to primary production of phytoplankton, *Hydrobiologia* **42:**85–95.
Drews, G., 1985, Structure and functional organization of light-harvesting complexes and photochemical reaction centers in membranes of phototrophic bacteria, *Microbiol. Rev.* **49:**59–70.
Eichler, B., and Pfennig, N., 1990, Seasonal development of anoxygenic phototrophic bacteria in a holomictic drumlin lake (Schleinsee, F. R. G.), *Arch. Hydrobiol.* **119:**369–392.
Fowler, C. F., Nugent, N. A., and Fuller, R. C., 1971, The isolation and characterization of a photochemically active complex from *Chloropseudomonas ethylica*, *Proc. Natl. Acad. Sci. USA* **68:**2278–2282.
Fründ, C., and Cohen, Y., 1992, Diurnal cycles of sulfate reduction under oxic conditions in cyanobacterial mats, *Appl. Environ. Microbiol.* **58:**70–77.
Fry, B., 1986, Sources of carbon and sulfur nutrition for consumers in three meromictic lakes of New York State, *Limnol. Oceanogr.* **31:**79–88.
Gorlenko, V. M., 1988, Ecological niches of green sulfur and gliding bacteria, in: *Green Photosynthetic Bacteria* (J. M. Olson, J. G. Ormerod, J. Amesz, E. Stakebrandt, and H. G. Trüper, eds.), Plenum Press, New York, pp. 257–267.
Gorlenko, V. M., Dubinina, G. A., and Kuznetsov, S. I., 1983, The ecology of aquatic microorganisms, in: *Die Binnengewässer,* Bd. 28, Schweizerbart'sche Verlagsbuchhandlung, Stuttgart.
Grant, W. D., and Tindall, B. J., 1986, The alkaline saline environment, in: *Microbes in Extreme Environments,* (R. A. Herbert, and G. A. Codd, eds.) Academic Press, London, pp. 25–54.
Guerrero, R., Montesinos, E., Pedrós-Alió, C., Esteve, I., Mas, J., van Gemerden, H., Hofman P. A. G., and Bakker, J. F., 1985, Phototrophic sulfur bacteria in two Spanish lakes: Vertical distribution and limiting factors, *Limnol. Oceanogr.* **30:**919–931.
Guerrero, R., Pedrós-Alió, C., Esteve, I., and Mas, J., 1987, Communities of phototrophic sulfur bacteria in lakes of the Spanish Mediterranean region, *Acta Academiae Aboensis* **47:**125–151.
Hall, K. J., and Northcote, T. G., 1986, A novel terrestrial–freshwater linkage: Robin predation on damselfly nymphs, *Discovery* (Canada) **15:**107–109.
Hall, K. J., and Northcote, T. G., 1990, Production and decomposition processes in a saline meromictic lake, *Hydrobiologia* **197:**115–128.
Hammer, T. U., 1978, The saline lakes of Saskatchewan III. chemical characterization, *Int. Rev. Ges. Hydrobiol.* **63:**311–335.
Hammer, T. U., 1986, Saline lake ecosystems of the world, *Monographiae Biologicae,* Vol. 59, W. Junk Publishers, Dordrecht.
Heinrichs, M. L., 1995, Chironomid-based paleosalinity reconstruction of three lakes in the south-central interior of British Columbia, Canada. M. Sc. thesis, Simon Fraser University, Vancouver, Canada.
Javor, B., 1983, Planktonic standing crop and nutrients in a saltern ecosystem, *Limnol. Oceanogr.* **28:**153–159.
Javor, B., 1984, Growth potential of halophilic bacteria isolated from solar salt environments: carbon sources and salt requirements, *Appl. Environ. Microbiol.* **48:**352–360.

Javor, B., 1989, Hypersaline environments, *Brock/Springer Series in Contemporary Bioscience.* Springer, Berlin.

Jørgensen, B. B., 1978, A comparison of methods for the quantification of bacterial sulfate reduction in coastal marine sediments, *Geomicrobiol. J.* **1**:49–64.

Jørgensen, B. B., 1982, Ecology of the bacteria of the sulphur cycle with special reference to anoxic-oxic interface environments, *Phil. Trans. R. Soc. Lond. B* **298**:543–561.

Jørgensen, B. B., and Cohen, Y., 1977, Solar Lake (Sinai). 5. The sulfur cycle of the benthic microbial mats, *Limnol. Oceanogr.* **22**:657–666.

Jørgensen, B. B., and Revsbech, N.P., 1983, Colorless sulfur bacteria, *Beggiatoa* spp. and *Thiovulum* spp., in O_2 and H_2S microgradients, *Appl. Environ. Microbiol.* **45**:1261–1270.

Jørgensen, B. B., Kuenen, J. G., and Cohen, Y., 1979, Microbial transformations of sulfur compounds in a stratified lake (Solar Lake, Sinai), *Limnol. Oceanogr.* **24**:799–822.

Karl, D. M., and Knauer, G. A., 1991, Microbial production and particle flux in the upper 350 m of the Black Sea, *Deep-Sea Res.* **38**:S921–S942.

Lawrence, J. R., Haynes, R. C., and Hammer, U. T., 1978, Contribution of photosynthetic green sulfur bacteria to total primary production in a meromictic saline lake, *Verh. Int. Ver. Limnol.* **20**:201–207.

Leavitt, P. R., and Carpenter, S. R., 1990, Aphotic pigment degradation in the hypolimnion: Implications for sedimentation studies and paleolimnology, *Limnol. Oceanogr.* **35**:520–534.

Lowe, D. J., Green, J. D., Northcote, T. G., and Hall, K. J., 1997, Fluctuating levels of a meromictic lake: evidence for Holocene climate flickering, *J. Quaternary Sciences* (in review).

Mas, J., Pedrós-Alió, C., and Guerrero, R., 1990, In situ specific loss and growth rates of purple sulfur bacteria in Lake Cisó, *FEMS Microbiol. Ecol.* **73**:271–281.

Montesinos, E., Guerrero, R., Abella, C., and Esteve, I., 1983, Ecology and physiology of the competition for light between *Chlorobium limicola* and *Chlorobium phaeobacteroides* in natural habitats, *Appl. Environ. Microbiol.* **46**:1007–1016.

Morris, D. P., and Lewis, W. M., 1992, Nutrient limitation of bacterioplankton growth in Lake Dillon, Colorado, *Limnol. Oceanogr.* **37**:1179–1192.

Nicholson, J. A. M., Stolz, J. F., and Pierson, B. K., 1987, Structure of a microbial mat at Great Sippewissett Marsh, Cape Cod, Massachusetts, *FEMS Microbiol. Ecol.* **45**:343–364.

Northcote, T. G., and Hall, K. J., 1983, Limnological contrasts and anomalies in two adjacent saline lakes, *Hydrobiologia* **105**:179–194.

Northcote, T. G., and Hall, K. J., 1990, Vernal microstratification patterns in a meromictic saline lake: Their causes and biological significance, *Hydrobiologia* **197**:105–114.

Northcote, T. G., and Halsey, T. G., 1969, Seasonal changes in the limnology of some meromictic lakes in southern British Columbia, *J. Fish. Res. Bd. Canada* **26**:1763–1787.

Oren, A., 1983, Population dynamics of halobacteria in the Dead Sea water column, *Limnol. Oceanogr.* **28**:1094–1103.

Overmann, J., and Pfennig, N., 1992, Buoyancy regulation and aggregate formation in *Amoebobacter purpureus* from Mahoney Lake, *FEMS Microbiol. Ecol.* **101**:67–79.

Overmann, J. and Tilzer, M. M., 1989, Control of primary productivity and the significance of photosynthetic bacteria in a meromictic kettle lake (Mittlerer Buchensee, West Germany), *Aquat. Sci.* **51**:261–278.

Overmann, J., Beatty, J. T., Hall, K. J., Pfennig, N., and Northcote, T. G., 1991, Characterization of a dense, purple sulfur bacterial layer in a meromictic salt lake, *Limnol. Oceanogr.* **36**:846–859.

Overmann, J., Cypionka, H., and Pfennig, N., 1992a, An extremely low-light-adapted phototrophic sulfur bacterium from the Black Sea, *Limnol. Oceanogr.* **37**:150–155.

Overmann, J., Fischer, U., and Pfennig, N., 1992b, A new purple sulfur bacterium from saline littoral sediments, *Thiorhodovibrio winogradskyi* gen. nov. and spec. nov, *Arch. Microbiol.* **157**:329–335.

Overmann, J., Sandmann, G., Hall, K. J., and Northcote, T. G., 1993, Fossil carotenoids and paleolimnology of meromictic Mahoney Lake, British Columbia, Canada, *Aquatic Sciences* **55**:31–39.

Overmann, J., Beatty, J. T., and Hall, K. J., 1994, Photosynthetic activity and population dynamics of *Amoebobacter purpureus* in a meromictic saline lake, *FEMS Microbiol. Ecol.* **15**:309–320.

Overmann, J., Beatty, J. T., Krouse, H. R., and Hall, K. J., 1996a, The sulfur cycle in the chemocline of a meromictic salt lake, *Limnol. Oceanogr.* **41**:147–156.

Overmann, J., Beatty, J. T., and Hall, K. J., 1996b, Purple sulfur bacteria control the growth of aerobic heterotrophic bacterioplankton in a meromictic salt lake, *Appl. Environ. Microbiol.* **62**:3251–3258.

Parkin, T. B., and Brock, T. D., 1980a, The effects of light quality of phototrophic bacteria in lakes. *Arch. Microbiol.* **125**:19–27.

Parkin, T. B., and Brock, T. D., 1980b, Photosynthetic bacterial production in lakes: The effects of light intensity, *Limnol. Oceanogr* **25**:711–718.

Parkin, T. B., and Brock, T. D., 1981a, Photosynthetic bacterial production and carbon mineralization in a meromictic lake, *Arch. Hydrobiol.* **91**:366–382.

Parkin, T. B., and Brock, T. D., 1981b, The role of phototrophic bacteria in the sulfur cycle of a meromictic lake, *Limnol. Oceanogr.* **26**:880–890.

Pfennig, N., 1978, General physiology and ecology of photosynthetic bacteria, in: *The Photosynthetic Bacteria* (R. K. Clayton, and W. R. Sistrom, eds.) Plenum Press, New York, pp. 3–18.

Pfennig, N., and Biebl, H., 1976, *Desulfuromonas acetoxidans* gen. nov. and sp. nov., a new anaerobic, sulfur-reducing, acetate-oxidizing bacterium, *Arch. Microbiol.* **110**:3–12.

Scudder, G. G. E., 1969, The fauna of saline lakes on the Fraser Plateau in British Columbia, *Verh. Internat. Verein. Limnol.* **17**:430–439.

Scudder, G. G. E., 1983, A review of factors governing the distribution of two closely related corixids in the saline lakes of British Columbia, *Hydrobiologia* **105**:143–154.

Shiah, F.-K., and Ducklow, H. W., 1994, Temperature regulation of heterotrophic bacterioplankton abundance, production, and specific growth rate in Chesapeake Bay, *Limnol. Oceanogr.* **39**:1243–1258.

Skyring, G. W., and Bauld, J., 1990, Microbial mats in coastal environments, in: *Advances in Microbial Ecology* Vol. 11, (K. C. Marshall, ed.), Plenum Press, New York, pp. 461–498.

Smith, R. L., and Oremland, R. S., 1987, Big Soda Lake (Nevada). 2. Pelagic sulfate reduction, *Limnol. Oceanogr.* **32**:794–803.

Sorokin, Yu. I., 1970, Interrelations between sulphur and carbon turnover in meromictic lakes, *Arch Hydrobiol.* **66**:391–446.

Steenbergen, C. L. M., 1982, Contribution of photosynthetic sulfur bacteria to primary production in Lake Vechten, *Hydrobiologia* **95**:59–64.

Takahashi, M., and Ichimura, S., 1968, Vertical distribution of organic matter production of photosynthetic sulfur bacteria in Japanese lakes, *Limnol. Oceanogr.* **13**:644–655.

Thienemann, A., 1925, Die Binnengewässer Mitteleuropas, in: *Die Binnengewässer,* Vol. I, Schweizerbart'sche Verlagsbuchhandlung, Stuttgart.

van Gemerden, H., 1967, On the bacterial sulfur cycle of inland waters, Ph. D. thesis, Rijksuniversiteit, Leiden.

van Gemerden, H., 1968, Growth measurement of *Chromatium* cultures. *Arch. Mikrobiol.* **64**:103–110.

van Gemerden, H., and Mas, J., 1995, Ecology of purple sulfur bacteria. In: *Anoxygenic Photosynthetic Bacteria* (R. E. Blankenship, M. T. Madigan, and C. E. Bauer, eds.), Kluwer Academic Publishers, Boston, pp. 49–85.

van Gemerden, H., Montesinos, E., Mas, J., and Guerrero, R., 1985, Diel cycle of metabolism of phototrophic purple sulfur bacteria in Lake Cisó (Spain), *Limnol. Oceanogr.* **30**:932–943.

van Gemerden, H., Tughan C. S., de Wit, R., and Herbert R. A., 1989, Laminated microbial ecosystems on sheltered beaches in Scapa Flow, Orkney Islands, *FEMS Microbiol. Ecol.* **62**:87–102.

Visscher, P. T., Nijburg, J. W., and van Gemerden, H., 1990, Polysulfide utilization by *Thiocapsa roseopersicina*, *Arch. Microbiol.* **155**:75–81.

Waiser, M. J., and Robarts, R. D., 1995, Microbial nutrient limitation in prairie saline lakes with high sulfate concentrations, *Limnol. Oceanogr.* **40**:566–574.

Wetzel, R. G., 1983, *Limnology,* 2nd edition, Saunders, New York.

Widdel, F., 1988, Microbiology and ecology of sulfate- and sulfur-reducing bacteria. In: *Biology of Anaerobic Microorganisms* (A. J. B. Zehnder, ed.), John Wiley & Sons, New York, pp. 469–585.

Zehr, J. P., Harvey, R. W., Oremland, R. S., Cloern, J. E., and George, L. H., 1987, Big Soda Lake (Nevada). 1. Pelagic bacterial heterotrophy and biomass, *Limnol. Oceanogr.* **32**:781–793.

Züllig, H., 1985, Pigmente phototropher Bakterien in Seesedimenten und ihre Bedeutung für die Seenforschung, *Schweiz. Z. Hydrol.* **47**:87–126.

Ecology and Biogeochemistry of *in Situ* Groundwater Bioremediation

ROBERT T. ANDERSON and DEREK R. LOVLEY

1. Introduction

The activity of microorganisms has a significant impact on the chemical composition of groundwaters (Chapelle, 1993). Microbial processes in both shallow (Madsen, 1995) and deep (Lovley and Chapelle, 1995) pristine aquifers have recently been reviewed in detail. The purpose of this chapter is to summarize recent research on the microbial ecology and biogeochemistry of contaminated aquifers.

There is an urgent need to remediate many polluted groundwater supplies, and *in situ* microbial degradation or immobilization of contaminants may be one of the most effective and least costly alternatives (Bouwer, 1992). In fact, as reviewed below, microorganisms often naturally remove large quantities of contaminants from polluted groundwater at no cost to the property owner, either by consuming the contaminants as electron donors or acceptors or by inadvertent metabolic side reactions. This naturally occurring restoration of groundwater has been termed "intrinsic bioremediation" (Water Science and Technology Board, 1993). Thus, if time is not of the essence, and the long-term expense of documenting and monitoring the intrinsic bioremediation is not unduly burdensome, intrinsic bioremediation can be an appropriate treatment option. When intrinsic bioremediation does not remove contaminants at a sufficient rate, engineered bioremediation, in which the desired microbial metabolism is artificially stimulated, may in some instances be more economically feasible than other non-biological engineering options. Many studies in the field of groundwater bio-

---

ROBERT T. ANDERSON • Department of Civil and Environmental Engineering, University of Massachusetts, Amherst, Massachusetts 01003.   **DEREK R. LOVLEY** • Department of Microbiology, University of Massachusetts, Amherst, Massachusetts 01003.
Advances in Microbial Ecology, Volume 15, edited by Jones. Plenum Press, New York, 1997.

remediation have been empirical in nature, with little consideration of the underlying principles that might be applied to contaminated sites other than the one under investigation. However, as reviewed here, a large number of studies have begun to define practices and principles that can be generally applied to further understanding both intrinsic and engineered *in situ* bioremediation of contaminated groundwater. This review only includes references that were available to us by February 1996.

2. Diversity and Distribution of Microorganisms in Pristine Aquifers

The many studies that have been conducted on the microbiology of pristine aquifers provide insight into the factors that may control the number and distribution of microorganisms in contaminated subsurface environments. Comparison of the microbiology of pristine aquifers with contaminated aquifers gives an indication of the changes in microbial community structure and activity in response to contamination. Detailed reviews of the microbiology of pristine aquifers have been previously provided (Chapelle, 1993; Ghiorse and Wilson, 1988; Hirsch, 1992; Kuznetsov *et al.*, 1963; Lovley and Chapelle, 1995; Madsen, 1995; Madsen and Ghiorse, 1993; McNabb and Dunlap, 1975). Therefore, only basic concepts that have been developed in studies of pristine aquifers, but also apply to contaminated aquifers are summarized below.

2.1. Oligotrophic Nature of the Subsurface

Most pristine aquifers that have been examined contain detectable populations of metabolically active microorganisms (Chapelle, 1993; Ghiorse and Wilson, 1988; Lovley and Chapelle, 1995; Madsen and Ghiorse, 1993; McNabb and Dunlap, 1975; Pedersen, 1993; White *et al.*, 1983). However, the *in situ* rates of metabolism may be slow because, prior to contamination, most aquifers are relatively oligotrophic environments (Ghiorse and Wilson, 1988; Lee *et al.*, 1988). There is little input of organic substrates to many aquifers because the most easily degraded carbon sources are utilized before they reach the deeper layers within the subsurface (Ghiorse and Wilson, 1988). In fact, the rates of organic matter oxidation in deep pristine aquifers are as low or lower than in any other environments known to support microbial life (Chapelle and Lovley, 1990). Subsurface microbial growth rates have been estimated to be as much as 1000-fold slower than in surface soils (Thorn and Ventullo, 1988).

The finding that plate counts on dilute organic media are generally higher than counts on full strength media is evidence that the microbial community is adapted to these oligotrophic conditions (Balkwill *et al.*, 1989; Balkwill and Ghiorse, 1985). In some cases plate counts with sediment and soil extract as culture media provide the largest estimates of viable subsurface microbes (Bone and Balkwill, 1988). Isolates from the subsurface may be metabolically less diverse than isolates from surface soils (Balkwill *et al.*, 1989). These observations have implications for *in situ* bioremediation as they imply that the microorganisms in some subsurface environments may not rapidly adapt to high inputs of organic contaminants.

Lipid analyses indicate that oligotrophic growth is a stressful existence as subsurface microorganisms have higher levels of beta-hydroxyalkanoic storage polymers and uronic acid-containing exopolymers relative to surface populations (Sinclair and Ghiorse, 1989; Smith *et al.*, 1986; White *et al.*, 1983). Both of these fatty acid signatures are indicators of unbalanced growth and, thus, a measure of nutritional stress (Smith *et al.*, 1986; White *et al.*, 1983).

2.2. Types of Microorganisms in the Subsurface

Despite the relatively poor conditions for growth in the subsurface, there is a wide diversity of microbial life in many aquifers, as has been previously reviewed in detail (Balkwill *et al.*, 1989; Chapelle, 1993; Ghiorse and Wilson, 1988; Hirsch and Rades-Rohkohl, 1983; Lovley and Chapelle, 1995; Madsen, 1995; Madsen and Ghiorse, 1993; Pedersen, 1993; Wilson *et al.*, 1983). These reviews should be consulted for lists of specific microorganisms that have been isolated from subsurface environments. In general, all physiological types of microorganisms other than photosynthetic organisms that are found in surface environments may also be active in subsurface environments (Lovley and Chapelle, 1995). Bacteria, fungi, and protozoa have all been recovered from a variety of shallow and deep aquifers. Several lines of evidence suggest that microorganisms found in the subsurface represent a distinct community, rather than just organisms carried into the subsurface with the recharge water (Balkwill *et al.*, 1989; Beloin *et al.*, 1988; Bone and Balkwill, 1988; Smith *et al.*, 1986).

By far, the vast majority of microbes living in the subsurface are bacteria (Beloin *et al.*, 1988; Sinclair and Ghiorse, 1989). Protozoans and fungi are often detected when assays for these organisms are conducted (Beloin *et al.*, 1988; Frederickson and Hicks, 1987; Sinclair and Ghiorse, 1987, 1989; Sinclair *et al.*, 1990). Phototrophic microbes have even been recovered from subsurface samples, but it was possible that there was significant communication between groundwater and surface water at the sampling sites (Beloin *et al.*, 1988; Sinclair and Ghiorse, 1989).

2.3. Numbers and Distribution of Microorganisms

Total microbial counts and numbers of culturable organisms are generally highest at the surface and decrease with depth through the unsaturated zone

(Beloin *et al.*, 1988; Bone and Balkwill, 1988; Colwell, 1989; Federle *et al.*, 1986, 1990; Sinclair and Ghiorse, 1987). However, once the water table is reached, this decrease in microbial numbers with depth generally stops. Most microbes within the subsurface are associated with the sediment (Arvin *et al.*, 1988; Harvey *et al.*, 1984; Hirsch and Rades-Rohkohl, 1983; Holm *et al.*, 1992), but free-living microbes are also observed in the ground water (Aamand *et al.*, 1989; Arvin *et al.*, 1988; Harvey and George, 1987; Harvey *et al.*, 1984; Holm *et al.*, 1992). The distribution of the subsurface microorganisms is often heterogeneous (Balkwill *et al.*, 1989; Kolbel-Boelke *et al.*, 1988a, 1988b), which is an important consideration when developing sampling strategies.

Direct counts of subsurface bacteria with fluorescent microscopy range from around 10^5-10^8 cells per gram of dry material with an overall average of 10^6 (Balkwill, 1989; Balkwill and Ghiorse, 1985; Beloin *et al.*, 1988; Ghiorse and Balkwill, 1983; Ghiorse and Wilson, 1988; Lee *et al.*, 1988; Sinclair and Ghiorse, 1989; Sinclair *et al.*, 1990; Wilson *et al.*, 1983). Generally, viable counts are one to three orders of magnitude lower than direct counts within the same sample, but viable numbers of cells approaching 100% of the total count have been observed (Balkwill, 1989; Balkwill and Ghiorse, 1985; Sinclair and Ghiorse, 1989; Sinclair *et al.*, 1990). Protozoa are usually restricted to transmissive strata where the highest number of viable bacteria are found; even here their viable numbers are typically low (10^1-10^3 per gram dry weight) (Beloin *et al.*, 1988; Sinclair and Ghiorse, 1987; Sinclair and Ghiorse, 1989; Sinclair *et al.*, 1990). Numbers of fungi within the subsurface are also quite variable. In sediments from an aquifer in South Carolina, low numbers of fungi [1–40 colony forming units (CFU) per gram dry material] were found to be evenly distributed throughout the sample profiles (Sinclair and Ghiorse, 1989) whereas in sediments from Kansas no fungi could be found (Sinclair *et al.*, 1990).

Direct counts of free-living bacteria in groundwater have been reported to range from 10^4-10^6 cells/milliliter (Buchanan-Mappin *et al.*, 1985; Harvey *et al.*, 1984; Marxsen, 1988). In contaminated aquifers, numbers of bacteria in the groundwater may be correlated with proximity to the source of contamination. Increased numbers of free-living bacteria in contaminated groundwater close to a contaminant source have been observed while the dominant population remains attached to the sediment (Federle *et al.*, 1990; Harvey *et al.*, 1984; Madsen and Ghiorse, 1993). These observations suggest that increased levels of metabolizable carbon or nutrients in groundwater may promote the growth of free-living bacteria or stimulate the activity of an attached population that may have an unattached life cycle component (Hirsch and Rades-Rohkohl, 1988). Experimental evidence for this has been obtained in laboratory column studies (Bengtsson, 1989).

A number of physico-chemical factors influence the proportion of free-living bacteria in the subsurface. For example, iron coatings on sediments confer a greater positive charge on the mineral surfaces, which facilitates adsorption of negatively-charged bacteria (Mills et al., 1994; Scholl and Harvey, 1992), thus increasing the proportion of attached bacteria. Increased ionic strength leads to increased adsorption between like-charged bacteria and mineral surfaces due to compression of the electrical double layer (EDL), which also favors attachment (Fontes et al., 1991; Gannon et al., 1991; Mills et al., 1994). The EDL between ionizable functional groups on bacterial and mineral surfaces can also be altered by the pH of the surrounding medium (Scholl and Harvey, 1992). In the unsaturated zone and capillary fringe, mobility of bacteria may be retarded due to sorption at the gas–water interface (Wan et al., 1994).

Within the saturated zone, sediment composition is a major determinant of microbial numbers. Gram-negative bacteria are more frequently found in the sandy aquifer layers and in the groundwater whereas gram-positive bacteria are more frequently attached to the sediment and in the layers of higher clay content (Balkwill, 1989; Kolbel-Boelke et al., 1988a, 1988b; Sinclair and Ghiorse, 1989). Overall, sandy aquifer layers contain more bacteria than clayey layers (Beloin et al., 1988; Chapelle and Lovley, 1990; Frederickson et al., 1989; Hicks and Frederickson, 1989; Phelps et al., 1989; Sinclair and Ghiorse, 1987, 1989; Sinclair et al., 1990) suggesting that conditions of lower pH (Sinclair and Ghiorse, 1989), hydraulic conductivity (Frederickson et al., 1989) or water potential (Frederickson et al., 1989) associated with clay-rich areas may be negatively affecting growth. In addition, pore throat size may restrict transport of microbes into clay-rich sediments, thus preventing colonization of the clay layers and resulting in smaller populations relative to more sandy layers (Chapelle and Lovley, 1990; Harvey et al., 1995; Sharma and McInerney, 1994). If sufficient metabolizable carbon exists within the clay layers, starved cells of smaller size that are able to penetrate into the clay layers may reduce the porosity of the clay layer further when they begin to grow within the pore space, thus restricting further transport in the clay (MacLeod et al., 1988). Whatever the reason for the differences in microbial numbers between sandy and clayey layers, it is clear that the best predictor of microbial number in the subsurface below the water table is clay content (Frederickson et al., 1989).

Seasonal variations in total microbial numbers in aquifers are generally not observed (Beloin et al., 1988; Hirsch and Rades-Rohkohl, 1988; Wilson et al., 1983). However, some seasonal variation in microbial activity has been noted (Beloin et al., 1988). Viable counts and ATP measurements from the Lula aquifer, Oklahoma, increased during the summer months relative to the winter months, indicating an increase in microbial activity at the site. However, seasonal activity variation at this site does not seem to be large and the majority of microbial subsurface parameters examined did not show a seasonal trend (Beloin et al., 1988).

In summary, studies in pristine aquifers have demonstrated that despite their generally oligotrophic nature, there are diverse metabolically active populations of microorganisms living in many aquifers. Although their rates of metabolism may be slow these microorganisms do have an influence on the chemical composition (i.e., quality) of the groundwater (Lovley and Chapelle, 1995).

3. Microbial Ecology and Biogeochemistry of Contaminated Aquifers

A major distinction between pristine aquifers and many contaminated aquifers is that anaerobic microbial processes are often much more important in contaminated aquifers than in pristine aquifers. Although anaerobic processes sometimes dominate deep pristine aquifers (Lovley and Chapelle, 1995), most shallow pristine aquifers are aerobic. However, when shallow aquifers are contaminated with organic compounds, extensive anaerobic zones frequently develop (Christensen et al., 1994; Lovley, 1997) (Fig. 1). This is because aerobic microorganisms can readily degrade many of the constituents of common aquifer pollutants such as petroleum products and landfill leachate (Table I), and thus rapidly remove the dissolved oxygen from the groundwater. Because of oxygen's low solubility, recharge water entering aquifers can only carry small amounts

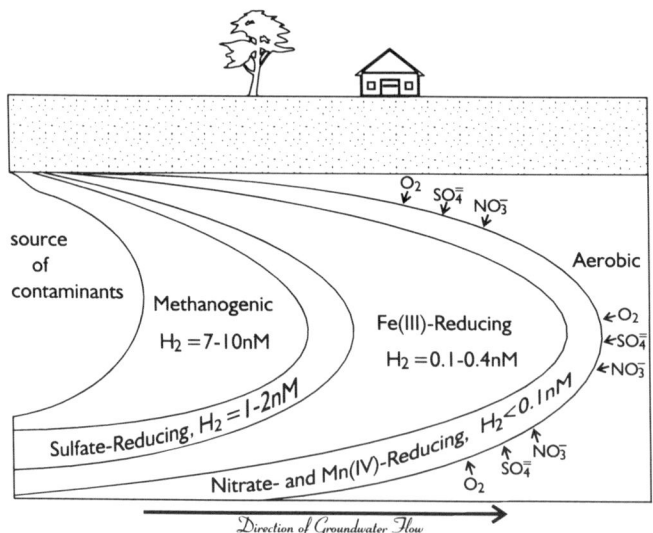

Figure 1. Idealized distribution of terminal electron accepting processes (TEAPs) and associated concentrations of dissolved H_2 in aquifers contaminated with organics such as petroleum or landfill leachate.

Table I. Isolated Microbial Species and Examples of Degradable Contaminants Found in Contaminated Areas

Compound	Species	Site	Reference
Aniline	Pseudomonas sp. AK20	Agricultural subsurface	(Konopka, 1993)
Benzene		Creosite site, ground water	(Thomas et al., 1987)
	Nocarida spp.	Gasoline contaminated well	(Ridgway et al., 1990)
	Pseudomonas putida	Gasoline contaminated well	(Ridgway et al., 1990)
	Alcaligenes denitrificans	Gasoline contaminated well	(Ridgway et al., 1990)
	Pseudomonas stutzeri	Gasoline contaminated well	(Ridgway et al., 1990)
	Pseudomonas maltophilia	Gasoline contaminated well	(Ridgway et al., 1990)
	Pseudomonas alcaligenes	Gasoline contaminated well	(Ridgway et al., 1990)
	Micrococcus	Gasoline contaminated well	(Ridgway et al., 1990)
	Pseudomonas aeruginosa	Gasoline contaminated well	(Ridgway et al., 1990)
	Arthrobacter	Natural gas facility	(Weber and Corseuil, 1994)
		Landfill leachate plume	(Holm et al., 1991)
Biphenyl		Oil spill, ground water	(Arvin et al., 1988)
		Fuel oil spill, ground water	(Aamand et al., 1989)
		Landfill leachate plume	(Holm et al., 1991)
Carbazole	Xanthobacter sp.	Coal gasification plant	(Grosser et al., 1991)
4-Chlorobenzoate	Acinetobacter 4CB1	PCB contaminated soil	(Adriaens et al., 1989)
Creosote	Pseudomonas cepacia	Creosote contaminated soil	(Ellis et al., 1991)
	Pseudomonas putida	Creosote contaminated soil	(Ellis et al., 1991)
	Pseudomonas flourescens	Creosote contaminated soil	(Ellis et al., 1991)
o-Cresol		Oil spill, ground water	(Arvin et al., 1988)
p-Cresol	Bacillus pumilus	Phenol contaminated aquifer	(Gunther et al., 1995)
Cyclohexane	Pseudomonas aeruginosa	Gasoline contaminated well	(Ridgway et al., 1990)
	Alcaligenes denitrificans	Gasoline contaminated well	(Ridgway et al., 1990)
	Pseudomonas maltophilia	Gasoline contaminated well	(Ridgway et al., 1990)

(continued)

Table I. (*Continued*)

Compound	Species	Site	Reference
	Pseudomonas putida	Gasoline contaminated well	(Ridgway et al., 1990)
	Micrococcus	Gasoline contaminated well	(Ridgway et al., 1990)
	Pseudomonas alcaligenes	Gasoline contaminated well	(Ridgway et al., 1990)
	Pseudomonas stutzeri	Gasoline contaminated well	(Ridgway et al., 1990)
Cycloheptane	*Pseudomonas putida*	Gasoline contaminated well	(Ridgway et al., 1990)
	Pseudomonas stutzeri	Gasoline contaminated well	(Ridgway et al., 1990)
	Pseudomonas maltophilia	Gasoline contaminated well	(Ridgway et al., 1990)
	Micrococcus	Gasoline contaminated well	(Ridgway et al., 1990)
Decane	*Nocardia spp.*	Gasoline contaminated well	(Ridgway et al., 1990)
	Alcaligenes denitrificans	Gasoline contaminated well	(Ridgway et al., 1990)
	Pseudomonas stutzeri	Gasoline contaminated well	(Ridgway et al., 1990)
	Pseudomonas maltophilia	Gasoline contaminated well	(Ridgway et al., 1990)
	Pseudomonas putida	Gasoline contaminated well	(Ridgway et al., 1990)
	Micrococcus	Gasoline contaminated well	(Ridgway et al., 1990)
	Pseudomonas alcaligenes	Gasoline contaminated well	(Ridgway et al., 1990)
Dibenzothiophene		Oil spill, ground water	(Arvin et al., 1988)
	F199 (CMN group)	Deep subsurface (410m)	(Frederickson et al., 1991)
Dibenzofurane		Oil spill, ground water	(Arvin et al., 1988)
Diesel fuel	*Pseudomonas vesicularis*	Diesel fuel spill	(Rainwater et al., 1993)
	Pseudomonas aeruginosa	Diesel fuel spill	(Rainwater et al., 1993)
1,4 Dimethylnaphthalene		Oil spill, ground water	(Arvin et al., 1988)
		Fuel oil spill, ground water	(Aamand et al., 1989)
Ethylbenzene	*Pseudomonas aeruginosa*	Gasoline contaminated well	(Ridgway et al., 1990)
	Pseudomonas putida	Gasoline contaminated well	(Ridgway et al., 1990)

Table I. (*Continued*)

Compound	Species	Site	Reference
	Alcaligenes denitrificans	Gasoline contaminated well	(Ridgway et al., 1990)
	Pseudomonas maltophilia	Gasoline contaminated well	(Ridgway et al., 1990)
	Micrococcus	Gasoline contaminated well	(Ridgway et al., 1990)
	Pseudomonas alcaligenes	Gasoline contaminated well	(Ridgway et al., 1990)
	Pseudomonas stutzeri	Gasoline contaminated well	(Ridgway et al., 1990)
2-Ethylnaphthalene		Fuel oil spill, ground water	(Aamand et al., 1989)
Fluorenone		Oil spill, ground water	(Arvin et al., 1988)
Hexahydro-1,3,5-Trinitro-1,3,5-Triazine	*Morganella morganii*	Explosive contaminated soil	(Kitts et al., 1994)
	Providenicia rettgeri	Explosive contaminated soil	(Kitts et al., 1994)
	Citrobacter freundii	Explosive contaminated soil	(Kitts et al., 1994)
Hexane	*Pseudomonas aeruginosa*	Gasoline contaminated well	(Ridgway et al., 1990)
	Nocardia spp.	Gasoline contaminated well	(Ridgway et al., 1990)
	Pseudomonas maltophilia	Gasoline contaminated well	(Ridgway et al., 1990)
	Pseudomonas putida	Gasoline contaminated well	(Ridgway et al., 1990)
	Micrococcus	Gasoline contaminated well	(Ridgway et al., 1990)
	Pseudomonas stutzeri	Gasoline contaminated well	(Ridgway et al., 1990)
Indole		Oil spill, ground water	(Arvin et al., 1988)
Methylaniline	*Pseudomonas sp.* AK20	Agricultural subsurface	(Konopka, 1993)
Methylbutane	*Nocardia spp.*	Gasoline contaminated well	(Ridgway et al., 1990)
	Pseudomonas stutzeri	Gasoline contaminated well	(Ridgway et al., 1990)
Methylbutane	*Pseudomonas putida*	Gasoline contaminated well	(Ridgway et al., 1990)
	Pseudomonas maltophilia	Gasoline contaminated well	(Ridgway et al., 1990)
	Micrococcus	Gasoline contaminated well	(Ridgway et al., 1990)
N-methyl-carbamates	ER2	Agricultural soil	(Topp et al., 1993)
Methylcyclohexane	*Rhodococcus* spp.	Oil refinery soil	(Lloyd-Jones and Trudgill, 1989)

(*continued*)

Table I. (*Continued*)

Compound	Species	Site	Reference
	Flavobacterium spp.	Oil refinery soil	(Lloyd-Jones and Trudgill, 1989)
	Pseudomonas spp.	Oil refinery soil	(Lloyd-Jones and Trudgill, 1989)
Methylcyclopentane	*Pseudomonas aeruginosa*	Gasoline contaminated well	(Ridgway et al., 1990)
	Nocardia spp.	Gasoline contaminated well	(Ridgway et al., 1990)
	Pseudomonas maltophilia	Gasoline contaminated well	(Ridgway et al., 1990)
	Pseudomonas stutzeri	Gasoline contaminated well	(Ridgway et al., 1990)
	Pseudomonas putida	Gasoline contaminated well	(Ridgway et al., 1990)
	Micrococcus	Gasoline contaminated well	(Ridgway et al., 1990)
1-Methylnaphthalene		Fuel oil spill, ground water	(Aamand et al., 1989)
Naphthaline		Creosote site, ground water	(Thomas et al., 1987)
		Creosote site contaminated soil	(Thomas et al., 1989)
	Pseudomonas putida	Hydrocarbon polluted soil	(Herbes and Schwall, 1978)
	Pseudomonas aeruginosa	Gasoline contaminated well	(Ridgway et al., 1990)
	Pseudomonas maltophilia	Gasoline contaminated well	(Ridgway et al., 1990)
	Pseudomonas putida	Gasoline contaminated well	(Ridgway et al., 1990)
	Micrococcus	Gasoline contaminated well	(Ridgway et al., 1990)
	Pseudomonas stutzeri	Gasoline contaminated well	(Ridgway et al., 1990)
		Landfill leachate plume	(Holm et al., 1991)
		Oil spill, ground water	(Arvin et al., 1988)
Naphthalene	F199 (CMN group)	Deep subsurface (410m)	(Frederickson et al., 1991)
p-Nitrophenol	*Pseudomonas putida*	Contaminated Pond	(Spain et al., 1984)
	Arthrobacter sp.	Contaminated Pond	(Spain et al., 1984)
Octane	*Pseudomonas aeruginosa*	Gasoline contaminated well	(Ridgway et al., 1990)
	Nocardia spp.	Gasoline contaminated well	(Ridgway et al., 1990)
	Pseudomonas stutzeri	Gasoline contaminated well	(Ridgway et al., 1990)

Table I. (*Continued*)

Compound	Species	Site	Reference
	Pseudomonas putida	Gasoline contaminated well	(Ridgway et al., 1990)
	Micrococcus	Gasoline contaminated well	(Ridgway et al., 1990)
Oil	*Pseudomonas* spp.	Oil refinery soil	(Ellis et al., 1990)
	Rhodococcus spp.	Oil refinery soil	(Ellis et al., 1990)
	Acinetobacter spp.	Oil refinery soil	(Ellis et al., 1990)
	Mycobacterium spp.	Oil refinery soil	(Ellis et al., 1990)
	Anthrobacter spp.	Oil refinery soil	(Ellis et al., 1990)
Pentachlorophenol	*Pseudomonas*	PCP contaminated soil	(Radehaus and Schmidt, 1992)
	Arthrobacter	PCP contaminated soil	(Stanlake and Finn, 1982)
Phenanthrene	*Flavobacterium*	Hydrocarbon polluted soil	(Herbes and Schwall, 1978)
Phenol	*Bacillus pumilus*	Phenol-contaminated aquifer	(Gunther et al., 1995)
Pyrene	*Mycobacterium sp.*	Coal gasification plant	(Grosser et al., 1991)
Pyridines		Chemical plant grounds	(Kaiser and Bollag, 1992)
Quinoline		Oil spill, ground water	(Arvin et al., 1988)
	Pseudomonas (presumed)	D.O.E. site S.C.	(Brockman et al., 1989)
	Pseudomonas putida	Creosote contaminated soil	(Bennett et al., 1985)
	Pseudomonas flourescens	Creosote contaminated soil	(Bennett et al., 1985)
	Pseudomonas chloraphis	Creosote contaminated soil	(Bennett et al., 1985)
	Pseudomonas pseudoalcaligenes	Creosote contaminated soil	(Bennett et al., 1985)
Quinoline	*Moraxella*	Garden soil	(Grant and Al-Najjar, 1976)
Toluene		Oil spill, ground water	(Arvin et al., 1988)
		Landfill leachate plume	(Holm et al., 1991)
		Fuel oil spill, ground water	(Aamand et al., 1989)
	Pseudomonas aeruginosa	Gasoline contaminated well	(Ridgway et al., 1990)
	Pseudomonas putida	Gasoline contaminated well	(Ridgway et al., 1990)
	Pseudomonas stutzeri	Gasoline contaminated well	(Ridgway et al., 1990)
	Alcaligenes denitrificans	Gasoline contaminated well	(Ridgway et al., 1990)
	Pseudomonas maltophilia	Gasoline contaminated well	(Ridgway et al., 1990)

(*continued*)

Table I. (*Continued*)

Compound	Species	Site	Reference
	Micrococcus	Gasoline contaminated well	(Ridgway *et al.*, 1990)
	F199 (CMN group)	Deep Subsurface (410m)	(Fredrickson *et al.*, 1990)
Trichloroethylene	*Pseudomonas cepacia* G4	TCE contaminated soil	(Nelson *et al.*, 1986)
	Strain 46-1	Solvent waste dump	(Little *et al.*, 1988)
	Methylomonas methanica	68-1TCE contaminated aquifer	(Koh *et al.*, 1993)
1,3,5 Trimethylbenzene		Oil spill, ground water	(Arvin *et al.*, 1988)
	Pseudomonas aeruginosa	Gasoline contaminated well	(Ridgway *et al.*, 1990)
	Nocardia spp.	Gasoline contaminated well	(Ridgway *et al.*, 1990)
	Pseudomonas stutzeri	Gasoline contaminated well	(Ridgway *et al.*, 1990)
	Pseudomonas maltophilia	Gasoline contaminated well	(Ridgway *et al.*, 1990)
	Pseudomonas putida	Gasoline contaminated well	(Ridgway *et al.*, 1990)
	Micrococcus	Gasoline contaminated well	(Ridgway *et al.*, 1990)
	Pseudomonas alcaligenes	Gasoline contaminated well	(Ridgway *et al.*, 1990)
		Fuel oil spill, ground water	(Aamand *et al.*, 1989)
Trimethylpentane	*Pseudomonas putida*	Gasoline contaminated well	(Ridgway *et al.*, 1990)
	Pseudomonas stutzeri	Gasoline contaminated well	(Ridgway *et al.*, 1990)
	Micrococcus	Gasoline contaminated well	(Ridgway *et al.*, 1990)
o-Xylene		Oil spill, ground water	(Arvin *et al.*, 1988)
	Pseudomonas aeruginosa	Gasoline contaminated well	(Ridgway *et al.*, 1990)
	Nocardia spp.	Gasoline contaminated well	(Ridgway *et al.*, 1990)
	Pseudomonas maltophilia	Gasoline contaminated well	(Ridgway *et al.*, 1990)
	Pseudomonas putida	Gasoline contaminated well	(Ridgway *et al.*, 1990)
	Pseudomonas stutzeri	Gasoline contaminated well	(Ridgway *et al.*, 1990)
	Micrococcus	Gasoline contaminated well	(Ridgway *et al.*, 1990)
	F199 (CMN group)	Deep subsurface (410m)	(Frederickson *et al.*, 1991)

Table I. (*Continued*)

Compound	Species	Site	Reference
p-Xylene	*Pseudomonas aeruginosa*	Gasoline contaminated well	(Ridgway *et al.*, 1990)
	Pseudomonas stutzeri	Gasoline contaminated well	(Ridgway *et al.*, 1990)
	Pseudomonas maltophilia	Gasoline contaminated well	(Ridgway *et al.*, 1990)
	Pseudomonas putida	Gasoline contaminated well	(Ridgway *et al.*, 1990)
	Micrococcus	Gasoline contaminated well	(Ridgway *et al.*, 1990)
p-Xylene	*Xanthomonas*	Natural gas facility	(Weber and Corseuil, 1994)
		Landfill leachate plume	(Holm *et al.*, 1991)
		Fuel oil spill, ground water	(Aamand *et al.*, 1989)
	F199 (CMN group)	Deep subsurface (410m)	(Frederickson *et al.*, 1991)

of oxygen and introduction of oxygen as the result of diffusion from the surface is slow. These small amounts of oxygen are rapidly consumed by further aerobic respiration and chemical oxidation of reduced products of anaerobic metabolism such as Fe(II), sulfide, and methane. Thus, aerobic biodegradation is most likely to be restricted to the fringes of many contaminant plumes. As summarized below, it is becoming increasingly apparent that anaerobic microorganisms also have a significant capacity for organic contaminant degradation. Given the importance of anaerobic metabolism in contaminated aquifers, the factors controlling the distribution of aerobic and the various potential anaerobic processes in contaminated aquifers will be reviewed first, followed by a discussion of *in situ* bioremediation of specific contaminants.

3.1. Terminal Electron-Accepting Processes in Contaminated Aquifers

The physiological characteristics of anaerobic microorganisms in aquifers have not been studied in great detail, but it is generally considered that the metabolic potential of aquifer microorganisms is similar to that of more well-studied microbial communities in surface waters, aquatic sediments, and soils (Lovley and Chapelle, 1995; Madsen, 1995). In aquatic sediments and submerged soils, organic matter is metabolized by a succession of microbial processes (Ponnamperuma, 1972; Reeburgh, 1983). When oxygen is available in the water overlying the sediment, organic matter in the most surficial layer is metabolized by aerobic microorganisms. With the depletion of oxygen, organic matter is then metabolized by a succession of anaerobic terminal-electron-accepting processes (TEAPs). Nitrate reduction is generally the first TEAP found with depth in aquatic sediments. Nitrate reduction is followed by zones of Mn(IV) reduction, Fe(III) reduction, sulfate reduction, and methanogenesis as each successive electron acceptor is depleted with depth (Froelich et al., 1979; Reeburgh, 1983). A similar succession of microbial processes is observed along the groundwater flow path of many deep pristine aquifers (Champ et al., 1979; Jackson and Patterson, 1982; Lovley and Chapelle, 1995; Lovley and Goodwin, 1988). This separation of processes is not absolute. Environmental heterogeneities may generate microzones such that several TEAPs take place within a small area, and when conditions are fluctuating under non-steady situations, several TEAPs may overlap (Lovley and Goodwin, 1988). However, in sediments that are at or approaching steady-state conditions, one TEAP generally dominates the geochemistry within each zone (Froelich et al., 1979; Reeburgh, 1983). The exception to this is the frequent finding that nitrate reduction and manganese reduction take place within the same sediment interval. Geochemical evidence indicates that there is also a segregation of processes within the anaerobic plumes of contaminated aquifers (Fig. 1) (Baedecker and Back, 1979; Baedecker et al., 1993; Bjerg et al., 1995; Champ et al., 1979; Chapelle et al., 1996; Chapelle and Lovley, 1992; Fish, 1993; Jackson and Patterson, 1982; Lovley et al., 1994a; Lovley and Goodwin, 1988; Lyngkilde and Christensen, 1992b; Matthess, 1992; Schwille, 1976; Williams et al., 1991).

The TEAP distribution in anaerobic aquifers and aquatic sediments follows the thermodynamic yield of the various processes, and thus TEAP distribution and/or succession is often explained as a function of thermodynamics, with microorganisms preferentially utilizing the electron acceptors that provide the maximum free energy for organic matter oxidation (Bouwer, 1992; Stumm and Morgan, 1981). However, on a purely thermodynamic basis, reactions yielding less energy should also take place as long as they are energetically favorable (McCarty, 1972). Thus, the thermodynamic rationale is inadequate as it does not explain why microorganisms that can only perform one TEAP (for example, methanogens) would not be active in other TEAP zones (in this example, the Fe(III)- or sulfate-reducing zone).

The segregation of TEAPs can be more accurately explained based on physiological controls on microbial metabolism and the competition between different types of microorganisms for electron donors. Although this competition has not been intensively studied in the anoxic zone of contaminated aquifers, in aquatic sediments a key factor in the segregation of Fe(III) reduction, sulfate reduction, and methane production into distinct zones is the competition between Fe(III) reducers, sulfate reducers, and methane producers for acetate and H_2, the central intermediates in much of the organic matter metabolism in anaerobic sedimentary environments (Caccavo et al., 1992; Chapelle and Lovley, 1992;

Coates et al., 1996b; Cord-Ruwisch et al., 1988; Lovley, 1985; Lovley and Chapelle, 1995a; Lovley et al., 1982, 1989b; Lovley and Goodwin, 1988; Lovley and Phillips, 1987). Fe(III) reducers have a higher affinity for H_2 and acetate than sulfate reducers or methanogens, and as long as a sufficient supply of Fe(III) is available, Fe(III) reducers are able to maintain the concentration of H_2 and acetate at a level too low for sulfate reducers or methanogens to utilize. Some sulfate reducers may also preferentially reduce Fe(III) when it is available (Coleman et al., 1993; Lovley et al., 1993a). Methane production is inhibited in the presence of sulfate concentrations greater than about 20 µM (Lovley and Klug, 1986) in freshwater environments because sulfate reducers have a higher affinity for H_2 and acetate than do methanogens. H_2 and acetate concentrations follow the pattern expected from such competition, with lowest concentrations in the Fe(III)-reducing zone, intermediate concentrations in the sulfate-reducing zone, and highest concentrations in the methanogenic zone.

There is a similar pattern of H_2 in the Fe(III)-reducing, sulfate-reducing and methanogenic zones of petroleum-contaminated aquifers (Lovley et al., 1994a; Vroblesky and Chapelle, 1994), which suggests that competition between these TEAPs may account for their distribution in contaminated aquifers. However, it should be kept in mind that, as discussed below, the aromatic hydrocarbons, which are the prime electron donors introduced into petroleum-contaminated groundwater, may be directly oxidized to carbon dioxide by Fe(III)- and sulfate-reducing microorganisms (Lovley et al., 1989a, 1995; Lovley and Lonergan, 1990; Rabus et al., 1993). Thus, in contrast to aquatic sediments where fermentative microorganisms are responsible for the initial metabolism of organic matter (primarily carbohydrates and proteins) to electron donors that Fe(III)- and sulfate-reducers can metabolize, competition in contaminated aquifers may be in the more initial stages of metabolism.

Inhibition of Fe(III) reduction, sulfate reduction, and methane production in the nitrate- and manganese-reducing zones of aquatic sediments may be explained, at least in part, by the finding that H_2 (Lovley and Goodwin, 1988) and, presumably, acetate concentrations are even lower in nitrate- and Mn(IV)-reducing sedimentary environments than they are in Fe(III)-reducing environments. Some Fe(III) reducers (Lovley, 1995) and sulfate reducers (Widdel, 1988) are capable of nitrate reduction and may preferentially reduce nitrate when it is available. In a similar manner, some sulfate reducers may reduce Mn(IV) (Lovley and Phillips, 1994). Net Fe(III) reduction is also prevented in the presence of Mn(IV) because Mn(IV) chemically reoxidizes Fe(II) produced during Fe(III) reduction back to Fe(III) (Lovley, 1991). These considerations presumably also apply to the anaerobic zones of contaminated aquifers.

Upon initial inspection, the distribution of TEAPs in anaerobic contaminated aquifers (Fig. 1), in which the TEAPs proceed from methanogenesis upgradient to oxygen respiration downgradient, may appear to be in the reverse order

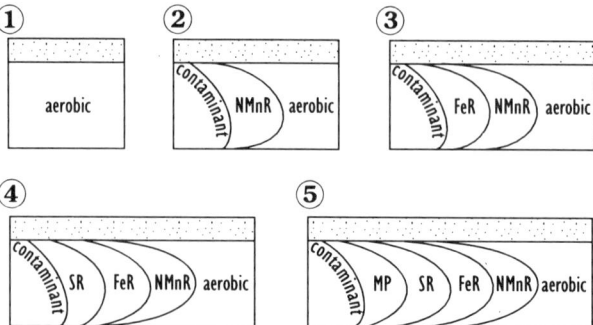

Figure 2. Hypothesized stages (2–5) in the succession of terminal electron accepting processes (TEAPs) after an aerobic pristine aquifer (1) is contaminated with organics such as petroleum or landfill leachate.

of the TEAP distribution in aquatic sediments where organic matter degradation starts with oxygen respiration at the top and proceeds through to methanogenesis with depth. However, this is merely due to the tendency to view diagrams from left to right and top to bottom. The distribution of TEAPs in anaerobic contaminated aquifers is in fact what would be expected from competition between the microorganisms catalyzing the different TEAPs. This can be visualized by considering the probable evolution of TEAPs in an aquifer following the introduction of organic contaminants (Fig. 2).

In this model, prior to contamination aerobic conditions predominate (Fig. 2). With the introduction of the contamination, oxygen is rapidly consumed in the groundwater closest to the contamination and anoxic conditions develop. Geochemical data indicates that there is then a succession of processes within this anaerobic zone (Bjerg et al., 1995; Christensen et al., 1994; Lovley, 1997; Lovley et al., 1989a; Lyngkilde and Christensen, 1992b). With the depletion of oxygen it is expected that there will be a brief period in which nitrate and Mn(IV) reduction will be the dominant TEAP, followed by Fe(III) reduction once the nitrate and Mn(IV) are depleted, and then sulfate reduction once the microbially reducible Fe(III) is depleted (Fig. 2). In aquifers that have been contaminated for a significant period of time, all of the alternative electron acceptors have been depleted from the sediments closest to the source of contamination and methanogenesis is the TEAP. Oxygen, nitrate, and sulfate entering with uncontaminated water from above or below the plume is consumed at the plume fringes, either by respiration or through abiotic oxidation of reduced endproducts. The TEAPs other than methanogenesis become reestablished further downgradient, where the appropriate electron acceptors are still available (Fig. 2).

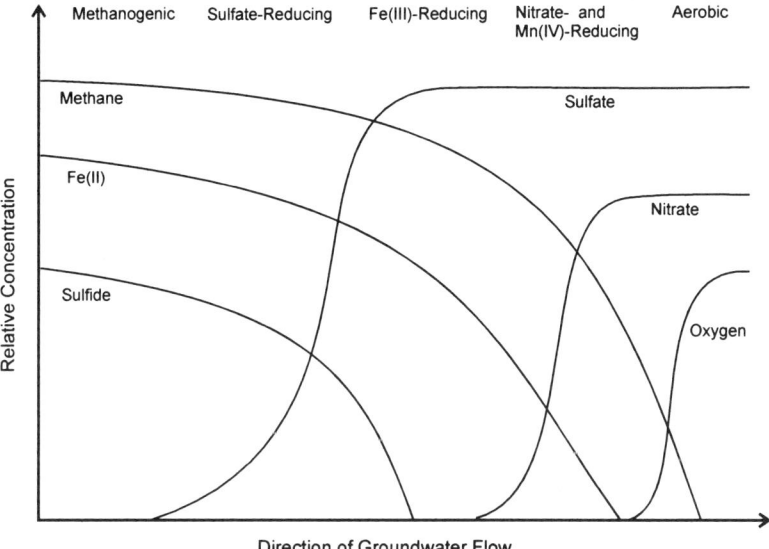

Figure 3. Idealized relative distribution of electron acceptors and reduced products of anaerobic respiration in aquifers contaminated with organics such as petroleum or landfill leachate.

Measurements of the depletion of electron acceptors and the accumulation of reduced endproducts can often be used to delineate the distribution of TEAPs in aquatic sediments and pristine aquifers. However, such geochemical signatures often do not accurately pinpoint the location of TEAPs in contaminated aquifers (Christensen *et al.*, 1994; Lovley *et al.*, 1994a). The reasons for this are evident from Figure 3. Reduced endproducts often persist in groundwater as it moves downgradient, blurring the geochemical signature. For example, methane produced upgradient is relatively stable under anoxic conditions and thus there may be high methane concentrations in zones of the aquifer where there is no methane production. Much of the Fe(II) produced during Fe(III) reduction goes into solid phases and then will continue to equilibrate with groundwater, giving high Fe(II) concentrations in zones where Fe(III) reduction is no longer important. The profiles of electron acceptor depletion may be complicated by the fact that oxygen, nitrate, and sulfate may be fluxing into the plume from top, bottom, and ends of the plume.

Measurements or calculations of redox potential are also not effective in delineating the TEAP distribution because there is no consistent relationship between redox potentials and the predominating TEAPs in sedimentary environments (Lovley and Goodwin, 1988). The TEAP can be determined in laboratory incubations of cored sediment material, typically by monitoring the fate of [2-^{14}C]-acetate and other geochemical considerations, such as the presence of reduced species (Lovley, 1997; Lovley *et al.*, 1994a). However, this is labor intensive and often expensive.

An alternative method for determining TEAPs in contaminated aquifers is based on the finding that, as discussed above, sedimentary environments with different TEAPs have different concentrations of dissolved H_2 (Fig. 1) (Lovley *et al.*, 1994a; Lovley and Goodwin, 1988). Thus, by measuring the concentration of dissolved H_2 in groundwater it is possible to determine the TEAP at the point of sampling. H_2 concentrations enable one to accurately predict the TEAPs in different zones of a variety of aquifers as verified by direct measurements of processes in collected sediments or other considerations (Chapelle *et al.*, 1995; Lovley *et al.*, 1994a). This was even true for petroleum-contaminated aquifers in which it was impossible to determine the TEAP by measuring concentrations of microbial end products or redox potential (Lovley *et al.*, 1994a).

Although measurements of dissolved H_2 are not yet commonly made in most studies of anaerobic groundwaters, one study in which H_2 was routinely monitored over an annual cycle demonstrated that the distribution of TEAPs in a shallow, petroleum-contaminated aquifer was not static and changed on a seasonal basis (Vroblesky and Chapelle, 1994). Increased rainfall during winter months introduced oxygen into the anaerobic groundwater and oxidized Fe(II) to Fe(III). The increased recharge of the aquifer with rainwater resulted in a rise in the water table so that previously aerobic, Fe(III)-containing soil was water saturated. This further increased the availability of Fe(III) for oxidation of contaminants in the aquifer. The combination of these two factors was sufficient to switch the TEAP to Fe(III) reduction at a site where Fe(III) reduction had previously become limited by Fe(III) availability. The rainwater also carried high concentrations of sulfate into the aquifer, which stimulated sulfate reduction in previously sulfate-depleted methanogenic zones. Determination of the TEAP with dissolved H_2 or other techniques is important because (as outlined in the following sections) which TEAP is involved in the degradation can affect the rate, extent, and endproducts of degradation. Not only is there often a major distinction between aerobic and anaerobic pathways for degradation, but different anaerobic processes are more effective than others in removing some contaminants.

3.2. *In Situ* Bioremediation of Specific Contaminants

Groundwaters have been contaminated with an incredible number of organic and inorganic pollutants, and the number of studies that have considered the potential for *in situ* bioremediation of aquifers is staggering. Therefore, it is well beyond the scope of this review to consider *in situ* bioremediation of all known groundwater contaminants. Discussion of organic contaminants will be restricted to aromatic hydrocarbons and chlorinated hydrocarbons. Not only are these some of the most prevalent organic groundwater contaminants, but studies on the *in situ* bioremediation of these contaminants illustrate general principles of bioremediation that can be applied to understanding degradation of other organic contaminants in groundwater. The potential for *in situ* bioremediation of inorganic contaminants will also be highlighted as many previous reviews on bioremediation have ignored this topic. Although an attempt has been made to focus solely on groundwater studies, in some instances pertinent findings on contaminant degradation in other environments that may have an impact on understanding *in situ* bioremediation in groundwater will be discussed.

3.2.1. Aerobic Degradation of Aromatic Hydrocarbons

Both monoaromatic hydrocarbons and polycyclic aromatic hydrocarbons are important groundwater contaminants. Aerobic and anaerobic degradation of these contaminants will be considered separately because of the great differences in the microorganisms involved and in the modes of degradation.

Monoaromatic hydrocarbon contamination typically results from petroleum pollution (terrestrial oil spills, leaking underground tanks or pipelines) and landfill leachate. The most common monoaromatic hydrocarbon contaminants are benzene, toluene, ethylbenzene, and xylenes, commonly abbreviated BTEX. Of the BTEX components, benzene is of greatest concern because of its high toxicity and because it is relatively soluble and thus mobile in aquifers. All of the BTEX components are degraded within the aerobic zone of contaminated aquifers (Barker *et al.*, 1987; Morgan *et al.*, 1993; Salanitro, 1993). In fact, the rate of aerobic BTEX degradation is typically fast enough that models designed to predict the transport of BTEX in groundwater generally model the reaction between BTEX and oxygen as an instantaneous reaction (Salanitro, 1993). However, the rates of degradation may slow markedly as the concentration of dissolved oxygen decreases to 1 mg/l or less (McAllister and Chiang, 1994; Salanitro, 1993).

The realization that BTEX can be rapidly degraded in aerobic groundwaters has been a major impetus to rely on intrinsic bioremediation in cases of BTEX contamination. Intrinsic bioremediation of BTEX contamination is typically evidenced by BTEX plumes much smaller than would be predicted from conservative transport of the BTEX with groundwater, or in some instances actual shrinkage of the plume over time (McAllister and Chiang, 1994)

Pathways of aerobic BTEX degradation have been extensively reviewed (Bouwer and Zehnder, 1993; Dagley, 1971; Gibson and Subramanian, 1984; Hoeppel and Hinchee, 1993; Smith, 1990) and thus will not be covered here. In general, pure culture studies on aerobic metabolism of BTEX have found that molecular oxygen is incorporated into the aromatic ring in order to activate it.

Presumably a similar form of metabolism takes place in aerobic BTEX-contaminated groundwater.

Polycyclic aromatic hydrocarbons (PAHs) consist of multiple fused benzene rings in either linear, cluster, or angular arrangements. Coal gasification processes, improper heavy waste oil disposal, coal tar wastes, and creosote waste pits are all potential sources of PAH contamination in local groundwater aquifers (Cerniglia, 1992; Luthy et al., 1994). Aerobic oxidation of PAHs in subsurface environments has been documented numerous times (Bouwer et al., 1994; Durant et al., 1995; Herbes and Schwall, 1978; Klecka et al., 1990; Lee et al., 1983; Madsen et al., 1991, 1992; Thomas et al., 1989, 1987; Werner, 1991). Napthalene and phenanthrene are the most commonly studied PAHs. Rates of aerobic naphthalene degradation in laboratory incubations of aquifer sediment microcosms range from 0–14% per day depending on whether the site has an adapted microbial population (Bouwer et al., 1994; Durant et al., 1995; Heitkamp et al., 1987; Klecka et al., 1990; Lee et al., 1983; Madsen et al., 1992; Thomas et al., 1989). Likewise rates of phenanthrene degradation in sediment microcosms vary from 0–4.5% per day (Durant et al., 1995; Madsen et al., 1992; Thomas et al., 1989). The fate of higher molecular weight compounds such as flourene or dibenzofuran have not been commonly determined in sediment incubations. However, there are reports of rates of degradation approaching 20–30% per week for these compounds (Lee et al., 1983).

The mechanisms for aerobic PAH degradation have been reviewed (Cerniglia, 1992; Gibson and Subramanian, 1984; Smith, 1990). As with the monoaromatic hydrocarbons, degradation is initiated by introduction of oxygen to produce *cis*-dihydrodiols at the aromatic ring of initial attack. Dehydrogenation of the dihydrodiol produces a catechol intermediate followed by *ortho* or *meta* ring cleavage (Cerniglia, 1992; Smith, 1990). The process is repeated for additional rings in the polyaromatic structure.

3.2.1.a. Measuring Rates of Aerobic Degradation. In order to prove that *in situ* bioremediation is effectively removing aromatic hydrocarbons from contaminated aquifers it is necessary to document hydrocarbon oxidation (Wilson et al., 1994). Unequivocal documentation of *in situ* biodegradation generally requires several convergent lines of indirect evidence, but such a complete assessment is seldom performed (Baker and Herson, 1990; Madsen, 1991; Madsen et al., 1991; Salanitro, 1993). Generally, demonstration of *in situ* biodegradation depends on the collection of three types of evidence (Water Science and Technology Board, 1993). First, a documented loss of contaminant from the site must be shown. Second, laboratory studies should be performed that indicate whether or not a microbial population exists within collected samples that is capable of transforming the contaminant of concern under the conditions expected at the site. Lastly, there must be some indication that degradation is occurring at the site. These data can be quite difficult to obtain but are essential in evaluating the efficacy of the process.

Several techniques have been employed to determine the biodegradation potential of BTEX compounds at contaminated sites (Aamand et al., 1989; Alvarez and Vogel, 1991; Armstrong et al., 1991; Arvin et al., 1988; Barker et al., 1987; Davis et al., 1994; Dobbins et al., 1992; Morgan et al., 1993; Patterson et al., 1993; Swindoll et al., 1988; Werner, 1985; Wilson et al., 1990a) and similar approaches apply to the study of PAH degradation. A common technique is to add a [^{14}C]-BTEX component to aquifer material and measure the evolution of $^{14}CO_2$ (Armstrong et al., 1991; Davis et al., 1994; Swindoll et al., 1988; Thomas et al., 1990, 1987). The ^{14}C should be in a ring position of the BTEX component so that production of $^{14}CO_2$ is definitive evidence for ring cleavage and oxidation. An alternative strategy, which is less satisfactory but also less technically difficult and expensive, is to follow the loss of BTEX components from sediment incubations over time with gas chromatography (Aamand et al., 1989; Anid et al., 1993; Arvin et al., 1988; Barker et al., 1987; Davis et al., 1994; Morgan et al., 1993; Nielson and Christensen, 1994).

It is preferable to use aquifer sediments and the associated groundwater for such measurements rather than just groundwater because, as discussed above, the majority of the aquifer microorganisms are associated with the sediment. However, when it is not possible to obtain cores, rate estimates with groundwater samples may be qualitatively useful to determine if the potential for contaminant degradation exists and if the microbial populations are better adapted for contaminant degradation than in uncontaminated sites (Arvin et al., 1988).

BTEX degradation rates measured in laboratory incubations can vary from 5–50% per day whereas estimates derived from modeling the disappearance of BTEX components in aquifers are generally lower (0.7–1.5% per day) (MacIntyre et al., 1993; Salanitro, 1993). Such results indicate that it is not advisable to assume that estimates from laboratory incubations can be used to predict rates of in situ bioremediation. When it is paramount to know in situ rates, a modeling approach to the study of microbial processes should be considered (Lovley and Chapelle, 1997). However, it should be pointed out that at least in one instance, laboratory incubations under anaerobic conditions did provide rate estimates comparable to those obtained from geochemical modeling (Chapelle et al., 1996).

Even when rate estimates of BTEX degradation obtained in laboratory incubations do not accurately reflect rates of in situ bioremediation, they are an indication of the metabolic potential in the aquifer. Comparison of rates of degradation from samples taken within a BTEX-contaminated area versus those from a site that is uncontaminated can effectively demonstrate that a change in the microbial metabolic potential, and therefore adaptation, has taken place within the contaminated sediments. Concurrent monitoring of the rate of degradation of contaminant compounds and other, more labile compounds, within both contaminated and uncontaminated sediment samples can illustrate rather distinct adaptational differences between the associated microbial populations (Armstrong *et al.*, 1991). For example, mineralization rates of labile compounds such as glucose and amino acids were as much as four times slower in BTEX-contaminated samples than in samples from an uncontaminated area (Armstrong *et al.*, 1991). In contrast, rates of toluene mineralization in uncontaminated samples were negligible relative to contaminated samples (Armstrong *et al.*, 1991).

An alternative to incubating sediments in the laboratory to estimate aerobic BTEX degradation rates is to measure the rate of degradation in *in situ* microcosms (Gillham *et al.*, 1990; Holm *et al.*, 1991, 1992). Briefly, an *in situ* microcosm is made up of a pipe that can be passed through the center of a hollow stem auger. The pipe is separated into two chambers. The lower chamber, known as the test chamber, is open at the bottom to allow entrance of aquifer material as the microcosm is driven into the subsurface. The upper chamber houses tubing ports that are connected to screens within the test chamber. The main screen is positioned at the top of the test chamber and is used to inject or pump groundwater in or out of the test chamber. Another screen is positioned at the end of a cylindrical rod extending down through the center of the test chamber and can be used for sampling (Gillham *et al.*, 1990). This technique causes less sediment disturbance than laboratory incubations and thus rate estimates may be more likely to reflect *in situ* rates.

3.2.1.b. Adaption of Aerobic Microbial Community to Aromatic Hydrocarbon Contamination. The contamination of groundwater with aromatic hydrocarbons generally stimulates the growth of aerobic aromatic hydrocarbon-degrading bacteria in the aquifer. Numerous aerobic microorganisms capable of degrading aromatic hydrocarbons have been recovered from contaminated aquifers (Table I). Most notable among these are various *Pseudomonas* species that can degrade a wide variety of contaminants commonly found at hydrocarbon spill sites, including benzene, toluene, ethylbenzene, naphthalene, methylnaphthalene, phenanthrene, flourene, quinoline, and others (Brockman *et al.*, 1989; Ellis *et al.*, 1991; Kampfer *et al.*, 1991; Madsen *et al.*, 1992; Ridgway *et al.*, 1990). *Pseudomonas* species accounted for 86.9% of the gasoline-degrading bacteria isolated from well water and core material from a gasoline-contaminated aquifer (Ridgway *et al.*, 1990). Most of the isolates were capable of degrading two or three of the hydrocarbons evaluated, with toluene being the hydrocarbon most frequently utilized. However, some of the isolates could use all 15 of the hydrocarbons evaluated. The predominance of *Pseudomonas* species in petroleum-contaminated groundwaters contrasts with the finding that in pristine groundwaters this genera accounted for less than 1% of the isolates (Stetzenbach *et al.*, 1986). Furthermore, the types of *Pseudomonas* species may differ significantly between petroleum-contaminated groundwaters and nearby uncontaminated sites (Kampfer *et al.*, 1991).

Identification of isolates from contaminated sites can be a useful part of a remedial assessment (Bouwer *et al.*, 1994). However, it can be difficult to interpret the results of broad-range isolation studies. This is evident from a large-scale study on various bioremediation options at a former waste oil refinery (Kampfer *et al.*, 1993). Enumeration of hydrocarbon degraders indicated an increase in numbers in response to bioremediation. However, more detailed studies in which 3400 isolates were characterized with 87 different physiological tests yielded less defined results. Although some of the isolates could be assigned to specific genera, many could not be identified and it was not clear how changes in the diversity of the isolates with time or with depth could be related to the bioremediation.

As previously reviewed (Dobbins *et al.*, 1992; Madsen *et al.*, 1992; Thomas *et al.*, 1989), the capacity for aerobic degradation of aromatic hydrocarbon contaminants typically increases in response to contamination. This is generally detected by a comparison of metabolic activity in the contaminated sediments and a nearby uncontaminated background site. The clearest case of adaption is when there is no detectable degradation of the hydrocarbon contaminants in the pristine site, but significant degradation in the contaminated site. However, other responses that have been observed include: contaminant degradation in the pristine sediments, but at a slower rate that in the contaminated sediments; a shorter lag period prior to degradation of added contaminants in the polluted sediments; or that the contaminant is degraded to a greater extent in the contaminated sediments (Madsen *et al.*, 1992).

Some of the most complete studies demonstrating adaption of the microbial community to contamination have been in cases of PAH contamination. A study employing a variety of indicators of microbial biomass and activity indicated that the adaption of the microbial population to petroleum contamination of an aquifer was very specific (Jensen, 1989). There was little change in the overall microbial biomass and activity, but contaminated sediments had the potential to degrade naphthalene, whereas uncontaminated sediments did not. In an aquifer contaminated with buried coal tar, PAHs (naphthalene and phenanthrene) were more rapidly oxidized to carbon dioxide in laboratory incubations of sediments from within the contaminant plume than in sediments from a nearby pristine portion of the aquifer (Madsen *et al.*, 1991). Viable counts of aerobic heterotrophic bacteria were highest closest to the source of contamination, and lowest in the pristine sediments. However, there was no significant difference in total bacteria as detected by microscopic counts. Numbers of protozoa were very high within the contaminant plume, reaching levels comparable to those found in activated sewage sludge in portions of the contaminant plume. In contrast, numbers of protozoans in the pristine groundwater were low. The high numbers of protozoans suggested that microorganisms were growing rapidly within the contaminant plume. The rapid PAH oxidation rates within the plume suggested that this rapid growth could be attributed to degradation of the contaminants.

In a study with PAH-contaminated soils, there was a direct correlation between levels of mRNA for the naphthalene degradative gene and the rate of [^{14}C]-naphthalene oxidation to $^{14}CO_2$ (Fleming *et al.*, 1993). Although not strictly a groundwater study, this investigation demonstrated that it is possible to document the adaption of microorganisms to aromatic hydrocarbon contamination at the molecular level.

3.2.1.c. Factors Influencing the Rate of Aerobic Hydrocarbon Degradation. The degradation of individual BTEX components may be greatly influenced by the presence of other BTEX compounds. Both synergistic and antagonistic effects have been observed in studies with pure cultures and slurries of aquifer sediment (Alvarez and Vogel, 1991; Arvin *et al.*, 1989; Chang *et al.*, 1993). Benzene and toluene have been observed to degrade at similar rates when present in aquifers slurries as a single substrate or together. The presence of *p*-xylene increases the lag period prior to any observed benzene or toluene degradation (Alvarez and Vogel, 1991). Others studies have indicated that toluene or *o*-xylene alone can stimulate benzene degradation, but if these species are present together the effect is inhibitory (Arvin *et al.*, 1989). Potential substrates, such as *p*-cresol commonly found at PAH-contaminatcd sites, may decrease the rates of degradation of more hydrophobic compounds such as phenanthrene and carbazole (Millette *et al.*, 1995).

Another factor limiting the rates of BTEX degradation may be the lack of a sufficient number of microorganisms with the capacity to degrade the BTEX components. In laboratory studies with sands from a petroleum-contaminated aquifer, lag times in the degradation of benzene, toluene, or xylene were increased as the population of microorganisms capable of degrading these components decreased (Corseuil and Weber, 1994). The longest lags were observed with xylene, which was consistent with field evidence indicating that the xylene plume had spread much further than that of toluene or benzene. These studies demonstrate that even when oxygen and nutrients are provided at optimal concentrations, the initial rate of BTEX degradation can be slow if the initial population of BTEX degraders is low (Corseuil and Weber, 1994).

Low populations of microorganisms with the appropriate degradative potential can be supplemented via inoculation. In a continuation of the studies outlined above in the previous paragraph, an inoculum of benzene- and xylene-degrading microorganisms was developed by cultivating a biofilm of these organisms on columns of activated carbon and then passing the effluent from the biofilm column through the aquifer sand (Weber and Corseuil, 1994). In studies with benzene, this treatment greatly increased the number of organisms in the sand by over 100-fold and eliminated the lag in benzene degradation that was observed in uninoculated aquifer sand. Qualitatively similar results were observed with xylene. Inoculation with organisms that had been grown in the biofilm on one BTEX component also stimulated degradation of other BTEX components. (Weber and Corseuil, 1994). The microorganisms sloughed from the biofilm did not significantly adhere to the aquifer sands and readily moved through the sand columns.

The rate of PAH degradation is typically limited by the desorption rate from the sediment surface (Bouwer et al., 1994; Mihelcic and Luthy, 1988a, 1988b; Volkering et al., 1995). PAHs are generally quite hydrophobic and sorb to mineral and clay surfaces (Cerniglia, 1992; Fu et al., 1994; Kishi et al., 1990; Podoll et al., 1989; Volkering et al., 1995). Surfactants can enhance the dissolution of PAHs and the degradation of PAHs in laboratory cultures (Volkering et al., 1995) and field remediation sites (Ellis et al., 1990, 1991). Even when the added surfactants do not solubilize the PAH, mineralization can be enhanced (Aronstein et al., 1991; Francy et al., 1991).

3.2.1.d. Stimulation of Aerobic Aromatic Hydrocarbon Degradation: Addition of Oxygen, Nutrients, and Microorganisms. As noted above, aerobic metabolism of aromatic hydrocarbons in groundwater can lead to the depletion of oxygen from the groundwater, with a resultant decrease in the rate of aromatic hydrocarbon degradation. However, aerobes can survive in the anaerobic zones of a contaminated aquifers and thus stimulation of aerobic activity can be a very effective method of engineered bioremediation of aromatic hydrocarbons and other contaminants (Bianchi-Mosquera et al., 1994; Brown et al., 1984; Flathman et al., 1994; Hoeppel and Hinchee, 1993; Norris et al., 1994; Raymond et al., 1986). In fact, addition of oxygen to groundwater is probably the most common engineering technique used for enhancing bioremediation of aquifers contaminated with petroleum-related products (Brown et al., 1984; Hoeppel and Hinchee, 1993; Lee et al., 1988; Raymond et al., 1986; Wilson et al., 1994). Several books dealing with the engineering aspects of aerobic bioremediation are available (Alexander, 1994; Flathman et al., 1994; King et al., 1992; Norris et al., 1994; Water Science and Technology Board, 1993) and only the microbiological aspects of this topic will be briefly covered here.

Oxygen may be introduced into the aquifer by injecting air or oxygen below the water table (Crocetti et al., 1992; Jenkins et al., 1993; Norris et al., 1994), but due to the low solubility of oxygen in water, additional oxygen sources are often used (Brown and Crosbie, 1994; Wilson et al., 1994). Hydrogen peroxide is completely soluble in water and can provide a much larger amount of oxygen to the subsurface than bubbling oxygen into the water (Anid et al., 1993; Bianchi-Mosquera et al., 1994; Brown and Crosbie, 1994; Brown et al., 1984; Morgan et al., 1993; Morgan and Watkinson, 1992; Pardieck et al., 1992; Spain et al., 1989). However, hydrogen peroxide is toxic to some microorganisms. For example, BTEX degradation at a gas-condensate-contaminated site was completely inhibited when hydrogen peroxide at a concentration of 200 mg/l was added to the groundwater (Morgan et al., 1993). Inhibition of degradation below 200mg/l has also been observed at bioremediation sites (Huling et al., 1990; Thomas and Ward, 1989). In contrast, hydrogen peroxide concentrations up to 400 mg/l stimulated contaminant degradation at an aviation gasoline-contaminated site and also in gasoline-contaminated sand columns (Brown et al., 1984; Wilson et al., 1994). These disparate results indicate that the impact of hydrogen peroxide within the subsurface is site specific.

The presence of metals such as iron and manganese can catalyze the decomposition of H_2O_2, potentially resulting in premature release of oxygen prior to entry into the desired zone (Pardieck et al., 1992). As with bioventing and air sparging systems, peroxide injection technologies must also take into account the redox nature of the site (Barcelona and Holm, 1991). If the site is a highly reduced environment, abiotic consumption of oxygen through the oxidation of reduced metals may prevent oxygenation of the subsurface.

Additionally, microorganisms are capable of decomposing H_2O_2 through enzymatic mechanisms. For example, when recirculated groundwater was used in a peroxide injection system the concentration of hydrogen peroxide at the point of injection was low and the effectiveness of the overall subsurface oxygen delivery system was diminished (Spain et al., 1989). This is because catalase-positive bacteria in the recirculated groundwater were growing on the amended nutrients and consuming the peroxide. The solution to this problem was to inhibit microbial activity within the infiltration gallery by using charcoal-filtered tap water as the injection fluid.

Inoculation of adapted bacteria into the subsurface has also been viewed as another way to enhance *in situ* biodegradation of contaminants (Alexander, 1994; Norris et al., 1994). The effluent from above-ground bioreactors treating contaminated groundwater may provide a supply of contaminant-degrading microorganisms when the effluent is pumped back into the subsurface (Ellis et al., 1991; Portier et al., 1990; von Wedel et al., 1988; Weber and Corseuil, 1994). Isolated microbes can be cultured to higher cell densities using contaminants as substrates and reinjected into the subsurface (von Wedel et al., 1988). However, the role of inoculated bacteria in the biodegradation of contaminants within the subsurface can rarely be distinguished from indigenous species (Norris et al., 1994). In order to determine whether such inoculation techniques are effective, a reliable method for evaluating the survival of the inoculated organisms must be devised.

Few studies have directly documented the survival of a microorganism inoculated into a contaminated aquifer. However, in one such test, a *Pseudomonas* sp. inoculated into an aquifer contaminated with sewage effluent was still viable after 14.5 months, as determined by detection of specific DNA sequences and isolation of the organism from the aquifer (Krumme et al., 1994;

Thiem et al., 1994). Newly emerging molecular techniques will aid in future studies on the survival of inoculated organisms in the subsurface. Of major interest is the detection of specific degradative genes introduced into contaminated environments and/or the products of these genes. However, probes must be specific enough to allow discrimination between inoculated and indigenous genes (Jain et al., 1987; Ka et al., 1994a, 1994b, 1994c; Sayler et al., 1985; Walia et al., 1990; Wyndham et al., 1994).

3.2.2. Anaerobic Degradation of Aromatic Hydrocarbons

3.2.2.a. TEAPs Involved. Although BTEX components are degraded faster under oxic conditions than under anaerobic conditions, a significant portion BTEX entering contaminated aquifers may be anaerobically degraded (Lovley, 1997). As recently reviewed (Lovley, 1997), anaerobic degradation of all BTEX components has been documented under Fe(III)-reducing (Anderson et al., 1997; Lovley and Lonergan, 1990; Lovley et al., 1989a, 1994b, 1996b), sulfate-reducing (Beller et al., 1992; Chapelle et al., 1996; Edwards and Grbic-Galic, 1992; Edwards et al., 1992; Flyvbjerg et al., 1993; Haag et al., 1991; Lovley et al., 1995; Patterson et al., 1993; Rabus et al., 1993), methanogenic (Acton and Barker, 1992; Edwards and Grbic-Galic, 1994; Grbic-Galic and Vogel, 1987; Van Beelen and Van Keulen, 1990; Wilson et al., 1986, 1990a), and nitrate-reducing (Altenschmidt and Fuchs, 1991; Anid et al., 1993; Barbaro et al., 1992; Dolfing et al., 1990; Evans et al., 1991a, 1991b; Flyvbjerg et al., 1993; Fries et al., 1994; Haner et al., 1995; Hutchins et al., 1991; Hutchins, 1991; Kuhn et al., 1988, 1985; Major et al., 1988; Morgan et al., 1993; Patterson et al., 1993; Rabus and Widdel, 1995; Schocher et al., 1991; Zeyer et al., 1986), conditions and the possibility of BTEX oxidation coupled to Mn(IV) reduction seems likely. However, not all BTEX components were degraded in all studies. Anaerobic benzene degradation is particularly sporadic, with the majority of studies concluding on the basis of laboratory incubations or the persistence of benzene in anaerobic groundwater that benzene is not degraded under anaerobic conditions (Acton and Barker, 1992; Anid et al., 1993; Barbaro et al., 1992; Barker et al., 1987; Flyvbjerg et al., 1993; Hutchins et al., 1991; Kuhn et al., 1988; Lee and Borden, 1988; Patterson et al., 1993; Thierrin et al., 1993). However, anaerobic benzene degradation as evidenced by loss of benzene over time or oxidation of [^{14}C]-benzene to $^{14}CO_2$ has been observed in anaerobic incubations of aquifer material (Anderson et al., 1997; Edwards and Grbic-Galic, 1992; Lovley et al., 1994b, 1996b; Major et al., 1988; Morgan et al., 1993; Wilson et al., 1986).

With the development of anaerobic conditions in contaminated aquifers, Fe(III) is in general the most abundant potential electron acceptor for organic matter oxidation (Lovley, 1991). Although no BTEX-degrading Fe(III) reducers have as yet been isolated from petroleum-contaminated aquifers, such organisms are likely to be present as toluene-oxidizing Fe(III) reducers have been isolated from aquatic sediments (Coates *et al.*, 1996b; Lovley and Lonergan, 1990). Loss of BTEX components as contaminated groundwater moved through the Fe(III)-reducing zone of contaminated aquifers suggests that Fe(III) reduction can be important in removing BTEX from groundwater (Borden *et al.*, 1995; Lyngkilde and Christensen, 1992a; Rugge *et al.*, 1995).

Several lines of evidence indicated that BTEX was being oxidized with the reduction of Fe(III) in detailed studies of an aquifer in Bemidji, Minnesota contaminated with crude oil (Baedecker *et al.*, 1993, 1989; Lovley *et al.*, 1989a). There was a loss of BTEX components from the groundwater that was accompanied by the accumulation of Fe(II) in the groundwater, depletion of Fe(III) from the sediments, and the accumulation of isotopically light carbon dioxide that originated from organic matter oxidation. The finding that [ring-^{14}C]-toluene and [^{14}C]-benzene injected into sediments from the Fe(III)-reducing zone was rapidly oxidized to $^{14}CO_2$ without a lag provided further evidence for anaerobic BTEX oxidation in the Fe(III)-reducing zone (Anderson *et al.*, 1997).

Extensive sulfate-reducing zones may be found in some contaminated aquifers (Beeman and Suflita, 1987; Thierrin *et al.*, 1993; Vroblesky and Chapelle, 1994) and there is evidence for natural attenuation of BTEX components under sulfate-reducing conditions. In a field study, there was clear loss of toluene from groundwater as it moved through the sulfate reduction zone of a petroleum-contaminated aquifer (Chapelle *et al.*, 1996). However, there was no detectable loss of benzene. These results were consistent with relatively rapid oxidation of [ring-^{14}C]-toluene to $^{14}CO_2$ but very slow oxidation of [^{14}C] benzene in laboratory incubations of the sediments. The slow metabolism of benzene in these sediments may be attributable to a lack of the appropriate benzene-oxidizing microorganisms in the sediments (see below). In the sulfate reduction zone of an aquifer in Australia in which fertilizer and septage inputs had artificially increased the sulfate levels, there was preferential loss of all of the BTEX components other than benzene over that of bromide and benzene when deuterated BTEX tracers were added to the groundwater (Thierrin *et al.*, 1993). Benzene, toluene, and xylene degradation under sulfate-reducing conditions have been documented with either pure cultures, enrichments or laboratory incubations of sediments (Beller *et al.*, 1992; Chapelle *et al.*, 1996; Davis *et al.*, 1994; Edwards and Grbic-Galic, 1992; Edwards *et al.*, 1992; Flyvbjerg *et al.*, 1993; Haag *et al.*, 1991; Lovley *et al.*, 1995; Patterson *et al.*, 1993; Thierrin *et al.*, 1993).

BTEX consumption has also been documented under methanogenic conditions. In fact, it was the study of benzene and toluene degradation under methanogenic conditions by Grbic-Galic and Vogel (Grbic-Galic and Vogel, 1987; Vogel and Grbic-Galic, 1986) that greatly stimulated research in the field of anaerobic BTEX degradation. BTEX uptake has been observed in laboratory incubations of methanogenic sediments from contaminated aquifers (Acton and Barker, 1992; Edwards and Grbic-Galic, 1994; Wilson et al., 1986, 1990a). Toluene and o-xylene can be completely converted to carbon dioxide and methane, (Edwards and Grbic-Galic, 1994). Although benzene disappearance from some methanogenic aquifer sediments has been documented (Wilson et al., 1986, 1990a), other studies have found that benzene persists under methanogenic conditions (Acton and Barker, 1992; Edwards and Grbic-Galic, 1994; Lovley et al., 1994b).

Except in instances of nitrate contamination (see below), nitrate is not generally an abundant electron acceptor in groundwater. However, nitrate is highly soluble and can readily be added to groundwater for engineered bioremediation strategies. Studies with pure cultures, enrichments, and aquifer sediments have definitively demonstrated that toluene, ethylbenzene, and xylenes can be degraded under nitrate-reducing conditions (Altenschmidt and Fuchs, 1991; Anid et al., 1993; Barbaro et al., 1992; Dolfing et al., 1990; Evans et al., 1991a, 1991b; Flyvbjerg et al., 1993; Fries et al., 1994; Haner et al., 1995; Hutchins et al., 1991; Hutchins, 1991; Kuhn et al., 1988, 1985; Major et al., 1988; Morgan et al., 1993; Patterson et al., 1993; Rabus and Widdel, 1995; Schocher et al., 1991; Zeyer et al., 1986). Most studies have found that benzene is not degraded under nitrate-reducing conditions (Anid et al., 1993; Barbaro et al., 1992; Flyvbjerg et al., 1993; Hutchins et al., 1991; Hutchins, 1991; Kuhn et al., 1988), but there have been instances in which nitrate-dependent loss of benzene has been observed in laboratory incubations of aquifer material (Davis et al., 1994; Major et al., 1988; Morgan et al., 1993).

3.2.2.b. Factors Limiting Anaerobic Benzene Degradation. Given the necessity to effectively remove benzene in order for BTEX bioremediation to be considered successful, it would be helpful if there was more information on anaerobic benzene degraders. No anaerobic benzene degraders are available in pure culture. However, studies with sediments have indicated that some factors limit anaerobic benzene degradation.

A major factor influencing benzene degradation may be whether the appropriate benzene-degrading organisms are present in the aquifer. Although it is commonly regarded that aerobic hydrocarbon-degrading microorganisms are ubiquitous (Dobbins et al., 1992), the appropriate anaerobic benzene-degrading organisms may not always be found where they are needed. This was evident from studies in the sulfate reduction zone of a petroleum-contaminated aquifer in Hanahan, South Carolina (Weiner and Lovley, 1997). There was no detectable uptake of benzene in the sediments, even after long laboratory incubations. However, benzene was rapidly consumed when the aquifer sediments were inoculated with freshwater sediments that had been adapted for benzene degradation coupled to sulfate reduction. Benzene degradation continued in the aquifer sediments with successive inoculations of new aquifer sediments from adapted aquifer sediments. This indicated that the stimulation of benzene degradation was not due to the addition of components in the aquatic sediments such as organic matter or nutrients, but rather was something that could multiply (i.e., microorganisms).

An important factor controlling benzene degradation in the Fe(III) reduction zone of aquifers is the ability of the Fe(III)-reducing bacteria to access the insoluble Fe(III) oxides that are the primary source of Fe(III) in aquifers. Addition of synthetic chelators such as nitrilotriacetic acid (NTA) or ethylenediaminetetraacetic acid (EDTA) greatly stimulates the rate of benzene degradation in aquifer sediments in which Fe(III) reduction is the TEAP (Lovley *et al.*, 1994b, 1996b). This stimulation is associated with the solubilization of Fe(III), which makes the Fe(III) more available for microbial reduction (Lovley and Woodward, 1996). In the presence of chelated Fe(III), adapted Fe(III)-reducing sediments oxidized benzene nearly as rapidly as it was oxidized in aerobic aquifer material. Thus, differences in the Fe(III) chelation capacity of dissolved organic matter in aquifers might influence rates of benzene degradation within the Fe(III) reduction zone.

The addition of Fe(III) chelators does not stimulate benzene degradation in the upgradient zones of petroleum-contaminated aquifers from which Fe(III) has been depleted and sulfate reduction or methanogenesis is the TEAP (Lovley *et al.*, 1994b). However, if instead of just the chelator, chelated Fe(III) is added, then benzene metabolism is stimulated (Lovley *et al.*, 1996b). These studies have suggested that chelated Fe(III) might be used as an alternative electron acceptor for bioremediation of petroleum-contaminated aquifers. Chelated Fe(III) would have the advantage over oxygen addition in that chelated Fe(III) is much more soluble than oxygen and that chelated Fe(III) will not result in the plugging of injection wells and aquifers that is associated with oxygen additions due to the formation of insoluble Fe(III) oxides.

Benzene degradation in the Fe(III)-reducing zone may also be influenced by the presence of naturally occurring organic matter in groundwater. The addition of humic acids to aquifer sediments in which Fe(III) reduction was the TEAP stimulated benzene degradation better than any of the synthetic chelators evaluated (Lovley *et al.*, 1996b). Subsequent studies (Lovley *et al.*, 1996a) revealed that humic acids are able to accept electrons from Fe(III)-reducing bacteria and then pass the electrons to Fe(III). This electron shuttling between Fe(III)-reducing bacteria and Fe(III) greatly accelerates the rate of Fe(III) oxide reduction and this is presumably the reason that benzene degradation proceeds more rapidly. However, it is also possible that humic acids are a better electron acceptor for anaerobic benzene degradation than Fe(III). It is yet to be determined if humic acids will stimulate benzene degradation in aquifer sediments that are depleted of

Fe(III). Bioremediation of benzene with the addition of naturally occurring humic substances may be preferable to the use of synthetic Fe(III) chelators.

Further evidence that the availability of anaerobic electron acceptors influences the rate of benzene degradation was the finding that benzene persisted as landfill leachate-contaminated groundwater moved through the methanogenic zone of shallow aquifers but was removed from the groundwater as it moved through the Fe(III) reduction zone (Lyngkilde and Christensen, 1992a; Rugge *et al.*, 1995).

Thus, it is clear from the rudimentary studies conducted to date that there are likely to be many factors controlling the activity of anaerobic BTEX-degrading microorganisms in aquifers. Further study of anaerobic BTEX degradation seems warranted in order to better understand the natural attenuation of BTEX under anaerobic conditions. Such studies might also lead to improved BTEX remediation strategies using anaerobic processes.

3.2.2.c. Anaerobic Degradation of PAHs. Anaerobic PAH oxidation in aquifers has yet to be investigated in detail. However, studies with sediments from other environments have indicated that at least some PAHs can be degraded under nitrate-reducing (Al-Bashir *et al.*, 1990; Leduc *et al.*, 1992; Mihelcic and Luthy, 1988a, 1988b) and sulfate-reducing conditions (Coates *et al.*, 1996a).

In one groundwater study, there was no naphthalene degradation under denitrifying or sulfate-reducing conditions (Flyvbjerg *et al.*, 1993). In a study on contaminated groundwater at a manufactured gas plant (Durant *et al.*, 1995), naphthalene was degraded under denitrifying conditions in 2 of the 11 samples evaluated and there was no phenanthrene degradation in any samples. There was significant loss of deuterated naphthalene injected into the sulfate-reducing zone of an aquifer with elevated sulfate concentrations (Thierrin *et al.*, 1993). This suggested that naphthalene was degraded with the reduction of sulfate. Modeling of naphthalene concentrations along the groundwater flow path further supported the possibility of degradation. Although there was evidence for naphthalene degradation, it appeared that there was no benzene degradation. These studies suggest that anaerobic PAH oxidation is possible in groundwater, but further investigation on this process is clearly warranted.

3.2.3. Halogenated Solvents and Related Compounds

Halogenated aliphatic hydrocarbons, which are widely used as solvents and degreasers, are among the most prevalent organic ground water contaminants (Hardman, 1991; McCarty and Semprini, 1994; Strand *et al.*, 1990; Westrick *et al.*, 1984). Some of the most common halogenated solvents found in groundwater are trichloroethene (TCE) and related compounds such as tetrachloroethene (PCE), 1,1,1-trichloroethane (TCA), isomers of dichloroethene (DCE), isomers of dichloroethane (DCA), and vinyl chloride (VC). Carbon tetrachloride (CT) and chloroform (CF) are also found in contaminated groundwater. Numerous studies have investigated the *in situ* bioremediation potential of these compounds (Bowman *et al.*, 1993; Fliermans *et al.*, 1988; Hopkins and McCarty, 1995; Hopkins *et al.*, 1993a, 1993b; McCarty *et al.*, 1991; Phelps *et al.*, 1988; Roberts *et al.*, 1990; Semprini *et al.*, 1992, 1991, 1990; Semprini and McCarty, 1991, 1992; Wilson and Wilson, 1985).

Within the subsurface, transformation of halogenated solvents can occur as a result of abiotic and microbially catalyzed reactions. Abiotic transformations occur primarily through substitution and elimination reactions (Vogel, 1994; Vogel *et al.*, 1987). Substitution reactions involve the replacement of a halogen substituent with a hydroxyl group (-OH) while elimination reactions involve the removal of hydrogen halide from the compound forming an alkene. Some halogenated compounds can undergo both types of reactions each producing different transformation products (Vogel, 1994). As an example, the halogenated solvent 1,1,1-trichloroethane (TCA) is known to be abiotically transformed to acetic acid through a series of substitution reactions yet TCA can also be transformed to 1,1-dichloroethylene (1,1-DCE) through an elimination reaction (Vogel *et al.*, 1987; McCarty and Semprini, 1994). Which type of reaction a given compound will undergo is largely dependent on the number of halogen substituents present in the compound. Highly halogenated organics tend to be more subject to elimination reactions whereas substitution reactions are more likely with less halogenated compounds (Vogel, 1994; Vogel *et al.*, 1987).

3.2.3.a. Aerobic Processes. In aerobic environments, microbial oxidation of halogenated organics is more likely to occur with the less halogenated compounds. Highly halogenated compounds are already quite oxidized and are rather poor sources of energy (poor electron donors) to support growth. Hence, most of the research on the aerobic oxidation of halogenated organics has centered around the mono- and dihalogenated organics which are better electron donors and more easily oxidized (Brunner *et al.*, 1980; Rittmann and McCarty, 1980; Stucki *et al.*, 1983; Janssen *et al.*, 1985). However, microbial oxidation of trichloroethylene (TCE) and chloroform (CF) have also been investigated with results indicating that these compounds are also potentially transformed by microbial oxidation (Fogel *et al.*, 1986; Nelson *et al.*, 1986; Little *et al.*, 1988; Strand and Shippert, 1986). While some microbes have been shown to grow on less halogenated organics when the compound is present in sufficient concentration (Brunner *et al.*, 1980), recent research has focused on the co-metabolic transformations of halogenated organics as these reactions are not related to growth and can theoretically result in the complete removal of the halogenated organic assuming that it is bioavailable (Vogel, 1994). During co-metabolism, the halogenated organic is fortuitously transformed by enzymes catalyzing a similar, growth related reaction.

The discovery that natural gas stimulated aerobic TCE degradation (Wilson and Wilson, 1985) was a major breakthrough in demonstrating the potential for *in situ* bioremediation of chlorinated solvents, as this suggested a way to promote the degradation of these contaminants with a relatively inexpensive additive. TCE degradation was the result of the co-metabolism of TCE by methylotrophic bacteria (Little *et al.*, 1988), The methane monooxygenase (MMO) of methane-oxidizing bacteria can fortuitously oxidize TCE and a variety of other halogenated aliphatics including VC and the isomers of DCE (Tsien *et al.*, 1989). TCE-epoxide, a transformation product of MMO, has been observed in laboratory preparations of soluble MMO and in field studies where co-metabolic removal of TCE was occurring (Henry and Grbic-Galic, 1991b; Semprini *et al.*, 1990). TCE-epoxide can decompose to form a variety of nonhalogenated intermediates including carbon monoxide (CO), formate, glyoxylate, and dichloroacetic acid. At neutral to basic pH, carbon monoxide and formate formation are favored, while neutral to acidic pH ranges favor dichloroacetic acid and glyoxylate formation. (Henry and Grbic-Galic, 1991b; Vogel *et al.*, 1987). In pure culture experiments, carbon monoxide has been shown to inhibit TCE oxidation. However, in mixed cultures this may not be as pronounced due to the presence of other CO-degrading microbes (Henry and Grbic-Galic, 1991b).

In practice, the addition of methane to groundwater serves as a primary carbon and energy source for methylotrophic microorganisms that fortuitously degrade TCE (Henry and Grbic-Galic, 1991a; Koh *et al.*, 1993; Little *et al.*, 1988; Oldenhius *et al.*, 1991; Semprini *et al.*, 1990; Semprini and McCarty, 1991). Other halogenated organics that are degraded under biostimulated, methylotrophic conditions within aquifers include VC and the *cis-* and *trans-* isomers of DCE (Semprini *et al.*, 1991, 1990). Removal percentages for the *cis-* and *trans-* isomers of DCE have been shown to be 50% and 90%, respectively, whereas VC was 95% transformed (Semprini *et al.*, 1991, 1990). TCE removal was 20–30% under these conditions. Additionally, pulsed injections of methane into the subsurface caused pulsed responses in the concentrations of VC and *t*-DCE, which were added at a steady rate indicating a dependence on the presence of methane within the treatment area (Semprini *et al.*, 1991). There was no pulsed response in the VC or *t*-DCE concentration if formate or methanol were used as the primary substrate, indicating that methane may be a competitive substrate in this system. Formate and methanol are intermediates in the methane oxidation pathway and are oxidized by enzymes other than MMO (Hardman, 1991; Semprini *et al.*, 1991). Their addition presumably allowed the expression of MMO and therefore transformation of haloorganics without substrate competition.

An alternative aerobic mechanism for stimulating TCE removal is to enhance the activity of toluene- and phenol-oxidizing microorganisms, as these organisms will fortuitously oxidize TCE during toluene and phenol oxidation

(Hopkins et al., 1993a, 1993b; Nelson et al., 1987, 1986; Shields et al., 1989; Wackett and Gibson, 1988). *Pseudomonas putida* F1 degrades TCE through the action of toluene dioxygenase during toluene oxidation (Wackett and Gibson, 1988). *Pseudomonas cepacia* G4 biotransforms TCE only during expression of the *meta* fission pathway for phenol and toluene oxidation (Nelson et al., 1987). The intermediates in toluene oxidation, *o*-cresol and *m*-cresol, are inducers of TCE oxidation in this organism (Shields et al., 1989). Additionally, a mutant strain (PR1) of this species possessing a constitutive oxygenase, toluene *ortho*-monooxygenase, may be able to degrade TCE in the absence of an inducer (Krumme et al., 1993). Such a trait would be ideal for situations where bioaugmentation was warranted.

Field pilot studies using biostimulation of phenol and toluene oxidizers to degrade TCE have been conducted in aquifers with the result that larger removal efficiencies were noted for TCE (>90%) than in similar tests conducted under methylotrophic conditions (Hopkins and McCarty, 1995; Hopkins et al., 1993a, 1993b). Improved efficiency was also observed for *cis*-DCE (>90%); however, the effectiveness of removal for *trans*-DCE efficiency was slightly reduced (>74%) relative to methylotrophic conditions. Additionally, in this study the presence of 1,1-DCE in the test solutions inhibited TCE oxidation, reducing the amount of TCE transformed from 90% to 50% (Hopkins and McCarty, 1995). This is of concern because 1,1-DCE is an abiotic transformation product of another widely used degreaser, 1,1,1-trichloroethane (TCA) (Vogel et al., 1987) and is likely to be found where TCA is present (McNabb and Narasimhan, 1994).

Other microorganisms that may be important halogenated solvent degraders under co-metabolic conditions include propane and ammonia oxidizers (Vannelli et al., 1990; Wackett et al., 1989). Propane oxidizers have been isolated from both soil and water and include members of the *Rhodococcus* (Woods and Murrell, 1989), *Arthrobacter* (Stephens and Dalton, 1986), *Pseudomonas, Nocardia, Acinetobacter, Brevibacter*, and *Actinomyces* genera (Hou et al., 1983). Propane monooxygenase is thought to be responsible for the transformation of halogenated solvents. As an example, the propane oxygenase of *Mycobacterium* catalyzes the transformation of TCE, *cis*-DCE, *trans*-DCE and VC (Wackett et al., 1989). Trichloroethane (TCA) has also been observed to be transformed under propane oxidizing conditions, however propane was observed to be a powerful inhibitor of TCA transformation at concentrations above 1% (Keenan et al., 1994). The ammonia oxidase of the ubiquitous soil microbe *Nitrosomonas europaea* transforms TCE under ammonia-oxidizing conditions (Hyman et al., 1995; Vannelli et al., 1990). *Nitrosomonas* also transformed a host of other halogenated organics including dichloromethane, dibromomethane, chloroform, ethylene dibromide, vinyl chloride, isomers of dichloroethylene, 1,1,1-trichloroethane and others. However, TCE is a potent competitive inhibitor ($K_i = 30\mu M$) of ammonia oxidase ($K_m = 40\mu M$) (Hyman et al., 1995) and at increased TCE concentrations (1.15 mM), the activity of ammonia oxidase decreases, eventually leading to inactivation (Hyman *et al.*, 1995). Thus, stimulation of propane or ammonia oxidizers within the subsurface to affect halogenated solvent transformation will require strict control over the delivery of a growth substrate in order to balance growth with solvent degradation.

3.2.3.b. Anaerobic Processes. While less halogenated organics can potentially serve as electron donors under aerobic conditions, highly halogenated organics can serve as electron acceptors under anaerobic conditions (Vogel *et al.*, 1987; Kuhn and Suflita, 1989; Mohn and Tiedje, 1992; Holliger and Schraa, 1994). Anaerobic treatment of halogenated organics can potentially be very advantageous in contaminated aquifers as the reduction of halogenated aliphatics as terminal electron acceptors is faster than fortuitous dehalogenations since electron flow is more directed towards the halogenated organic (Holliger and Schraa, 1994). For anaerobic bioremediation to be useful however, halogenated organics must be shown to be transformed to nontoxic, dehalogenated products.

Anaerobic reduction of aliphatic hydrocarbons results in dehalogenation, the replacement of halogen(s) with hydrogen(s) (Mohn and Tiedje, 1992). Reductive dehalogenation of chlorinated solvents has been observed under methanogenic conditions (Bouwer and McCarty, 1983; Galli and McCarty, 1989a, 1989b; Gibson and Sewell, 1992; Holliger *et al.*, 1993; Vogel and McCarty, 1985; Freedman and Gossett, 1989) and a series of reductive dehalogenations is generally observed. As an example PCE has been observed to be sequentially reduced to TCE, DCE, and VC with VC being a dominant product of this reaction (Vogel and McCarty, 1985; Parsons *et al.*, 1985, 1984). The production of VC is problematic in terms of the bioremediation potential of this process because of the carcinogeneity of VC. While VC has been shown to transform under methanogenic conditions in the presence of mixed nutrients (Barrio-Lage *et al.*, 1990), an ideal process would be transformation of PCE or TCE to nonhalogenated intermediates. Such a process has been observed in column studies of anaerobic river sediment and in bioreactors where PCE was reductively transformed to ethane (De Bruin *et al.*, 1992) and ethylene (Freedman and Gossett, 1989) demonstrating complete biological removal of this solvent in natural samples. Radiolabel pathway experiments indicate that there are two PCE degradation pathways under methanogenic conditions (Vogel and McCarty, 1985). One pathway results in the formation of reduced products while the other results in the production of carbon dioxide. Earlier studies with CT also indicated transformation and mineralization (Bouwer and McCarty, 1983). CT was reduced to chloroform and ultimately oxidized to carbon dioxide. Other studies under acetogenic conditions with *Acetobacterium woodii* have shown reductive dehalogenation presumably preceding mineralization (Egli *et al.*, 1988). Dichloromethane and chloromethane intermediates were observed in the cultures over time, however, radiolabel experiments indicated mineralization of PCE (Egli *et al.*, 1988). The mechanism by which these highly halogenated species are oxidized is still speculative. However, there is evidence that the presence of corrinoids (metallo-prophyrins) in PCE-transforming cells may mediate some of these reactions (Egli *et al.*, 1990). A summary of other mechanisms involving transition metal complexes relating to halogenated solvent transformation can be found in Mohn and Tiedje (1992).

Isolation of organisms capable of coupling growth to the reduction of PCE, is an important step in the development of stable cultures for potential bioremediation uses (Holliger and Schumacher, 1994; Holliger *et al.*, 1993; Scholz-Muramatsu *et al.*, 1995). *Dehalobacter restrictus* grows with hydrogen as an electron donor and PCE as an electron acceptor and could not grow with other donors or acceptors such as oxygen, nitrate, sulfate, or carbon dioxide (Holliger and Schumacher, 1994). *Dehalospirillum multivorans* is also capable of coupling growth to PCE reduction but has a much wider range of potential donors and acceptors than *Dehalobacter restrictus* (Holliger and Schumacher, 1994; Scholz-Muramatsu *et al.*, 1995). In addition to hydrogen, *D. multivorans* can use pyruvate and formate with PCE as an acceptor. In the absence of PCE, fumarate or sulfur can be used as an electron acceptor in contrast to *D. restrictus* which requires PCE for growth (Holliger and Schumacher, 1994; Scholz-Muramatsu *et al.*, 1995). Both organisms transform PCE to *cis*-1,2-DCE indicating that complete dehalogenation in the environment may be the result of a dehalogenating consortium and that other organisms specifically adapted to using transformation products of reductive dehalogenation as electron acceptors may be present in contaminated areas (Holliger and Schraa, 1994). Such organisms could be useful in bioaugmentation applications for the treatment of halogenated organics at contaminated sites.

Denitrification conditions may also be important in reductive dehalogenation of halogenated solvents (Bouwer and McCarty, 1983). *Pseudomonas* sp. strain KC was isolated from groundwater aquifer sediments (Criddle *et al.*, 1990; Lewis and Crawford, 1993; Tatara *et al.*, 1993). Strain KC is able to degrade CT to carbon dioxide without production of intermediates such as chloroform (CF)—a trait ideally suited for bioremediation purposes. However, this activity is limited by the presence of reduced iron species (<10 μM) in the groundwater. In a unique pilot study of a potential *in situ* bioremediation system, CT was introduced into an aquifer and monitored over time (Semprini *et al.*, 1992). The transformation of carbon tetrachloride *in situ* was facilitated by the injection of acetate and nitrate into the groundwater (Semprini *et al.*, 1992). Denitrification conditions were induced within the subsurface and CT transformation was observed. The appearance of the transformation product, chloroform, in the groundwater suggested that *Pseudomonas* KC type microorganisms were not mediating the transformation (Semprini *et al.*, 1992). In fact, rates of CT transformation increased when nitrate was removed from the injection fluid. The presence of sulfate in the groundwater indicated the possibility of sulfate reduction within the aquifer, however, what impact sulfate reduction had on the transformation of CT in the pilot study could not be determined from the available data (Semprini et al., 1992).

Degradation of halogenated organics under dissimilatory Fe(III)- and Mn(IV)-reducing conditions has not been extensively investigated with the exception of observed transformation of chlorofluorocarbons (CFCs) under Fe(III)-reducing conditions in aquatic sediment (Lovley and Woodward, 1992). Given the widespread occurrence of Fe(III) in aquifer sediments and the fact that halogenated organics have been found to biotransform under nitrate and methanogenic conditions it seems likely that transformation of halogenated solvents is at least possible if not probable under Fe(III)- and Mn(IV)-reducing conditions.

Transformation of highly chlorinated organics under sulfate-reducing conditions in the groundwater from a deep aquifer contaminated with chlorofluorocarbons and TCE has been reported (Sonier et al., 1994). Transformation of trichlorofluoromethane (CFC-11) was found to occur without a lag period, in the absence of nitrate and in the presence of 2 mM bromoethane sulfonic acid (Sonier et al., 1994) producing fluorodichloromethane (HCFC-21). The reaction was shown to be dependent on the presence of acetate and, after adaptation, butyrate (Sonier et al., 1994). Lactate, a common electron donor for sulfate reducers, did not support CFC-11 transformation and sulfide did not mediate dechlorination activity. Other studies have also reported transformation of halogenated organics under sulfate-reducing conditions (Suflita et al., 1988; Bagley and Gossett, 1990) in aquifer material including the transformation of PCE to ethene and ethane at a contaminated landfill site (Beeman et al., 1994).

Recently, a rapidly growing facultative aerobe has been isolated from a contaminated aquifer that reductively dechlorinates PCE to cis-DCE (Sharma and McCarty, 1996). This organism is able to carry out this reaction in the presence of near saturation levels of PCE (10 mM) and may be an ideal organism for the bioaugmentation of heavily contaminated areas (Sharma and McCarty, 1996). Potential advantages of this organism over previously described (Holliger and Schumacher, 1994; Maymo-Gatell et al., 1995; Scholz-Muramatsu et al., 1995) organisms that respire PCE is that it is a facultative aerobe and can be grown on a variety of carbon sources. As with other PCE respirers, PCE dehalogenation required anaerobic conditions and the end product of PCE reduction was cis-DCE, which is less toxic than PCE or TCE, but is still a concern (Sharma and McCarty, 1996). Complete dechlorination of PCE to ethene or ethane has been observed in enrichment cultures and could lead to the identification of a microbial consortium that is responsible for the complete removal of PCE from contaminated sediments (De Bruin et al., 1992; Maymo-Gatell et al., 1995; Sharma and McInerney, 1994; Freedman and Gossett, 1989). The future use of halorespiration for bioremediation of aquifers contaminated with halogenated organics appears promising given the isolation of strains demonstrating a capacity to couple growth to the reduction of these compounds.

3.2.4. Bioremediation of Nitrate

Nitrate is the most common groundwater contaminant and causes the most frequently reported health-related problems associated with groundwater pollution (Korom, 1992; Spalding and Exnter, 1993). Ingestion of nitrate in drinking water is known to cause methemoglobinemia in infants and may also be related to various other health problems (Spalding and Exnter, 1993). Agricultural practices account for most of the nitrate inputs to groundwater (Cole, 1993; Korom, 1992). Microbial reduction of nitrate to N_2 under anaerobic conditions, commonly referred to as denitrification, is a potential mechanism for removal of nitrate from contaminated aquifers. A large number of studies have indicated that when anaerobic conditions develop in aquifers, nitrate can be denitrified with either organic carbon, Fe(II), or sulfide acting as the electron donor (Gayle *et al.*, 1989; Korom, 1992).

Sediments from a variety of aquifers have been shown to have the potential for denitrification in laboratory incubations (Bengtsson and Annadotter, 1989; Bradley *et al.*, 1992; Francis *et al.*, 1989; Morris *et al.*, 1988; Smith *et al.*, 1988, 1991; Starr and Gillham, 1993) or in *in situ* microcosms (Starr and Gillham, 1993). Field evidence for *in situ* denitrification includes the loss of nitrate accompanied with the accumulation of excess dissolved N_2 in the groundwater (Vogel *et al.*, 1981; Wilson *et al.*, 1990b) as well as nitrous oxide, an intermediate of denitrification (Spalding and Parrott, 1994). Isotopic fractionation as the result of denitrification in ground water is evident from isotopically heavy nitrogen and oxygen in the unreacted nitrate as well as in the production of isotopically light N_2 (Bottcher *et al.*, 1990; Mariotti *et al.*, 1988; Smith *et al.*, 1991; Spalding and Parrott, 1994; Vogel *et al.*, 1981; Wilson *et al.*, 1990b). Depletion of nitrate that has been injected into groundwater along groundwater flow paths (Korom, 1992; Trudell *et al.*, 1986) has provided further evidence for *in situ* denitrification. Organic carbon is often the electron donor for denitrification in groundwater but inorganic electron donors, most notably pyrite, are also important in some instances (Korom, 1992; Postma *et al.*, 1991)

Naturally occurring denitrification in aquifers can be viewed as a form of intrinsic bioremediation in cases where nitrate concentrations in portions of the aquifer are above regulatory standards. Engineered denitrification is employed in above-ground bioreactors to treat some nitrate-contaminated drinking waters in France, Great Britain, and Germany (Mateju *et al.*, 1992). It is also possible to engineer *in situ* bioremediation strategies for nitrate-contaminated groundwater. The accumulation of nitrate in groundwater is indicative of a shortage of electron donors for denitrification. Sucrose, ethanol, acetate, and methanol have been used as electron donors for *in situ* stimulation of denitrification (Mateju *et al.*, 1992). Stimulation of denitrification by the addition of electron donors has resulted in significant nitrate removal from groundwater in both laboratory and field studies (Dahab, 1993; Hamon and Fustec, 1991; Janda *et al.*, 1988; Mateju *et al.*, 1992; Mercado *et al.*, 1988). However, the addition of organic electron donors also can result in undesirable growth of bacteria. For example, microbial growth can be so extensive that the pumped groundwater is visibly turbid with high numbers of coliform bacteria (Mercado *et al.*, 1988). If surplus electron donor remains in the treated groundwater this may stimulate the growth of biofilms on the walls of the pipes in the water distribution system (Janda *et al.*, 1988). Accumulation of N_2 and carbon dioxide bubbles as the result of denitrification can also impede groundwater flow (Mercado *et al.*, 1988; Soares *et al.*, 1988).

H_2 is an alternative electron donor that might result in less growth of undesirable bacteria and less aquifer plugging than organic electron donors (Smith *et al.*, 1994). Addition of H_2 or formate greatly stimulated denitrification in sediments from a nitrate-contaminated sand and gravel aquifer (Smith *et al.*, 1994). This was true even though the autotrophic denitrifiers that could be recovered in culture media were less than 10% of the culturable heterotrophic denitrifiers. Hydrogen-oxidizing denitrifiers were isolated from the aquifer, and their affinity for H_2 was comparable to the H_2 affinity of the denitrifying sediments. The isolates could also reduce O_2 with H_2, which is an important consideration for *in situ* bioremediation because most nitrate-contaminated aquifers are aerobic and O_2 must be removed prior to denitrification. The use of H_2 as an electron donor might result in less plugging of aquifers than is observed with organic electron donors because the non-specific growth of heterotrophic microorganisms can be avoided (Smith *et al.*, 1994). As summarized in a European patent application, when nitrate-contaminated groundwater was pumped to the surface, amended with H_2, carbon dioxide, and phosphate, and then reinjected into the aquifer, nearly 90% of the nitrate contamination was removed within four weeks (Mateju *et al.*, 1992).

There does not appear to have been any detailed investigation into the microorganisms involved in denitrification in contaminated groundwater. One study has indicated that high nitrate concentrations in groundwater may select for denitrifying microorganisms that can reduce nitrate faster than denitrifiers in less contaminated aquifers (Bengtsson and Bergwall, 1995). Given the importance of the nitrate contamination problem and the need to develop effective, inexpensive remediation strategies, further investigation into the microbial ecology of denitrification in groundwater seems warranted.

3.2.5. Bioremediation of Metal and Metalloid Contaminants

In situ bioremediation of metal contamination of groundwater has not received as much investigation as bioremediation of organics. Microbial activity can only remove metals from contaminated groundwater or convert them to less toxic forms, it cannot remove the metals from the aquifer. This is unlike the bioremediation of organics in which the contaminants are generally destroyed and converted to harmless carbon dioxide. Therefore, most bioremediation strategies for metal contamination have focused on *ex situ* methods using aboveground reactors. However, there may be instances where pumping costs are prohibitive or other considerations may favor an *in situ* approach for metals remediation.

The most common adverse effect of metals on groundwater quality is undesirably high concentrations of dissolved Fe(II) and Mn(II). In general, this cannot be considered a pollution problem as it most often results from naturally occurring microbial oxidation of organic compounds coupled to the reduction of Fe(III) and Mn(IV) in groundwaters (Lovley, 1991). Dissolved Fe(II) and Mn(II) in groundwater become a problem when the groundwater is pumped to the surface and contacts oxygen. The Fe(II) and Mn(II) are then oxidized to insoluble oxides that plug the plumbing system or discolor the water and cause staining problems.

One strategy for bioremediation of high dissolved Fe(II) and Mn(II) in groundwater is to stimulate the activity of Fe(II)- and Mn(II)-oxidizing bacteria within the aquifer (Braester and Martinell, 1988; Gounot, 1994; Jaudon *et al.*, 1989). This is done by pumping aerated groundwater into the aquifer. Introduction of oxygen has two effects (Jaudon *et al.*, 1989). One is that the dissolved oxygen inhibits microbial Fe(III) and Mn(IV) reduction. Furthermore, the oxygen stimulates the activity of Fe(II)- and Mn(II)-oxidizing bacteria such as *Leptothrix* and *Gallionella* species (Braester and Martinell, 1988), which convert the Fe(II) and Mn(II) to insoluble Fe(III) and Mn(IV) oxides that precipitate out of the groundwater.

A much explored avenue for removing metals from waste streams and contaminated surface waters is biosorption (Gadd, 1988; Macaskie, 1991). In biosorption, metals are adsorbed onto microbial biomass, often using dead microorganisms. Biosorption is probably not a feasible option for *in situ* metal remediation because the biomass to which the metals are adsorbed will eventually be degraded leading to the re-release of the metals.

Microbial processes that convert soluble metals to insoluble forms via redox reactions might be used for *in situ* bioremediation. For example, highly soluble U(VI) is reduced to highly insoluble U(IV) by dissimilatory Fe(III)-reducing microorganisms that substitute U(VI) for Fe(III) in their metabolism (Lovley *et al.*, 1991). Some sulfate-reducing microorganisms can also reduce U(VI), but unlike the Fe(III) reducers, they do not conserve energy to support growth from this reaction (Lovley and Phillips, 1992b; Lovley *et al.*, 1993b). Microbial U(VI) reduction has been shown to effectively remove uranium from uranium-contaminated groundwaters in laboratory incubations (Lovley and Phillips, 1992a). Furthermore, it has been demonstrated that inoculation of aquifer sediments with

U(VI)-reducing microorganisms can remove uranium from groundwater moving through columns of the aquifer sediments in laboratory incubations. These studies suggest that stimulation of the activity of U(VI)-reducing microorganisms in subsurface environments could be used to form a barrier that would prevent further migration of uranium contamination. When the upgradient groundwater no longer posed a uranium threat, the uranium that was precipitated in the U(VI)-reduction zone could be extracted using techniques that are commonly employed for *in situ* uranium mining. The formation of uranium deposits known as rollfront or sandstone-type deposits, in which U(VI) in solution in aerobic groundwater is precipitated as U(IV) as the groundwater enters an anaerobic portion of the aquifer may represent an example of natural attenuation of uranium (Lovley, 1991; Lovley *et al.*, 1991).

Microbial reduction of highly toxic, highly soluble Cr(VI) to less toxic, less soluble Cr(III) has been widely studied as a possible bioremediation strategy for chromium contamination (Bopp and Ehrlich, 1988; Hardoyo and Ohtake, 1991; Horitsu *et al.*, 1987; Ishibashi *et al.*, 1990; Lovley and Phillips, 1994). Both anaerobic and aerobic Cr(VI) reduction have been documented (Lovley, 1993). The usefulness of anaerobic Cr(VI) reduction as a bioremediation tool has been questioned because constituents such as Fe(II) and sulfide, which are typically present in anaerobic environments, will rapidly reduce Cr(VI) via nonenzymatic reactions. However, some microorganisms can reduce Cr(VI) in the presence of oxygen and the Cr(III) is stable under aerobic conditions (Bopp and Ehrlich, 1988; Horitsu *et al.*, 1987; Ishibashi *et al.*, 1990). Thus, if a technique could be devised to specifically stimulate the activity of aerobic Cr(VI)-reducing microorganisms in groundwater, this might be a useful strategy for *in situ* immobilization.

Radioactive cobalt chelated to EDTA that was generated as a byproduct of processing of nuclear materials is a contaminant of concern in some groundwaters (Caccavo *et al.*, 1994). Co(III)-EDTA is soluble and does not tend to adsorb onto sediment particles. In contrast, Co(II) strongly adsorbs to aluminum oxides in soils (Girvin *et al.*, 1993). Although Co(III)-EDTA is resistant to chemical reduction, the dissimilatory Fe(III)-reducer *Geobacter sulfurreducens* can grow with Co(III)-EDTA as the sole electron acceptor, reducing the Co(III) to Co(II) (Caccavo *et al.*, 1994). Thus, stimulation of microbial Co(III)-EDTA reduction might be able to prevent the migration of this contaminant in the subsurface.

Microbial reduction of selenate, Se(VI), and selenite, Se(IV), to insoluble elemental selenium, Se°, in anaerobic aquatic sediments is a natural attenuation mechanism for selenium contamination in ponds used to collect agricultural drainage water in the Western United States, and a similar process may naturally remove selenate from ground waters (Oremland, 1994; Oremland *et al.*, 1990, 1991). Studies with soils have indicated that establishment of anaerobic conditions is sufficient to stimulate selenium removal from surface water (Long *et al.*,

1990). It seems likely that microbial Se(VI) reduction could be stimulated in aquifers in similar manner.

The possibility of microorganisms volatilizing mercury as the result of reducing ionic mercury, Hg(II), to metallic mercury, Hg°, was investigated with sediments from deep pristine aquifers (Radosevich and Klein, 1993). The extent of mercury reduction in sediments containing indigenous microorganisms was not significantly greater than in sterile controls. However, numerous studies have demonstrated intrinsic bioremediation of mercury as the result of microbial Hg(II) reduction volatilizing mercury from contaminated soils and surface water environments (Barkay et al., 1991; Goldstein et al., 1988; Ogunseitan and Olson, 1991). The aerobic microbial community typically responds to mercury contamination with an enhanced capacity for Hg(II) reduction as a means of mercury detoxification (Baldi et al., 1992, 1991; Barkay, 1987; Barkay et al., 1989a, 1989b, 1991; Barkay and Olson, 1986; Olson et al., 1991; Rochelle et al., 1991). It may be possible to stimulate rates of microbial Hg(II) reduction as an engineering approach to *in situ* bioremediation of contaminated soils and water (Barkay et al., 1991; Goldstein et al., 1988; Ogunseitan and Olson, 1991), but a major limitation is that much of the Hg(II) is typically unavailable for microbial reduction (Barkay et al., 1991; Regnell, 1990; Silver, 1991).

Although microbial reduction of some contaminant metals and metalloids can remove them from groundwater, microbial arsenate reduction is likely to lead to arsenic contamination. It has recently been discovered that some microorganisms can use arsenate As(V) as an electron acceptor for the oxidation of organic matter and H_2 (Ahmann et al., 1994; Laverman et al., 1995). As(V) is reduced to As(III). As(V) reduction may be detrimental in aquifers because As(V) is less toxic than As(III) and tends to adsorb on Fe(III) oxides in the sediments whereas As(III) is more soluble. Thus, microbial As(V) reduction could release As(III) into groundwaters. Although microbial As(V) reduction does not appear to have been studied in aquifers, a report on groundwaters in the Midwestern United States that have elevated arsenic levels (Korte, 1991) provides geochemical data that is consistent with microbial As(V) reduction generating groundwaters with high As(III) concentrations.

4. Conclusions

This chapter has covered just a small fraction of the published studies that have considered either intrinsic or engineered bioremediation of organic and inorganic contaminants in groundwater. However, even from this partial overview it is readily apparent that microorganisms living in subsurface environments have the potential to remove significant quantities of groundwater contaminants. Most research on groundwater has focused on aerobic processes for organic contaminant degradation. However, it is becoming increasingly apparent that anaerobic processes naturally play an important role in degrading organic contaminants in groundwater and that it may be possible to stimulate this metabolism in order to enhance contaminant removal. Anaerobic processes also show potential for the removal of significant inorganic contaminants such as nitrate and some metals. Although a large number of aerobic microorganisms capable of degrading hydrocarbon contaminants have been isolated from aquifers, it is not clear whether these organisms are the ones that are important for *in situ* contaminant degradation. Even less is known about the anaerobic microorganisms involved in contaminant degradation. With the introduction of improved molecular techniques, it should be possible to begin elucidating the structure of the microbial community in contaminated aquifers. However, true understanding of the factors controlling the rate of intrinsic bioremediation and improved design of engineered bioremediation strategies will require detailed interdisciplinary studies that include consideration of the geochemistry and hydrology of aquifers as well as the microbial ecology.

References

Aamand, J., Jorgensen, C., Arvin, E., and Jensen, B. K., 1989, Microbial adaptation to degradation of hydrocarbons in polluted and unpolluted groundwater, *J. Contam. Hydrol.* **4**:299–312.

Acton, D. W., and Barker, J. F., 1992, In situ biodegradation potential of aromatic hydrocarbons in anaerobic groundwaters, *J. Contam. Hydrol.* **9**:325–352.

Adriaens, P., Kohler, H.-P. E., Kohler-Staub, D., and Focht, D. D., 1989, Bacterial dehalogenation of chlorobenzoates and coculture biodegradation of 4,4'-dichlorobiphenyl, *Appl. Environ. Microbiol.* **55**:887–892.

Ahmann, D., Roberts, A. L., Krumholtz, L. R. and Morel, F. M. M., 1994, Microbe grows by reducing arsenic, *Nature* **317**:750.

Al-Bashir, B., Cseh, T., Leduc, R., and Samson, R., 1990, Effect of soil/contaminant interactions on the biodegradation of naphthalene in flooded soil under denitrifying conditions, *Appl. Microbiol. Biotechnol.* **34**:414–419.

Alexander, M., 1994, *Biodegradation and Bioremediation*, Academic Press, London.

Altenschmidt, U., and Fuchs, G., 1991, Anaerobic degradation of toluene in denitrifying Pseudomonas sp.: indication for toluene methylhydroxylation and benzoyl-co A as central intermediate, *Arch. Microbiol.* **156**:152–158.

Alvarez, P. J. J., and Vogel, T. M., 1991, Substrate interactions of benzene, toluene and para-xylene during microbial degradation by pure cultures and mixed culture aquifers slurries, *Appl. Environ. Microbiol.* **57**:2981–2985.

Anderson, R. T., Gaw, C. V., and Lovley, D. R., 1997, Benzene oxidation in the Fe(III) reduction zone of a petroleum-contaminated aquifer, *Environ. Sci. Technol.* submitted.

Anid, P. J., Alvarez, P. J. J., and Vogel, T. M., 1993, Biodegradation of monoaromatic hyrocarbons in aquifer columns amended with hydrogen peroxide and nitrate, *Water Res.* **27**:685–691.

Armstrong, A. Q., Hodson, R. E., Hwang, H.-M., and Lewis, D. L., 1991, Environmental factors affecting toluene degradation in ground water at a hazardous waste site, *Environ. Toxicol. Chem.* **10**:147–158.

Aronstein, B. N., Calvillo, Y. M., and Alexander, M., 1991, Effect of surfactants at low concentrations on the desorption and biodegradation of sorbed aromatic compounds in soil, *Environ. Sci. Technol.* **25**:1728–1731.

Arvin, E., Jensen, B., Aamand, J., and Jorgensen, C., 1988, The potential of free-living ground water bacteria to degrade aromatic hydrocarbons and heterocyclic compounds, *Wat. Sci. Tech.* **20**:109–118.

Arvin, E., Jensen, B. K., and Gundersen, A. T., 1989, Substrate interactions during aerobic biodegradation of benzene, *Appl. Environ. Microbiol.* **55**:3221–3225.

Baedecker, M. J., and Back, W., 1979, Hydrogeological processes and chemical reactions at a landfill, *Ground Water* **17**:429–437.

Baedecker, M. J., Siegel, D. I., Bennett, P., and Cozzarelli, I. M., 1989, The fate and effects of crude oil in a shallow aquifer I. The distribution of chemical species and geochemical facies, in: *U. S. Geological Survey Water Resources Division Report 88–4220*, (G. E. Mallard and S. E. Ragone, eds., U. S. Geological Survey, Reston, VA., pp. 13–20.

Baedecker, M. J., Cozzarelli, I. M., Siegel, D. I., Bennett, P. C., and Eganhouse, R. P., 1993, Crude oil in a shallow sand and gravel aquifer: 3. Biogeochemical reactions and mass balance modeling in anoxic ground water, *Appl. Geochem.* **8**:569–586.

Bagley, D. M., and Gossett, J. M., 1990, Tetrachloroethene transformation to trichloroethene and *cis*-1,2-dichloroethene by sulfate-reducing enrichment cultures, *Appl. Environ. Microbiol.* **56**: 2511–2516.

Baker, K. H., and Herson, D. S., 1990, In situ bioremediation of contaminated aquifers and subsurface soils, *Geomicrobiology Journal* **8**:133–146.

Baldi, F., Semplici, F., and Filippelli, M., 1991, Environmental applications of mercury resistant bacteria, *Water, Air, Soil Pollut.* **56**:465–475.

Baldi, F., Boudou, A., and Ribeyre, F., 1992, Response of a freshwater bacterial community to mercury contamination ($HgCl_2$ and CH_3HgCl) in a controlled system, *Arch. Environ. Contam. Toxicol.* **22**:439–444.

Balkwill, D. L., 1989, Numbers, diversity, and morphological characteristics of aerobic, chemoheterotrophic bacteria in deep subsurface sediments from a site in South Carolina, *Geomicrobiol. J.* **7**:33–52.

Balkwill, D. L., and Ghiorse, W. C., 1985, Characterization of subsurface bacteria associated with two shallow aquifers in Oklahoma, *Appl. Environ. Microbiol.* **50**:580–588.

Balkwill, D. L., Frederickson, J. K., and Thomas, J. M., 1989, Vertical and horizontal variations in the physiological diversity of the aerobic chemoheterotrophic bacterial microflora in deep southeast coastal plain subsurface sediments, *Appl. Environ. Microbiol.* **55**:1058–1065.

Barbaro, J. R., Barker, J. F., Lemon, L. A., and Mayfield, C. I., 1992, Biotransformation of BTEX under anaerobic, denitrifying conditions: field and laboratory observations, *J. Contam. Hydrol.* **11**:245–272.

Barcelona, M. J., and Holm, T. R., 1991, Oxidation-reduction capacities of aquifers solids, *Environ. Sci. Technol.* **25**:1565–1572.

Barkay, T., 1987, Adaptation of aquatic microbial communities to Hg^{2+} stress, *Appl. Environ. Microbiol.* **53**:2725–2732.

Barkay, T., and Olson, B. H., 1986, Phenotypic and genotypic adaptation of aerobic heterotrophic sediment bacterial communities to mercury stress, *Appl. Environ. Microbiol.* **52**:403–406.

Barkay, T., Liebert, C., and Gillman, M., 1989a, Environmental significance of the potential for *mer*(Tn21)-mediated reduction of Hg^{2+} to Hg^0 in natural waters, *Appl. Environ. Microbiol.* **55**:1196–1202.

Barkay, T., Liebert, C., and Gillman, M., 1989b, Hybridization of DNA probes with whole-community genome for detection of genes that encode microbial responses to pollutants: *mer* genes and Hg^{2+} resistance, *Appl. Environ. Microbiol.* **55**:1574–1577.

Barkay, T., Turner, R. R., VandenBrook, A., and Liebert, C., 1991, The relationships of Hg(II)

volatilization from a freshwater pond to the abundance of *mer* genes in the gene pool of the indigenous microbial community, *Microb. Ecol.* **21:**151–161.
Barker, J. F., Patrick, G. C., and Major, D., 1987, Natural attenuation of aromatic hydrocarbons in a shallow sand aquifer, *Ground Water Monitoring Review,* **7:**64–71.
Barrio-Lage, G. A., Parsons, F. Z., Narbaitz, R. M., and Lorenzo, P. A., 1990, Enhanced anaerobic biodegradation of vinyl chloride in ground water, *Environ. Toxicol. Chem.* **9:**430–415.
Beeman, R. E., and Suflita, J. M., 1987, Microbial ecology of a shallow unconfined groundwater aquifer polluted by municipal landfill leachate, *Microb. Ecol.* **14:**39–54.
Beeman, R. E., Howell, J. E., Shoemaker, S. H., Salazar, E. A., and Buttram, J. R., 1994) A field evaluation of in situ microbial reductive dehalogenation by the biotransformation of chlorinated ethenes, in *Bioremediation of Chlorinated and Polycyclic Aromatic Hydrocarbon Compounds,* (Hinchee, R. E., Leeson, A., Semprini, L., and Ong, S. K. eds., Lewis Publishers, Boca Raton, pp. 14–27.
Beller, H. R., Grbic-Galic, D., and Reinhard, M., 1992, Microbial degradation of toluene under sulfate-reducing conditions and the influence of iron on the process, *Appl. Environ. Microbiol.* **58:**786–793.
Beloin, R. M., Sinclair, J. L., and Ghiorse, W. C., 1988, Distribution and activity of microorganisms in subsurface sediments of a pristine site in Oklahoma, *Microbial Ecology* **16:**85–97.
Bengtsson, G., 1989, Growth and metabolic flexibility in groundwater bacteria, *Microb. Ecol.* **18:**235–248.
Bengtsson, G., and Annadotter, H., 1989, Nitrate reduction in a groundwater microcosm determined by ^{15}N gas chromatography-mass spectrometry, *Appl. Environ. Microb.* **55:**2861–2870.
Bengtsson, G., and Bergwall, C., 1995, Heterotrophic denitrification potential as an adaptive response in groundwater bacteria, *FEMS Microbiol. Ecol.* **16:**307–318.
Bennett, J. L., Updegraff, J. M., Pereira, W. E., and Rostad, C. E., 1985, Isolation and degradation of four species of quinoline-degrading *Pseudomonas* from a creosote-contaminated site at Pensacola, Florida, *Microbios Lett.* **29:**147–154.
Bianchi-Mosquera, G. C., Allen-King, R. M., and Mackay, D. M., 1994, Enhanced degradation of dissolved benzene and toluene using a solid oxygen-releasing compound, *Ground Water Monitoring Review,* **14:**120–128.
Bjerg, P. L., Rugge, K., Pedersen, J. K., and Christensen, T. H., 1995, Distribution of redox-sensitive groundwater quality parameters downgradient of a landfill (Grindsted, Denmark, *Environ. Sci. Technol.* **29:**1387–1394.
Bone, T. L., and Balkwill, D. L., 1988, Morphological and cultural comparison of microorganisms in subsurface soil and subsurface sediments at a pristine study site in Oklahoma, *Microbial Ecology* **16:**49–64.
Bopp, L. H., and Ehrlich, H. L., 1988, Chromate resistance and reduction in *Pseudomonas fluorescens* strain LB300, *Arch. Microbiol.* **150:**426–431.
Borden, R. C., Gomez, C. A., and Becker, M. T., 1995, Geochemical indicators of intrinsic bioremediation, *Ground Water* **33:**180–189.
Bottcher, J., Strebel, O., Voerkelius, S., and Schmidt, H. L., 1990, Using isotope fractionation of nitrate-nitrogen and nitrate-oxygen for evaluation of microbial denitrification in a sandy aquifer, *J. Hydrol.* **114:**413–424.
Bouwer, E. J., 1992, Bioremediation of organic contaminants in the subsurface, in: *Environmental Microbiology* (R. Mitchell, ed.), John Wiley, New York, pp. 287–318.
Bouwer, E. J., and McCarty, P. L., 1983, Transformation of halogenated organic compounds under denitrification conditions, *Appl. Environ. Microbiol.* **45:**1295–1299.
Bouwer, E. J., and Zehnder, A. J. B., 1993, Bioremediation of organic compounds—putting microbial metabolism to work, *TIBTECH* **11:**360–367.

Bouwer, E., Durant, N., Wilson, L., Zhang, W., and Cunningham, A., 1994, Degradation of xenobiotic compounds in situ: capabilities and limits, *FEMS Microbiol. Rev.* **15**:307–317.
Bowman, J. P., Jimenez, L., Rosario, I., Hazen, T. C., and Sayler, G. S., 1993, Characterization of the methanotrophic bacterial community present in a trichloroethylene-contaminated subsurface groundwater site, *Appl. Environ. Microbiol.* **59**:2380–2387.
Bradley, P. M., Aelion, C. M., and Vroblesky, D. A., 1992, Influence of environmental factors on denitrification in sediment contaminated with JP-4 jet fuel, *Ground Water* **30**:843–848.
Braester, C., and Martinell, R., 1988, The vyredox and nitredox methods of *in situ* treatment of groundwater, *Wat. Sci. Tech.* **20**:149–163.
Brockman, F. J., Denovan, B. A., Hicks, R. J., and Frederickson, J. K., 1989, Isolation and characterization of quinoline-degrading bacteria from subsurface sediments, *Appl. Environ. Microbiol.* **55**:1029–1032.
Brown, R. A., and Crosbie, J. R., 1994, Oxygen sources for *in situ* bioremediation, in: *Bioremediation Field Experience* (E. P. Flathman, E. D. Jerger, and H. J. Exner, eds., Lewis Publishers, Boca Raton, pp. 311–332.
Brown, R. A., Norris, R. D., and Raymond, R. L., 1984, Oxygen transport in contaminated aquifers with hydrogen peroxide. *Petroleum Hydrocarbons and Organic Chemicals in Ground Water—Prevention, Detection, and Restoration*, National Water Well Association, Worthington, OH, pp. 441–450.
Brunner, W., Staub, D., and Leisinger, T., 1980, Bacterial degradation of dichloroethane, *Appl. Environ. Microbiol.*, **40**: 950–958.
Buchanan-Mappin, J. M., Wallis, P. M., and Buchanan, A. G., 1985, Enumeration and identification of heterotrophic bacteria in groundwater and in a mountain stream, *Can. J. Microbiol.* **32**:93–98.
Caccavo, F., Jr., Blakemore, R. P., and Lovley, D. R., 1992, A hydrogen-oxidizing, Fe(III)-reducing microorganism from the Great Bay Estuary, New Hampshire, *Appl. Environ. Microbiol.* **58**:3211–3216.
Caccavo, F., Lonergan, D. J., Lovley, D. R., Davis, M., Stolz, J. F., and McInerney, M. J., 1994, *Geobacter sulfurreducens* sp. nov., a hydrogen- and acetate-oxidizing dissimilatory metal-reducing microorganism, *Appl. Environ. Microbiol.* **60**:3752–3759.
Cerniglia, C., E., 1992, Biodegradation of polycyclic aromatic hydrocarbons, *Biodegradation* **3**:351–368.
Champ, D. R., Gulens, J., and Jackson, R. E., 1979, Oxidation-reduction sequences in ground water flow systems, *Can. J. Earth Sci.* **16**:12–23.
Chang, M.-K., Voice, T. C., and Criddle, C. S., 1993, Kinetics of competitive inhibition and cometabolism in the biodegradation of benzene, toluene, and p-xylene by two *Pseudomonas* isolates, *Biotechnol. Bioeng.* **41**:1057–1065.
Chapelle, F. H., 1993, *Ground-water Microbiology and Geochemistry*, John Wiley, New York.
Chapelle, F. H., and Lovley, D. R., 1990, Rates of microbial metabolism in deep coastal plain aquifers, *Appl. Environ. Microbiol.* **56**:1865–1874.
Chapelle, F. H., and Lovley, D. R., 1992, Competitive exclusion of sulfate reduction by Fe(III)-reducing bacteria: a mechanism for producing discrete zones of high-iron ground water, *Ground Water* **30**:29–36.
Chapelle, F. H., McMahon, P. B., Dubrovsky, N. M., Fujii, R. F., Oaksford, E. T., and Vroblesky, D. A., 1995, Deducing the distribution of terminal electron-accepting processes in hydrologically diverse groundwater systems, *Water Resour. Res.* **31**:359–371.
Chapelle, F. H., Bradley, P. M., Vroblesky, D. A., and Lovley, D. R., 1996, Measuring rates of biodegradation in a petroleum hydrocarbon-contaminated aquifer, *Ground Water*, **34**:691–698.
Christensen, T., H., Kjeldsen, P., Albrechtsen, H.-J., and Heron, G., 1994, Attenuation of pollu-

tants in landfill leachate polluted aquifers, *Critical Reviews in Environ. Sci. Technol.* **24:**119–202.
Coates, J. D., Anderson, R. T., Woodward, J. C., Phillips, E. J. P., and Lovley, D. R., 1996a, Anaerobic hydrocarbon degradation in petroleum contaminated harbor sediments under sulfate- and artificially imposed iron-reducing conditions, *Environ. Sci. Technol.,* **30:**2784–2789.
Coates, J. D., Lonergan, D. J., Jenter, H., and Lovley, D. R., 1996b, Isolation of *Geobacter* species from a variety of sedimentary environments, *Appl. Environ. Microbiol.* **62:**1531–1536.
Cole, J., 1993, Controlling environmental nitrogen through microbial metabolism, *Trends Biotechnol.* **11:**368–372.
Coleman, M. L., Hedrick, D. B., Lovley, D. R., White, D. C., and Pye, K., 1993, Reduction of Fe(III) in sediments by sulphate-reducing bacteria, *Nature* **361:**436–438.
Colwell, F. S., 1989, Microbiological comparison of surface soil and unsaturated subsurface soil from a semiarid high desert, *Appl. Environ. Microbiol.* **55:**2420–2423.
Cord-Ruwisch, R., Seitz, H., and Conrad, R., 1988, The capacity of hydrogenotrophic anaerobic bacteria to compete for traces of hydrogen depends on the redox potential of the terminal electron acceptor, *Arch. Microbiol.* **149:**350–357.
Corseuil, H. X., and Weber, W. J. J., 1994, Potential biomass limitations on rates of degradation of monoaromatic hydrocarbons by indigenous microbes in subsurface soils, *Wat. Res.* **28:**1415–1423.
Criddle, C. S., DeWitt, J. T., Grbic-Galic, D., and McCarty, P., 1990, Transformation of carbon tetrachloride by *Pseudomonas* sp. strain KC under denitrification conditions, *Appl. Environ. Microbiol.* **56:**3240–3246.
Crocetti, C. A., Head, C. L., and Ricciardelli, A. J., 1992, Aeration-enhanced bioremediation of oil- contaminated soils: A laboratory treatability study, in: *Proceedings of the Conference entitled Petroleum Hydrocarbons and Organic Chemicals in Ground Water: Prevention, Detection, and Restoration,* Houston, pp. 427–440.
Dagley, S., 1971, Catabolism of aromatic compounds by microorganisms, *Adv. Microb. Physiol.* **6:**1–46.
Dahab, M. F., 1993, Comparison and evaluation of *in-situ* bio-denitrification systems for nitrate reduction in groundwater, *Wat. Sci. Tech.* **28:**359–368.
Davis, J. W., Klier, N. J., and Carpenter, C. L., 1994, Natural biological attenuation of benzene in ground water beneath a manufacturing facility, *Ground Water* **32:**215–226.
De Bruin, W. P., Kotterman, M. J. J., Posthumus, M. A., Schraa, G., and Zehnder, A. J. B., 1992, Complete biological reductive transformation of tetrachloroethene to ethane, *Appl. Environ. Microbiol.* **58:**1996–2000.
Dobbins, D. C., Aelion, C. M., and Pfaender, F., 1992, Subsurface terrestial microbial ecology and biodegradation of organic chemicals: a review, *Critical Reviews in Environmental Control* **22:**67–136.
Dolfing, J., Zeyer, J., Binder-Eicher, P., and Schwarzenbach, R. P., 1990, Isolation and characterization of a bacterium that mineralizes toluene in the absence of molecular oxygen, *Arch Microbiol.* **134:**336–341.
Durant, N. D., Wilson, L. P., and Bouwer, E. J., 1995, Microcosm studies of subsurface PAH- degrading bacteria from a former manufactured gas plant, *J. Contam. Hydrol.* **17:**213–237.
Edwards, E. A., and Grbic-Galic, D., 1992, Complete mineralization of benzene by aquifer microorganisms under strictly anaerobic conditions, *Appl. Environ. Microbiol.* **58:**2663–2666.
Edwards, E. A., and Grbic-Galic, D., 1994, Anaerobic degradation of toluene and *o*-xylene by a methanogenic consortium, *Appl. Environ. Microbiol.* **60:**313–322.
Edwards, E. A., Wills, L. E., Reinhard, M., and Grbic-Galic, D., 1992, Anaerobic degradation of toluene and xylene by aquifer microorganisms under sulfate-reducing conditions, *Appl. Environ. Microbiol.* **58:**794–800.

Egli, C., Tschan, T., Scholtz, R., Cook, A. M., and Leisinger, T., 1988, Transforamtion of tetrachloromethane to dichloromethane and carbon dioxide by *Acetobacterium woodii*, *Appl. Environ. Microbiol.* **54:**2819–2824.

Egli, C., Stromeyer, S., Cook, A. M., and Leisinger, T., 1990, Transformation of tetra- and trichloromethane to CO_2 by an anaerobic bacteria is a non-enzymic process, *FEMS Microbiol. Lett.* **68:**207–212.

Ellis, B., Balba, M. T., and Theile, P., 1990, Bioremediation of oil contaminated land, *Environ. Technol.* **11:**443–455.

Ellis, B., Harold, P., and Kronberg, H., 1991, Bioremediation of a creosote contaminated site, *Environ. Technol.* **12:**447–459.

Evans, P. J., Mang, D. T., and Young, L. Y., 1991a, Degradation of toluene and m-xylene and transformation of o-xylene by denitrifying enrichment cultures, *Appl. Environ. Microbiol.* **57:**450–454.

Evans, P. J., Mang, D. T., Kim, K. S., and Young, L. Y., 1991b, Anaerobic degradation of toluene by a denitrifying bacterium, *Appl. Environ. Microbiol.* **57:**1139–1145.

Federle, T. W., Dobbins, D. C., and Thornton-Manning, J. R., 1986, Microbial biomass, activity, and community structure in subsurface soils, *Ground Water* **24:**365–374.

Federle, T. W., Ventullo, R. M., and White, D. C., 1990, Spatial distribution of microbial biomass, activity, community structure, and the biodegradation of linear alkylbenzene sulfonate (LAS) and linear alcohol ethoxylate (LAE) in the subsurface, *Microbial Ecology* **20:**297–313.

Fish, W., 1993, Sub-surface redox chemistry: a comparison of equilibrium and reaction-based approaches, in: *Metals in Groundwater* (H. E. Allen, E. M.Perdue and D. S. Brown, eds.), Lewis Publishers, Ann Arbor, MI, pp. 73–101.

Flathman, P. E., Jerger, D. E., and Exner, J. H., 1994, *Bioremediation Field Experience*, CRC Press, Boca Raton.

Fleming, J. T., Sanseverino, J., and Sayler, G. S., 1993, Quantitative relationship between naphthalene catabolic gene frequency and expression in predicting PAH degradation in soils at town gas manufacturing sites, *Environ. Sci. Technol.* **27:**1068–1074.

Fliermans, C. B., Phelps, T. J., Ringelberg, D., Mikell, A. T., and White, D. C., 1988, Mineralization of trichloroethylene by heterotrophic enrichment cultures, *Appl. Environ. Microbiol.* **54:**1709–1714.

Flyvbjerg, J., Arvin, E., Jensen, B. K., and Olsen, S. K., 1993, Microbial degradation of phenols and aromatic hydrocarbons in creosote-contaminated groundwater under nitrate-reducing conditions, *J. Contam. Hydrol.* **12:**133–150.

Fogel, M. M., Taddeo, A. R., and Fogel, S., 1986, Biodegradation of chlorinated ethenes by a methane utilizing mixed culture, *Appl. Environ. Microbiol.* **51:**720–724.

Fontes, D. E., Mills, A. L., Hornberger, G. M., and Herman, J. S., 1991, Physical and chemical factors influencing transport of microorganisms through porous media, *Appl. Environ. Microbiol.* **57:**2473–2481.

Francis, A. J., Slater, J. M., and Dodge, C. J., 1989, Denitrification in deep subsurface sediments, *Geomicrobiol. J.* **7:**103–116.

Francy, D. S., Thomas, J. M., Raymond, R. L., and Ward, C. H., 1991, Emulsification of hydrocarbons by subsurface bacteria, *J. Indus. Microbiol.* **8:**237–246.

Frederickson, J. K., and Hicks, R. J., 1987, Probing reveals many microbes beneath Earth's surface, *ASM News* **53:**78–79.

Frederickson, J. K., Garland, T. R., Hicks, R. J., Thomas, J. M., Li, S. W., and McFadden, K. M., 1989, Lithotrophic and heterotrophic bacteria in deep subsurface sediments and their relation to sediment properties, *Geomicrobiol. J.* **7:**53–66.

Frederickson, J. K., Brockman, F. J., Workman, D. J., Li, S. W., and Stevens, T. O., 1991, Isolation

and characterization of a subsurface bacterium capable of growth on toluene, naphthalene, and other aromatic compounds. *Appl. Environ. Microbiol.* **57**:796–803.
Freedman, D. L., and Gossett, J. M., 1989, Biological reductive dechlorination of tetrachloroethylene and trichloroethylene to ethylene under methanogenic conditions, *Appl. Environ. Microbiol.*, **55**: 2144–2151.
Fries, M. R., Zhou, J., Chee-Sanford, J., and Tiedje, J. M., 1994, Isolation, characterization, and distribution of denitrifying toluene degraders from a variety of habitats, *Appl. Environ. Microbiol.* **60**:2802–2810.
Froelich, P. N., Klinkhammer, G. P., Bender, M. L., Luedtke, N. A., Heath, G. R., Cullen, D., Dauphin, P., Hammond, D., Hartman, B., and Maynard, V., 1979, Early oxidation of organic matter in pelagic sediments of the eastern equatorial Atlantic: suboxic diagenesis, *Geochim. Cosmochim. Acta.* **43**:1075–1090.
Fu, G., Kan, A. T., and Tomson, M., 1994, Adsorption and hysteresis of PAHs in surface sediment, *Environ. Toxicol. Chem.* **13**:1559–1567.
Gadd, G. M., 1988, Accumulation of metals by microorganisms and algae, *Biotechnology* **6b**:403–433.
Galli, R., and McCarty, P. L., 1989a, Biotransformation of 1,1,1-trichloroethane, trichloromethane, and tetrachloromethane by a *Clostridium* sp., *Appl. Environ. Microbiol.* **55**:837–844.
Galli, R., and McCarty, P. L., 1989b, Kinetics of biotransformation of 1,1,1-trichloroethane by *Clostridium* sp. strain TCAIIB, *Appl. Environ. Microbiol.* **55**:845–851.
Gannon, J., Tan, Y., Baveye, P., and Alexander, M., 1991, Effect of sodium chloride on transport of bacteria in a saturated aquifer material, *Appl. Environ. Microbiol.* **57**:2497–2501.
Gayle, B. P., Boardman, G. E., Sherrard, J. H., and Benoit, R. E., 1989, Biological denitrification of water, *J. Environ. Engineer.* **115**:930–943.
Ghiorse, W. C., and Balkwill, D. L., 1983, Enumeration and morphological characterization of bacteria indigenous to subsurface environments, *Dev. Ind. Microbiol.* **24**:213–224.
Ghiorse, W. C., and Wilson, J. T., 1988, Microbial ecology of the terrestial subsurface, *Adv. Appl. Microbiol.* **33**:107–172.
Gibson, D. T., and Subramanian, V., 1984, Microbial degradation of aromatic hydrocarbons, in: *Microbial Degradation of Organic Compounds* (D. T. Gibson, ed.), Marcel Dekker, New York, pp. 181–252.
Gibson, S. A., and Sewell, G. W., 1992, Stimulation of reductive dechlorination of tetrachloroethene in anaerobic aquifer microcosms by addition of short-chain organic acids or alcohols, *Appl. Environ. Microbiol.* **58**:1392–1393.
Gillham, R. W., Starr, R. C., and Miller, D. J., 1990, A device for *in situ* determination of geochemical transport parameters 2. biochemical reactions, *Ground Water* **28**:858–862.
Girvin, D. C., Gassman, P. L., and Bolton, H., Jr., 1993, Adsorption of aqueous cobalt ethylenediaminetetraacetate by Al_2O_3: effects of oxidation state, ionic strength and sorbent concentration, *Soil Sci. Soc. Am. J.* **57**:47.
Goldstein, R. A., Olson, B. H., and Porcella, D. B., 1988, Conceptual model of genetic regulation of mercury biogeochemical cycling, *Environ. Technol. Let.* **9**:957–964.
Gounot, A.-M., 1994, Microbial oxidation and reduction of manganese: consequences in groundwater applications, *FEMS Microbiol. Rev.* **14**:339–350.
Grant, D. J. W., and Al-Najjar, T. R., 1976, Degradation of quinoline by a soil bacterium, *Microbios* **15**:177–189.
Grbic-Galic, D., and Vogel, T., 1987, Transformation of toluene and benzene by mixed methanogenic cultures, *Appl. Environ. Microbiol.* **53**:254–260.
Grosser, R. J., Warshawsky, D., and Vestal, J. R., 1991, Indigenous and enhanced mineralization of pyrene, benzo[a]pyrene, and carbazole in soils, *Appl. Environ. Microbiol.* **57**:3462–3469.

Gunther, K., Schlosser, D., and Fritsche, W., 1995, Phenol and cresol metabolism in *Bacillus pumilus* isolated from contaminated groundwater. *J. Basic Microbiol.* **35**:83–92.

Haag, F., Reinhard, M., and McCarty, P. L., 1991, Degradation of toluene and *p*-xylene in anaerobic microcosms: evidence for sulfate as a terminal electron acceptor, *Environ. Toxicol. Chem.* **10**:1379–1389.

Hamon, M., and Fustec, E., 1991, Laboratory and field study of an in situ groundwater denitrification reactor, *J. Wat. Pollut. Control Fed.* **63**:942–949.

Haner, A., Hohener, P., and Zeyer, J., 1995, Degradation of *p*-xylene by a denitrifying enrichment culture, *Appl. Environ. Microbiol.* **61**:3185–3188.

Hardman, D. J., 1991, Biotransformation of halogenated compounds, *Critical Reviews in Biotechnology* **11**:1–40.

Hardoyo, J. K., and Ohtake, H., 1991, Effects of heavy metal cations on chromate reduction by *Enterobacter cloacae* strain HO1, *J. Gen. Appl. Microbiol.* **37**:519–522.

Harvey, R. W., and George, L. H., 1987, Growth determinations for unattached bacteria in a contaminated aquifer, *Appl. Environ. Microbiol.* **53**:2992–2996.

Harvey, R. W., Smith, R. L., and George, L., 1984, Effect of organic contamination upon the microbial distributions and heterotrophic uptake in a Cape Cod, Mass., aquifer, *Appl. Environ. Microbiol.* **48**:1197–1202.

Harvey, R. W., Kinner, N. E., Bunn, A., MacDonald, D., and Metge, D., 1995, Transport behavior of groundwater protozoa and protozoan-sized microspheres in sandy aquifer sediments, *Appl. Environ. Microbiol.* **61**:209–217.

Heitkamp, M. A., Freeman, J. P., and Cerniglia, C. E., 1987, Naphthalene biodegradation in environmental microcosms: estimates of degradation rates and charcterization of metabolites, *Appl. Environ. Microbiol.* **53**:129–136.

Henry, S. M., and Grbic-Galic, D., 1991a, Influence of endogenous and exogenous electron donors and trichloroethylene oxidation toxicity on trichloroethylene oxidation by methanotrophic cultures from a groundwater aquifer, *Appl. Environ. Microbiol.* **57**:236–244.

Henry, S. M., and Grbic-Galic, D., 1991b, Inhibition of trichloroethylene oxidation by the transformation intermediate carbon monoxide, *Appl. Environ. Microbiol.* **57**:1770–1776.

Herbes, S. E., and Schwall, L. R., 1978, Microbial transformation of polycyclic aromatic hydrocarbons in pristine and petroleum-contaminated sediments, *Appl. Environ. Microbiol.* **35**:306–316.

Hicks, R. J., and Frederickson, J. K., 1989, Aerobic metabolic potential of microbial populations indigenous to deep subsurface environments, *Geomicrobiol. J.* **7**:67–77.

Hirsch, P., 1992, Microbiology, in: *Progress in Hydrogeochemistry* (G. Matthess, F.Frimmel, P. Hirsch, H. D. Schulz, and H.-E. Usdowski, eds.), Springer-Verlag, New York, pp. 308–311.

Hirsch, P., and Rades-Rohkohl, E., 1983, Microbial diversity in a groundwater aquifer in northern Germany, *Dev. Ind. Microbiol.* **24**:183–200.

Hirsch, P., and Rades-Rohkohl, E., 1988, Some special problems in the determination of viable counts of groundwater microorganisms, *Microbial Ecology* **16**:99–113.

Hoeppel, R. E., and Hinchee, R. E., 1993, Enhanced biodegradation for on-site remediation of contaminated soils and groundwater, in: *Hazardous Waste Site Soil Remediation* (J. D. Wilson, and N. A. Clarke, eds.), Marcel Dekker, New York, pp. 311–431.

Holliger, C., and Schraa, G., 1994, Physiological meaning and potential for application of reductive dechlorination by anaerobic bacteria, *FEMS Microbiol. Rev.* **15**:297–305.

Holliger, C., and Schumacher, W., 1994, Reductive dehalogenation as a respiratory process, *Antoine van Leeuwenhoek* **66**:239–246.

Holliger, C., Schraa, G., Stams, A. J. M., and Zehnder, A. J. B., 1993, A highly purified enrichment culture couples the reductive dechlorination of tetrachloroethene to growth, *Appl. Environ. Microbiol.* **59**:2991–2997.

Holm, P. E., Nielsen, P. H., and Christensen, T. H., 1991, Aerobic groundwater and groundwater

sediment degradation potential for xenobiotic compounds measured *in situ*, in: *In Situ Bioreclamation*, (E. R. Hinchee, and F. R. Olfenbuttel, eds.), Butterworth-Heinemann, Stoneham, MA, pp. 413–419.
Holm, P. E., Nielson, P. H., Albrechtson, H.-J., and Christensen, T. H., 1992, Importance of unattached bacteria and bacteria attached to sediment in determining potentials for degradation of xenobiotic organic contaminants in an aerobic aquifer, *Appl. Environ. Microbiol.* **58**:3020–3026.
Hopkins, G. D., and McCarty, P. L., 1995, Field evaluation of *in situ* aerobic cometabolism of trichloroethylene and three dichloroethylene isomers using phenol and toluene as the primary substrates, *Environ. Sci. Technol.* **29**:1628–1637.
Hopkins, G. D., Munakata, J., Semprini, L., and McCarty, P. L., 1993a, Trichloroethylene concentration effects on pilot field-scale in-situ groundwater bioremediation by phenol-oxidizing microorganisms, *Environ. Sci. Technol.* **27**:2542–2547.
Hopkins, G. D., Semprini, L., and McCarty, P. L., 1993b, Microcosm and in situ field studies of enhanced biotransformation of trichloroethylene by phenol-utilizing microorganisms, *Appl. Environ. Microbiol.* **59**:2277–2285.
Horitsu, H., Futo, S., Miyazawa, Y., Ogai, S., and Kawai, K., 1987, Enzymatic reduction of hexavalent chromium by hexavalent chromium tolerant *Pseudomonas ambigua* G-1, *Agric. Biol. Chem.* **51**:2417–2420.
Hou, C. T., Patel, R., Laskin, A. I., Barnabe, N., Barist, I., 1983, Epoxidation of short-chain alkenes by resting-cell suspensions of propane-grown bacteria, *Appl. Environ. Microbiol.* **46**:171–177.
Huling, S. G., Bledsoe, B. E., and White, M. V., 1990, Enhanced bioremediation using hydrogen peroxide as a supplemental source of oxygen: a laboratory and field study, US Environmental Protection Agency Report, EPA/600/52–90/006
Hutchins, S. R., 1991, Optimizing BTEX biodegradation under denitrifying conditions, *Environ. Toxicol. Chem.* **10**:1437–1448.
Hutchins, S. R., Sewell, G. W., Kovacs, D. A., and Smith, G. A., 1991, Biodegradation of aromatic hydrocarbons by aquifer microorganisms under denitrifying conditions, *Environ. Sci. Technol.* **25**:68–76.
Hyman, M. R., Russel, S. A., Ely, R. L., Williamson, K. J., and Arp, D. J., 1995, Inhibition, inactivation and recovery of ammonia-oxidizing activity in cometabolism of trichloroethylene by *Nitrosomonas europaea*, *Appl. Environ. Microbiol.* **61**:1480–1487.
Ishibashi, Y., Cervantes, C., and Silver, S., 1990, Chromium reduction in *Pseudomonas putida*, *Appl. Environ. Microbiol.* **56**:2268–2270.
Jackson, R. E., and Patterson, R. J., 1982, Interpretation of pH and Eh trends in a fluvial-sand aquifer system, *Wat. Resour. Res.* **18**:1255–1268.
Jain, R. K., Sayler, G. S., Wilson, J. T., Houston, L., and Pacia, D., 1987, Maintenance and stability of introduced genotypes in groundwater aquifer material, *Appl. Environ. Microbiol.* **53**:996–1002.
Janda, V., Rudovsky, J., Wanner, J., and Marha, K., 1988, *In situ* denitrification of drinking water, *Wat. Sci. Tech.* **20**:215–219.
Janssen, D. B., Scheper, A., Dijkhuizen, L., and Witholt, B., 1985, Degradation of halogenated aliphatic compounds by *Xanthobacter autotrophicus* GJ10, *Appl. Environ. Microbiol.*, **49**: 673–677.
Jaudon, P., Massiani, J. G., Rey, J., and Vacelet, E., 1989, Groundwater pollution by manganese. manganese speciation: application to the selection and discussion of an in situ groundwater treatment, *Sci. Tot. Environ.* **84**:169–183.
Jenkins, K. B., Michelsen, D. L., and Novak, J. T., 1993, Application of oxygen microbubbles for in situ biodegradation of p-xylene-contaminated groundwater in a soil column, *Biotechnol. Prog.* **9**:394–400.

Jensen, B. K., 1989, ATP-related specific heterotrophic activity in petroleum contaminated and uncontaminated groundwaters, *Can. J. Microbiol.* **35**:814–818.

Ka, J. O., Holben, W. E., and Tiedje, J. M., 1994a, Analysis of competition in soil among 2,4-dichlorophenoxyacetic acid-degrading bacteria, *Appl. Environ. Microbiol.* **60**:1121–1128.

Ka, J. O., Holben, W. E., and Tiedje, J. M., 1994b, Use of gene probes to aid in the recovery and identification of functionally dominant 2,4-dichlorophenoxyacetic acid-degrading populations in soil, *Appl. Environ. Microbiol.* **60**:1116–1120.

Ka, J. O., Holben, W. E., and Tiedji, J. M., 1994c, Genetic and phenotypic diversity of 2,4-dichlorophenoxyacetic acid (2,4-D)-degrading bacteria isolated from 2,4-D-treated field soils, *Appl. Environ. Microbiol.* **60**:1106–1115.

Kaiser, J.-P., and Bollag, J.-M., 1992, Influence of soil inoculum and redox potential on the degradation of several pyridine derivatives, *Soil Biology and Biochemistry,* **24**:351–357.

Kampfer, P., Steiof, M., and Dott, W., 1991, Microbiological characterization of a fuel-oil contaminated site including numerical identification of heterotrophic water and soil bacteria, *Microb. Ecol.* **21**:227–251.

Kampfer, P., Steiof, M., Becker, P. M., and Dott, W., 1993, Characterization of chemoheterotrophic bacteria associated with the in situ bioremediation of a waste-oil contaminated site, *Microb. Ecol.* **26**:161–188.

Keenan, J. E., Strand, S. E., and Stensel, H. D., 1994, Degradation kinetics of chlorinated solvents by a propane-oxidizing culture, in *Bioremediation of Chlorinated and Polycyclic Aromatic Hydrocarbon Compounds,* (R. E. Hinchee, A. Leeson, L. Semprini, and S.K. Ong, eds., Lewis Publishers, Boca Raton, pp. 1–13.

King, R., Barry, Long, G. M., and Sheldon, J. K., 1992, *Practical Environmental Bioremediation,* CRC Press, Inc., Boca Raton.

Kishi, H., Kogure, N., and Hashimoto, Y., 1990, Contribution of soil constituents in adsorption coefficient of aromatic compounds, halogenated alicyclic and aromatic compounds to soil, *Chemosphere* **21**:867–876.

Kitts, C. L., Cunningham, D. P., and Unkefer, P. J., 1994, Isolation of three hexahydro-1,3,5-trinitro-1,3,5-triazine-degrading species of the family *Enterobacteriaceae* from nitramine explosive-contaminated soil, *Appl. Environ. Microbiol.* **60**:4608–4711.

Klecka, G. M., Davis, J. W., Gray, D. R., and Madsen, S. S., 1990, Natural bioremediation of organic contaminants in ground water: cliffs-dow superfund site, *Ground Water* **28**:534–543.

Koh, S.-C., Bowman, J. P., and Sayler, G. S., 1993, Soluble methane monooxygenase production and trichloroethylene degradation by a type I methanotroph *Methylomonas methanica* 68-1, *Appl. Environ. Microbiol.* **59**:960–967.

Kolbel-Boelke, J., Anders, E.-M., and Nehrkorn, A., 1988b, Microbial communities in the saturated groundwater environment II: diversity of bacterial communities in a pleistocene sand aquifer and their in vitro activities, *Microbial. Ecol.* **16**:31–48.

Kolbel-Boelke, J., Teinken, B., and Nehrkorn, A., 1988a, Microbial communities in the saturated groundwater environment I: methods of isolation and characterization of heterotrophic bacteria, *Microbial. Ecol.* **16**:17–29.

Konopka, A., 1993, Isolation and characterization of a subsurface bacterium that degrades aniline and methylanilines, *FEMS Microbiology Letters,* **111**:93–100.

Korom, S. F., 1992, Natural denitrification in the saturated zone: a review, *Wat. Resour. Res.* **28**:1657–1668.

Korte, N., 1991, Naturally occurring arsenic in groundwaters of the midwestern United States, *Environ. Geol. Water. Sci.* **18**:137–141.

Krumme, M. L., Timmis, K. N., and Dwyer, D. F., 1993, Degradation of trichloroethylene by *Pseudomonas cepacia* G4 and the constitutive mutant strain G4 5223 PR1 in aquifer microcosms, *Appl. Environ. Microbiol.* **59**:2746–2749.

Krumme, M. L., Smith, R. L., Egestorff, J., Thiem, S. M., Tiedje, J. M., Timmis, K. N. and Dwyer, D. F., 1994, Behavior of pollutant-degrading microorganisms in aquifers: predictions for genetically engineered organisms, *Environ. Sci. Technol.* **28:**1134–1138.

Kuhn, E. P., and Suflita, J. M., 1989, Dehalogenation of pesticides by anaerobic microorganisms in soils and groundwater—a review, in *Reactions and Movements of Organic Chemicals in Soils*, (B. L. Sawhney, and K. Brown, eds.), Soil Science Society of America and American Society of Agronomy, Madison, WI., pp. 111–180.

Kuhn, E. P., Colberg, P. J., Schnoor, J. L., Wanner, O., Zehnder, A. J. B., and Schwarzenbach, R. P., 1985, Microbial transformations of substituted benzenes during infiltration of river water to groundwater: laboratory column studies, *Environ. Sci. Technol.* **19:**961–968.

Kuhn, E. P., Zeyer, J., Eicher, P., and Schwarzenbach, R. P., 1988, Anaerobic degradation of alkylated benzenes in denitrifying laboratory aquifer columns, *Appl. Environ. Microbiol.* **54:**490–496.

Kuznetsov, S. I., Ivanov, M. V., and Lyalikova, N. N., 1963, *Introduction to Geological Microbiology*, McGraw-Hill, New York.

Laverman, A. M., Blum, J., Switzer, Schaeffer, J. K., Phillips, E. J. P., Lovley, D. R., and Oremland, R. S., 1995, Growth of SES-3 with arsenate and other diverse electron acceptors, *Appl. Environ. Microbiol.* **61:**3556–3561.

Leduc, R., Samson, R., Al-Bashir, B., Al-Hawari, J., and Cseh, T., 1992, Biotic and abiotic disappearance of four PAH compounds from flooded soil under various redox conditions, *Wat. Sci. Tech.* **26:**51–60.

Lee, M. D., Thomas, J. M., Borden, R. C., Bedient, P. B., Ward, C. H., and Wilson, J. T., 1988, Biorestoration of aquifers contaminated with organic compounds, *CRC Critical Reviews in Environmental Control* **18:**29–89.

Lee, M. D., Wilson, J. T., and Ward, C. H., 1983, Microbial degradation of selected aromatics in a harzardous waste site, *Dev. Ind. Microbiol.* **44:**557–565.

Lee, W. E., and Borden, R. C., 1988, Anaerobic biotransformation of hydrocarbons in the subsurface: field observations, *EOS* **69:**368.

Lewis, T. A., and Crawford, R. L., 1993, Physiological factors affecting carbon tetrachloride dehalogenation by the denitrifying bacterium *Pseudomonas* sp. strain KC, *Appl. Environ. Microbiol.* **59:**1635–1641.

Little, C. D., Palumbo, A. V., Herbes, S. E., Lidstrom, M. E., Tyndall, R. L., and Gilmer, P. J., 1988, Trichloroethylene biodegradation by a methane-oxidizing bacterium, *Appl. Environ. Microbiol.* **54:**951–956.

Lloyd-Jones, G., and Trudgill, P. W., 1989, The degradation of alicyclic hydrocarbons by a microbial consortium, *International Biodeterioration*, **25:**197–206.

Long, R. H. B., Benson, S. M., Tokunaga, T. K., and Yee, A., 1990, Selenium immobilization in a pond sediment at Kesterson Reservoir, *J. Environ. Qual.* **19:**302–311.

Lovley, D. R., 1985, Minimum threshold for hydrogen metabolism in methanogenic bacteria, *Appl. Environ. Microbiol.* **49:**1530–1531.

Lovley, D. R., 1991, Dissimilatory Fe(III) and Mn(IV) reduction, *Microbiol. Rev.* **55:**259–287.

Lovley, D. R., 1993, Dissimilatory metal reduction, *Annu. Rev. Microbiol.* **47:**263–90.

Lovley, D. R., 1995, Microbial reduction of iron, manganese, and other metals, *Adv. Agron.* **54:**175–231.

Lovley, D. R., 1997, Potential for anaerobic bioremediation of BTEX in petroleum-contaminated aquifers, *J. Industr. Microbiol.* **18:**75–81.

Lovley, D. R., and Chapelle, F. H., 1995a, Deep subsurface microbial processes, *Rev. Geophys.* **33:**365–381.

Lovley, D. R., and Chapelle, F. H., 1997, A modeling approach to elucidating the distribution and rates of microbially catalyzed redox reactions in anoxic groundwater, in: *Mathematical Models in Microbial Ecology*, (J. A. Robinson, ed.), Chapman and Hall, New York, (in press).

Lovley, D. R., and Goodwin, S., 1988, Hydrogen concentrations as an indicator of the predominant terminal electron accepting reactions in aquatic sediments, *Geochim. Cosmochim. Acta.* **52**:2993–3003.

Lovley, D. R., and Klug, M. J., 1986, Model for the distribution of sulfate reduction and methanogenesis in freshwater sediments, *Geochim. Cosmochim. Acta.* **50**:11–18.

Lovley, D. R., and Lonergan, D. J., 1990, Anaerobic oxidation of toluene, phenol, and p-cresol by the dissimilatory iron-reducing organism, GS-15, *Appl. Environ. Microbiol.* **56**:1858–1864.

Lovley, D. R., and Phillips, E. J. P., 1987, Competitive mechanisms for inhibition of sulfate reduction and methane production in the zone of ferric iron reduction in sediments, *Appl. Environ. Microbiol.* **53**:2636–2641.

Lovley, D. R., and Phillips, E. J. P., 1992a, Bioremediation of uranium contamination with enzymatic uranium reduction, *Environ. Sci. Technol.* **26**:2228–2234.

Lovley, D. R., and Phillips, E. J. P., 1992b, Reduction of uranium by *Desulfovibrio desulfuricans*, *Appl. Environ. Microbiol.* **58**:850–856.

Lovley, D. R., and Phillips, E. J. P., 1994a, Novel processes for anoxic sulfate production from elemental sulfur by sulfate-reducing bacteria, *Appl. Environ. Microbiol* **60**:2394–2399.

Lovley, D. R., and Phillips, E. J. P., 1994b, Reduction of chromate by *Desulfovibrio vulgaris* (Hildenborough) and its c_3 cytochrome, *Appl. Environ. Microbiol.* **60**:726–728.

Lovley, D. R., and Woodward, J. C., 1992, Consumption of CFC-11 and CF-12 by anaerobic sediments and soils, *Environ. Sci. Technol.* **26**:925–929.

Lovley, D. R., and Woodward, J. C., 1996, Mechanism for chelator stimulation of microbial Fe(III) oxide reduction, *Chem. Geol.*, **132**: 19–24.

Lovley, D. R., Dwyer, D. F., and Klug, M. J., 1982, Kinetic analysis of competition between sulfate reducers and methanogens for hydrogen in sediments, *Appl. Environ. Microbiol.* **43**:1373–1379.

Lovley, D. R., Baedecker, M. J., Lonergan, D. J., Cozzarelli, I. M., Phillips, E. J. P., and Siegel, D. I., 1989a, Oxidation of aromatic contaminants coupled to microbial iron reduction, *Nature* **339**:297–299.

Lovley, D. R., Phillips, E. J. P., and Lonergan, D. J., 1989b, Hydrogen and formate oxidation coupled to dissimilatory reduction of iron or manganese by *Alteromonas putrefaciens*, *Appl. Environ. Microbiol.* **55**:700–706.

Lovley, D. R., Phillips, E. J. P., Gorby, Y. A., and Landa, E. R., 1991, Microbial reduction of uranium, *Nature* **350**:413–416.

Lovley, D. R., Roden, E. E., Phillips, E. J. P., and Woodward, J. C., 1993a, Enzymatic iron and uranium reduction by sulfate-reducing bacteria, *Marine Geol.* **113**:41–53.

Lovley, D. R., Widman, P. K., Woodward, J. C., and Phillips, E. J. P., 1993b, Reduction of uranium by cytochrome c_3 of *Desulfovibrio vulgaris*, *Appl. Environ. Microbiol.* **59**:3572–3576.

Lovley, D. R., Chapelle, F. H., and Woodward, J. C., 1994a, Use of dissolved H_2 concentrations to determine the distribution of microbially catalyzed redox reactions in anoxic groundwater, *Environ. Sci. Technol.* **28**:1205–1210.

Lovley, D. R., Woodward, J. C., and Chapelle, F. H., 1994b, Stimulated anoxic biodegradation of aromatic hydrocarbons using Fe(III) ligands, *Nature* **370**:128–131.

Lovley, D. R., Coates, J. D., Woodward, J. C., and Phillips, E. J. P., 1995, Benzene oxidation coupled to sulfate reduction, *Appl. Environ. Microbiol.* **61**:953–958.

Lovley, D. R., Coates, J. D., Blunt-Harris, E. L., Phillips, E. J. P., and Woodward, J. C., 1996a, Humic substances as electron acceptors for microbial respiration, *Nature*, **382**: 445–448.

Lovley, D. R., Woodward, J. C., and Chapelle, F. H., 1996b, Rapid anaerobic benzene oxidation with a variety of chelated Fe(III) forms, *Appl. Environ. Microbiol.* **62**:288–291.

Luthy, R. G., Dzombak, D. A., Peters, C. A., Roy, S. B., Ramaswami, A., Nakles, D. V., and Nott, B. R., 1994, Remediating tar-contaminated soils at manufactured gas plant sites, *Environ. Sci. Technol.* **28**:266A–276A.

Lyngkilde, J., and Christensen, T. H., 1992a, Fate of organic contaminants in the redox zones of a landfill leachate pollution plume (Vejen, Denmark), *J. Contam. Hydrol.* **10**:291–307.

Lyngkilde, J., and Christensen, T., H., 1992b, Redox zones of a landfill leachate pollution plume (Vejen, Denmark), *J. Contam. Hydrol.* **10**:273–289.

Macaskie, L. E., 1991, The application of biotechnology to the treatment of wastes produced from the nuclear fuel cycle: Biodegradation and bioaccumulation as a means of treating radionuclide-containing streams, *Crit. Rev. in Biotech.* **11**:41–112.

MacIntyre, W., G., Boggs, M., Antworth, C., P., and Stauffer, T., B., 1993, Degradation kinetics of aromatic organic solutes introduced into a heterogenous aquifer, *Water Resources Research* **20**:4045–4051.

MacLeod, F. A., Lappin-Scott, H. M., and Costerton, J. W., 1988, Plugging of a model rock system by using starved bacteria, *Appl. Environ. Microbiol.* **54**:1365–1372.

Madsen, E. L., 1991, Determining in situ biodegradation, *Environ. Sci. Technol.* **25**:1663–1673.

Madsen, E. L., 1995, Impacts of agricultural practices on subsurface microbial ecology, in: *Advances in Agronomy,* 54 (D. L. Sparks, ed.), Academic Press, San Diego, pp. 1–67.

Madsen, E. L., and Ghiorse, W. C., 1993, Groundwater microbiology: subsurface ecosystem processes, in: *Aquatic Microbiology, An Ecological Approach* (T. E. Ford, ed.), Blackwell Scientific, Boston, pp. 167–213.

Madsen, E. L., Sinclair, J. L., and Ghiorse, W. C., 1991, In situ biodegradation: microbiological patterns in a contaminated aquifer, *Science* **252**:830–833.

Madsen, E. L., Winding, A., Malachowsky, K., Thomas, C. T., and Ghiorse, W. C., 1992, Contrasts between subsurface microbial communities and their metabolic adaptation to polycyclic aromatic hydrocarbons at a forested and an urban coal-tar site, *Microb. Ecol.* **24**:199–213.

Major, D. W., Mayfield, C. I., and Barker, J. F., 1988, Biotransformation of benzene by denitrification in aquifer sand, *Ground Water* **26**:8–14.

Mariotti, A. A., Landreau, A., and Simon, B., 1988, ^{15}N isotope biogeochemistry and natural denitrification process in groundwater: application to the chalk aquifer of northern France, *Geochim. Cosmochim. Acta* **52**:1869–1878.

Marxsen, J., 1988, Investigations into the number of respiring bacteria in groundwater from sandy and gravelly deposits, *Microb. Ecol.* **16**:65–72.

Mateju, V., Cizinska, S., Krejci, J., and Janoch, T., 1992, Biological water denitrification-a review, *Enzyme Microb. Technol.* **14**:170–183.

Matthess, G., 1992, Silicate systems, in: *Progress in Hydrogeochemistry,* (G. Matthess, F. Frimmel, P. Hirsch, H. D. Schulz, and H.-E. Usdowski, eds.), Springer-Verlag, New York, pp. 199–201.

Maymo-Gatell, X., Tandoi, V., Gossett, J. M., and Zinder, S. H., 1995, Characterization of an H_2-utilizing enrichment culture that reductively dechlorinates tetrachloroethene to vinyl chloride and ethene in the absence of methanogenesis and acetogenesis, *Appl. Environ. Microbiol.* **61**:3928–3933.

McAllister, P. M., and Chiang, C. Y., 1994, A practical approach to evaluating natural attenuation of contaminants in ground water, *Ground Wat. Monit. Remed.* **14**:.

McCarty, P. L., 1972, Energetics of organic matter degradation, in: *Water Pollution Microbiology* (R. Mitchell, ed.), John Wiley, New York, pp. 91–118.

McCarty, P. L., and Semprini, L., 1994, Ground-water treatment for chlorinated solvents, in: *Handbook of Bioremediation* (R D. Norris, R. E. Hirchee, R. Brown, P. L. McCarty, L. Semprimi, J. T. Wilson, D. H . Kampbell, M. Reinhard, E. J. Bouwer, R. C., Borden, T. M. Vogel, J. M. Thomas, C. H. Ward, eds.), Lewis Publishers, Boca Raton, pp. 87–116.

McCarty, P. L., Semprini, L., Dolan, M. E., Harmon, T. C., Tiedeman, C., and Gorelick, S. M., 1991, *In situ* methanogenic bioremediation for contaminated groundwater at St. Joseph, Michigan, in: *On-Site Bioreclamation: Processes for Xenobiotic and HydrocarbonTreatment* (E. R. Hinchee, and F. R. Olfenbuttel, eds.), Butterworth-Heinemann, Boston, pp. 16–40.

McNabb, J. F., and Dunlap, W. J., 1975, Subsurface biological activity in relation to ground-water pollution, *Ground Water* **13**:33-44.

McNabb, W. W, Jr., and Narasimhan, T. N., 1994, Degradation of chlorinated hydrocarbons and groundwater geochemistry: A field study, *Environ. Sci. Technol.* **28**:769-775.

Mercado, A., Libhaber, M., and Soares, M. I. M., 1988, In situ biological groundwater denitrification: concepts and preliminary field tests, *Wat. Sci. Tech.* **20**:197-209.

Mihelcic, J. R., and Luthy, R. G., 1988a, Degradation of polycyclic aromatic hydrocarbon compounds under various redox conditions in soil-water systems, *Appl. Environ. Microbiol.* **54**:1182-1187.

Mihelcic, J. R., and Luthy, R. G., 1988b, Microbial degradation of acenaphthalene and naphthalene under denitrification conditions in soil-water systems, *Appl. Environ. Microbiol.* **54**:1188-1198.

Millette, D., Barker, J. F., Comeau, Y., Butler, B. J., Frind, E. O., Clement, B., and Samson, R., 1995, Substrate interaction during aerobic biodegradation of creosote-related compounds: a factorial batch experiment, *Environ. Sci. Technol.* **29**:1944-1952.

Mills, A. L., Herman, J. S., Hornberger, G. M., and DeJesus, T. H., 1994, Effect of solution ionic strength and iron coatings on mineral grains on the sorption of bacterial cells to quartz sand, *Appl. Environ. Microbiol.* **60**:3300-3306.

Mohn, W. W., and Tiedje, J. M., 1992, Microbial reductive dehalogenation, *Microbiol. Rev.* 482-507.

Morgan, P., and Watkinson, R. J., 1989, Microbiological methods for the cleanup of soil and ground water contaminated with halogenated organic compounds, *FEMS Microbiol. Rev.* **63**:277-300.

Morgan, P., and Watkinson, R. J., 1992, Factors limiting the supply and frequency of nutrient and oxygen supplements for the *in situ* biotreatment of contaminated soil and groundwater, *Water Research* **26**:73-78.

Morgan, P., Lewis, S. T., and Watkinson, R. J., 1993, Biodegradation of benzene, toluene, ethylbenzene and xylenes in gas-condenstate-contaminated ground-water, *Environ. Poll.* **82**:181-190.

Morris, J. T., Whiting, G. J., and Chapelle, F. H., 1988, Potential denitrification rates in deep sediments from the southeastern coastal plain, *Environ. Sci. Technol.* **22**:332-335.

Nelson, M. J. K., Montgomery, S. O., O'Neill, E. J., and Pritchard, P. H., 1986, Aerobic metabolism of trichloroethylene by a bacterial isolate, *Appl. Environ. Microbiol.* **52**:383-384.

Nelson, M. J. K., Montgomery, S. O., Mahaffey, W. R., and Pritchard, P. H., 1987, Biodegradation of trichloroethylene and involvement of an aromatic biodegradative pathway, *Appl. Environ. Microbiol.* **53**:949-954.

Nielson, P. H., and Christensen, T. H., 1994, Variability of biological degradation of aromatic hydrocarbons in an aerobic aquifer determined by laboratory batch experiments, *J. Contam. Hydrol.* **15**:305-320.

Norris, R. D., Hinchee, R. E., Brown, R., McCarty, P. L., Semprini, L., Wilson, J. T., Kampbell, D. H., Reinhard, M., Bouwer, E. J., Borden, R. C., Vogel, T. M., Thomas, J. M., and Ward, C. H., 1994, *Handbook of Bioremediation*, CRC Press, Boca Raton.

Ogunseitan, O. A., and Olson, B. H., 1991, Potential for genetic enhancement of bacterial detoxification of mercury waste in: Proceedings of the *Mineral Bioprocessing* Conference (R. W. Smith and M. Misra, eds.), The Minerals, Metals and Materials Society, Santa Barbara, pp. 325-337.

Oldenhius, R., Oedzes, J. Y., van der Waarde, J. J., and Janssen, D. B., 1991, Kinetics of chlorinated hydrocarbon degradation by *Methylosinus trichosporium* OB3b and toxicity of trichloroethylene, *Appl. Environ. Microbiol.* **57**:7-14.

Olson, B. H., Cayless, S. M., Ford, S., and Lester, J. N., 1991, Toxic element contamination and the occurrence of mercury-resistant bacteria in Hg-contaminated soil, sediments, and sludges, *Arch. Environ. Contam. Toxicol.* **20:**226–233.

Oremland, R. S., 1994, Biogeochemical transformations of selenium in anoxic environments, in: *Selenium in the Environment* (W. T. J. Frankenberger and S. N. Benson, eds.), Marcel Dekker, New York, pp. 389–419.

Oremland, R. S., Steinberg, N. A., Maest, A. S., Miller, L. G., and Hollibaugh, J. T., 1990, Measurement of in situ rates of selenate removal by dissimilatory bacterial reduction in sediments, *Environ. Sci. Technol.* **24:**1157–1164.

Oremland, R. S., Steinberg, N. A., Presser, T. S., and Miller, L. G., 1991, In situ bacterial selenate reduction in the agricultural drainage systems of western Nevada, *Appl. Environ. Microbiol.* **57:**615–617.

Pardieck, D. L., Bouwer, E. J., and Stone, A. T., 1992, Hydrogen peroxide use to increase oxidant capacity for in situ bioremediation of contaminated soils and aquifers: A Review, *J. Contam. Hydrol.* **9:**221–242.

Parsons, F., Barrio-Lage, G., and Rice, R., 1985, Biotransformation of chlorinated organic solvents in static microcosms, *Environ. Toxicol. Chem.*, **4:** 739–742.

Parsons, F., Wood, P. R., DeMarco, J., 1984, Transformations of tetrachloroethene and trichloroethene in microcosms and groundwater, *Am. Water Works, Assoc.* **71:** 56–59.

Patterson, B., M., Pribac, F., Barber, C., Davis, G., B., and Gibbs, R., 1993, Biodegradation and retardation of PCE and BTEX compounds in aquifer material from Western Australia using large-scale columns, *J. Contam. Hydrol.* **14:**261–278.

Pedersen, K., 1993, The deep subterranean biosphere, *Earth-Science Rev.* **34:**243–260.

Phelps, T. J., Ringleberg, D., Hedrick, D., Davis, J., Fliermans, C. B., and White, D. C., 1988, Microbial biomass and activities associated with subsurface environments contaminated with chlorinated hydrocarbons, *Geomicrobiol. J.* **6:**157–170.

Phelps, T. J., Raione, E. G., White, D. C., and Fliermans, C. B., 1989, Microbial activities in deep subsurface environments, *Geomicrobiol. J.* **7:**79–91.

Podoll, R. T., Irwin, K. C., and Parish, H. J., 1989, Dynamic studies of naphthalene sorption on soil from aqueous solution, *Chemosphere* **18:**2399–2412.

Ponnamperuma, F. N., 1972, The chemistry of submerged soils, *Adv. Agron.* **24:**29–96.

Portier, R. J., Zoeller, A. L., and Fujisaki, K., 1990, Bioremediation of pesticide-contaminated groundwater, *Remediation* **1:**41–60.

Postma, D., Boesen, C., Kristiansen, H., and Larsen, F., 1991, Nitrate reduction in an unconfined sandy aquifer: water chemistry, reduction processes, and geochemical modeling, *Wat. Resour. Res.* **27:**2027–2045.

Rabus, R., and Widdel, F., 1995, Anaerobic degradation of ethylbenzene and other aromatic hydrocarbons by a new denitrifying bacteria, *Arch. Mirobiol.* **163:**96–103.

Rabus, R., Nordhaus, R., Ludwig, W., and Widdel, F., 1993, Complete oxidation of toluene under strictly anoxic conditions by a new sulfate-reducing bacterium, *Appl. Environ. Microbiol.* **59:**1444–1451.

Radehaus, P. M., and Schmidt, S. K., 1992, Characterization of a novel *Pseudomomas* sp. that mineralizes high concentrations of pentachlorophenol, *Appl. Environ. Microbiol.* **58:**2879–2885.

Radosevich, M., and Klein, D. A., 1993, Bacterial enumeration and mercury volatilization in deep subsurface sediment samples, *Bull. Environ. Contam. Toxicol.* **51:**226–233.

Rainwater, K., Mayfield, M. P., Heintz, C., and Claborn, B. J., 1993, Enhanced *in situ* biodegradation of diesel fuel by cyclic vertical water table movement: preliminary studies, *Water Environ. Res.* **65:** 717–725.

Raymond, R. L., Brown, R. A., Norris, R. D., and O'Neill, E. T., 1986, Stimulation of biooxidation processes in subterranean formations, U.S. Patent 4,588,506 (5-13-86), FMC Corporation, Philadelphia.

Reeburgh, W. S., 1983, Rates of biogeochemical processes in anoxic sediments, *Ann. Rev. Earth Planet. Sci.* **11**:269–298.

Regnell, O., 1990, Conversion and partitioning of radio-labelled mercury chloride in aquatic model systems, *Can. J. Fish. Aquat. Sci.* **47**:548–553.

Ridgway, H. F., Safarik, J., Phipps, D., Carl, P., and Clark, D., 1990, Identification and catabolic activity of well-derived gasoline-degrading bacteria from a contaminated aquifer, *Appl. Environ. Microbiol.* **56**:3565–3575.

Rittman, B. E., and McCarty, P. L., 1980, Utilization of dichloromethane by suspended and fixed-film bacteria, *Appl. Environ. Microbiol.* **39**: 1225–1226.

Roberts, P. V., Hopkins, G. D., Mackay, D. M., and Semprini, L., 1990, A field evaluation of in-situ biodegradation of chlorinated ethenes: part 1, methodology and field site characterization, *Ground Water* **28**:591–604.

Rochelle, P. A., Wetherbee, M. K., and Olson, B. H., 1991, Distribution of DNA sequences encoding narrow- and broad-spectrum mercury resistance, *Appl. Environ. Microbiol.* **57**:1581–1589.

Rugge, K., Bjerg, P. L., and Christensen, T. H., 1995, Distribution of organic compounds from municipal solid waste in the groundwater downgradient of a landfill (Grindsted, Denmark, *Environ. Sci. Technol.* **29**:1395–1400.

Salanitro, J. P., 1993, The role of bioattenuation in the management of aromatic hydrocarbon plumes in aquifers, *Ground Water Monitoring Review* **13**:150–161.

Sayler, G. S., Shields, M. S., Tedford, E. T., Breen, A., Hooper, S. W., Sirotin, K. M., and Davis, J. W., 1985, Application of DNA-DNA colony hybridization to the dectection of catabolic genotypes in environmental samples, *Appl. Environ. Microbiol.* **49**:1295–1303.

Schocher, R. J., Seyfried, B., Vazquez, F., and Zeyer, J., 1991, Anaerobic degradation of toluene by pure cultures of dentrifying bacteria, *Arch. Microbiol.* **157**:7–12.

Scholl, M. A., and Harvey, R. W., 1992, Laboratory investigations on the role of sediment surface and groundwater chemistry in transport of bacteria through a contaminated sandy aquifer, *Environ. Sci. Technol.* **26**:1410–1417.

Scholz-Muramatsu, H., Neuman, A., Meßmer, M., Moore, E., and Diekert, G., 1995, Isolation and characterization of *Dehalospirillum multivorans* gen. nov., sp. nov., a tetrachloroethene-utilizing, strictly anaerobic bacterium, *Arch. Microbiol.* **163**:48–56.

Schwille, F., 1976, Anthropogenically reduced groundwaters, *Hydrol. Sci. Bull.* **21**:629–645.

Semprini, L., and McCarty, P. L., 1991, Comparison between model simulations and field results of in-situ biorestoration of chlorinated aliphatics: part 1. biostimulation of methanotrophs, *Ground Water* **29**:365–374.

Semprini, L., and McCarty, P. L., 1992, Comparison between model simulations and field results of in-situ biorestoration of chlorinated aliphatics: part 2. cometabolic transformations, *Ground Water* **30**:37–44.

Semprini, L., Hopkins, G. D., Roberts, P. V., and McCarty, P. L., 1990, In-situ biotransformation of carbon tetrachloride, 1,1,1-trichloroethane, Freon-11, and Freon-113 under anoxic conditions, *EOS, Trans. Amer. Geophys. Union* **71**:1324.

Semprini, L., Hopkins, G. D., Roberts, P. V., Grbic-Galic, D., and McCarty, P. L., 1991, A field evaluation of in-situ biodegradation of chlorinated ethenes: part 3. studies of competitive inhibition, *Ground Water* **29**:239–250.

Semprini, L., Hopkins, G. D., McCarty, P. L., and Roberts, P. V., 1992, In-situ transformation of carbon tetrachloride and other halogenated compounds resulting from biostimulation under anoxic conditions, *Environ. Sci. Technol.* **26**:2454–2461.

Sharma, P. K., and McCarty, P. L., 1996, Isolation and characterization of a facultatively aerobic bacterium that reductively dehalogenates tetrachloroethene to *cis*-1,2-dichloroethene, *Appl. Environ. Microbiol.* **62**:761–765.

Sharma, P. K., and McInerney, M. J., 1994, Effect of grain size on bacterial penetration, reproduction, and metabolic activity in porous glass bead chambers, *Appl. Environ. Microbiol.* **60**:1481–1486.

Shields, M. S., Montgomery, S. O., Chapman, P. J., Cuskey, S. M., and Pritchard, P. H., 1989, Novel pathway of toluene catabolism in the trichloroethylene-degrading bacterium G4, *Appl. Environ. Microbiol.* **55**:1624–1629.

Silver, S., 1991, *Proceedings to the Eighth International Biodeterioration and Biodegradation Symposium* (H. Rossmore, ed.), Elsevier, London, pp. 308–339.

Sinclair, J. L., and Ghiorse, W. C., 1987, Distribution of protozoa in subsurface sediments of a pristine groundwater site in Oklahoma, *Appl. Environ. Microbiol.* **53**:1157–1163.

Sinclair, J. L., and Ghiorse, W. C., 1989, Distribution of aerobic bacteria, protozoa, algae, and fungi in deep subsurface sediments, *Geomicrobiology* **7**:15–31.

Sinclair, J. L., Randtke, S. J., Denne, J. E., Hathaway, L. R., and Ghiorse, W. C., 1990, Survey of microbial populations in buried-valley aquifer sediments from northeastern Kansas, *Ground Water* **28**:369–377.

Smith, G. A., Nickels, J. S., Kerger, B. D., Davis, J. D., Collins, S. P., Wilson, J. T., McNabb, J. F., and White, D. C., 1986, Quantitative characterization of microbial biomass and community structure in subsurface material: a prokaryotic consortium responsive to organic contamination, *Can. J. Microbiol.* **32**:104–111.

Smith, J. A., Witkowski, P. J., and Chiou, C. T., 1988, Partition of nonionic organic compounds in aquatic systems, *Rev. Environ. Contam. Toxicol.* **103**:127–151.

Smith, M., R., 1990, The biodegradation of aromatic hydrocarbons by bacteria, *Biodegradation* **1**:191–206.

Smith, M. R., Ewing, M., and Ratledge, C., 1991, The interactions of various aromatic substrates degraded by *Pseudomomas* sp. NCIB 10643: synergistic inhibition of growth by two compounds that serve as growth substrates, *Appl. Microbiol. Biotechnol.* **34**:536–538.

Smith, R. L., Ceazan, M. L., and Brooks, M. H., 1994, Autotrophic, hydrogen-oxidizing, denitrifying bacteria in groundwater, potential agents for biotransformation of nitrate contamination, *Appl. Environ. Microbiol.* **60**:1949–1955.

Soares, M. I. M., Belkin, S., and Abeliovich, A., 1988, Biological groundwater denitrification: laboratory studies, *Wat. Sci. Tech.* **20**:189–195.

Sonier, D. N., Duran, N. L., and Smith, G. B., 1994, Dechlorination of trichlorofluoromethane (CFC-11) by sulfate-reducing bacteria from an aquifer contaminated with halogenated aliphatic compounds, *Appl. Environ. Microbiol.* **60**:4567–4572.

Spain, J. C., Van Veld, P. A., Monti, C. A., Pritchard, P. H., and Cripe, C. R., 1984, Comparison of p-nitrophenol biodegradation in field and laboratory test systems, *Appl. Environ. Microbol.* **48**:944–950.

Spain, J. C., Milligan, J. D., Downey, D. C., and Slaughter, J. K., 1989, Excessive bacterial decomposition of H_2O_2 during enhanced biodegradation, *Ground Water* **27**:163–167.

Spalding, R. F., and Exnter, M. E., 1993, Occurrence of nitrate in groundwater-a review, *J. Environ. Qual.* **22**:392–402.

Spalding, R. F., and Parrott, J. D., 1994, Shallow groundwater denitrification, *Sci. Tot. Environ.* **141**:17–25.

Stanlake, G. J., and Finn, R. K., 1982, Isolation and characterization of a pentachlorophenol-degrading bacterium, *Appl. Environ. Microbiol.* **44**:1421–1427.

Starr, R. C., and Gillham, R. W., 1993, Denitrification and organic carbon availability in two aquifers, *Groundwater* **31**:934–947.

Stephen, G. M., and Dalton, H., 1986, The role of the terminal and subterminal oxidation pathways in propane metabolism by bacteria, *J. Gen. Microbiol.* **132**: 2453–2462.

Stetzenbach, L. D., Kelley, L. M., and Sinclair, N. A., 1986, Isolation, identification, and growth of well-water bacteria, *Ground Water* **24**:6–10.

Strand, S. E., and Shippert, L., 1986, Oxidation of chloroform in an aerobic soil exposed to natural gas, *Appl. Environ. Microbiol.* **52**:203–205.

Strand, S. E., Bjelland, M. D., and Stensel, H. D., 1990, Kinetics of chlorinated hydrocarbon degradation by suspended cultures of methane-oxidizing bacteria, *Research Journal WPCF* **62**:124–129.

Stucki, G., Krebser, U., and Leisinger, T., 1983, Bacterial growth on 1,2-dichloroethane, *Experientia*, **39**: 366–371.

Stumm, W., and Morgan, J. J., 1981, *Aquatic Chemistry*, John Wiley & Sons, New York.

Suflita, J. M., Gibson, S. A., and Beeman, R. E., 1988, Anaerobic biotransformation of pollutant chemicals in aquifers, *J. Indust. Microbiol.* **3**: 179–194.

Swindoll, C. M., Aelion, M. C., Dobbins, D. C., Jiang, O., Long, S., and Pfaender, F. K., 1988, Aerobic biodegradation of natural and xenobiotic organic compounds by subsurface microbial communities, *Environ. Toxicol. Chem.* **7**:291–299.

Tatara, G. M., Dybas, M. J., and Criddle, C. S., 1993, Effects of medium and trace metals on kinetics of carbon tetrachloride transformation by *Pseudomonas* sp. strain KC, *Appl. Environ. Microbiol.* **59**:2126–2131.

Thiem, S. M., Krumme, M. L., Smith, R. L., and Tiedje, J. M., 1994, Use of molecular techniques to evaluate the survival of a microorganism injected into an aquifer, *Appl. Environ. Microbiol.* **60**:1059–1067.

Thierrin, J., Davis, G. B., Barber, C., Patterson, B. M., Pribac, F., Power, T. R., and Lambert, M., 1993, Natural degradation rates of BTEX compounds and naphthalene in a sulphate reducing groundwater environment, *Hydrol. Sci.* **38**:309–322.

Thomas, J. M., and Ward, C. H., 1989, In situ biorestoration of organic contaminants in the subsurface, *Environ. Sci. Tech.* **23**:760–766.

Thomas, J. M., Lee, M. D., and Ward, C. H., 1987, Use of ground water in an assessment of biodegradation potential in the subsurface, *Environ. Toxicol. Chem.* **6**:607–614.

Thomas, J. M., Lee, M. D., Scott, M. J., and Ward, C. H., 1989, Microbial ecology of the subsurface at an abandoned creosote waste site, *J. Indus. Microbiol.* **4**:109–120.

Thomas, J. M., Gordy, V. R., Fiorenza, S., and Ward, C. H., 1990, Biodegradation of BTEX in subsurface materials contaminated with gasoline: Granger, Indiana, *Wat. Sci. Tech.* **20**:53–62.

Thorn, P. M., and Ventullo, R. M., 1988, Measurement of bacterial growth rates in subsurface sediments using the incorporation of tritiated thymidine into DNA, *Microb. Ecol.* **16**:3–16.

Topp, E., Hanson, R. S., Ringelberg, D. B., White, D. C., and Wheatcroft, R., 1993, Isolation and characterization on an *n*-methylcarbamate insecticide-degrading methylotrophic bacterium, *Appl. Environ. Microbiol.* **59**:3339–3349.

Trudell, M. R., Gillham, R. W., and Cherry, J. A., 1986, An in-situ study of the occurrence and rate of denitrification in a shallow unconfined sand aquifer, *J. Hydrol.* **83**:251–268.

Tsien, H.-C., Brusseau, G. A., Hanson, R. S., and Wackett, L. P., 1989, Biodegradation of trichloroethylene by *Methylosinus trichosporium* OB3b, *Appl. Environ. Microbiol.* **55**:3155–3161.

Van Beelen, P., and Van Keulen, F., 1990, The kinetics of the degradation of chloroform and benzene in anaerobic sediment from the river Rhine, *Hydrobiol. Bull.* **24**:13–21.

Vannelli, T., Logan, M., Arciero, D. M., and Hooper, A. B., 1990, Degradation of halogenated aliphatic compounds by the ammonia-oxidizing bacterium *Nitrosomonas europaea*, *Appl. Environ. Microbiol.* **56**:1169–1171.

Vogel, T. M., 1994, Natural bioremediation of chlorinated solvents, in: *Handbook of Bioremediation*, (R. D. Norris, R. F. Hinchee, R. Brown, P. L. McCarty, L. Semprini, J. T. Wilson, D. H. Kampbell, M. Reinhard, E. J. Bouwer, R.C. Borden, T. M. Vogel, J. M. Thomas, C. H. Ward, eds.), Lewis Publishers, Boca Raton, pp. 201–225.

Vogel, T. M., and Grbic-Galic, D., 1986, Incorporation of oxygen from water into toluene and benzene during anaerobic fermentative transformation, *Appl. Environ. Microbiol.* **52**:200–202.

Vogel, T. M., and McCarty, P. L., 1985, Biotransformation of tetrachloroethylene to trichloroethylene, dichloroethylene, vinyl chloride, and carbon dioxide under methanogenic conditions, *Appl. Environ. Microbiol.* **49**:1080–1083.

Vogel, J. C., Talma, A. S., and Heaton, T. H. E., 1981, Gaseous nitrogen as evidence for denitrification in groundwater, *J. Hydrol.* **50**:191–200.

Vogel, T. M., Criddle, C. S., and McCarty, P. L., 1987, Transformations of halogenated aliphatic compounds, *Environ. Sci. Technol.* **21**:722–736.

Volkering, F., Breure, A. M., van Andel, J. G., and Rulkens, W. H., 1995, Influence of nonionic surfactants on bioavailability and biodegradation of polycyclic aromatic hydrocarbons, *Appl. Environ. Microbiol.* **61**:1699–1705.

von Wedel, R. J., Mosquera, J. F., Goldsmith, C. D., Hater, G. R., Wong, A., Fox, T. A., Hunt, W. T., Paules, M. S., Quiros, J. M., and Wiegand, J. W., 1988, Bacterial biodegradation of petroleum hydrocarbons in groundwater: *in situ* augmented bioreclamation with enrichment isolates in California, *Wat. Sci. Tech.* **20**:501–503.

Vroblesky, D. A., and Chapelle, F. H., 1994, Temporal and spatial changes of terminal electron-accepting processes in a petroleum hydrocarbon-contaminated aquifer and the significance for contaminant biodegradation, *Water Res. Res.* **30**:1561–1570.

Wackett, L. P., and Gibson, D. T., 1988, Degradation of trichloroethylene by toluene dioxygenase in whole-cell studies with *Pseudomonas putida* F1, *Appl. Environ. Microbiol.* **54**:1703–1708.

Wackett, L. P., Brusseau, G. A., Householder, S. R., and Hanson, R. S., 1989, Survey of microbial oxygenases: trichloroethylene degradation by propane-oxidizing bacteria, *Appl. Environ. Microbiol.* **55**:2960–2964.

Walia, S., Kahn, A., and Rosenthal, N., 1990, Construction and applications of DNA probes for detection of polychlorinated biphenyl-degrading genotypes in toxic organic-contaminated soil environments, *Appl. Environ. Microbiol.* **56**:254–259.

Wan, J., Wilson, J. L., and Kieft, T. L., 1994, Influence of the gas-water interface on transport of microorganisms through unsaturated porous media, *Appl. Environ. Microbiol.* **60**:509–516.

Water Science and Technology Board, Commission on Engineering and Technical Systems, National Research Council, 1993, *In Situ Bioremediation*, National Academy Press, Washington D.C.

Weber, W. J., and Corseuil, H. X., 1994, Inoculation of contaminated subsurface soils with enriched indigenous microbes to enhance bioremediation rates, *Wat. Res.* **28**:1407–1414.

Weiner, J., and Lovley, D. R., 1997, Stimulation of anaerobic benzene degradation in petroleum-contaminated aquifer sediments with a freshwater benzene-oxidizing, sulfate-reducing inoculum, *Appl. Environ. Microbiol.* (submitted).

Werner, P., 1985, A new way for the decontamination of polluted aquifers by biodegradation, *Wat. Supply* **3**:41–47.

Werner, P., 1991, German experiences in the biodegradation of creosote and gas work-specific substances, in: *In Situ Bioreclamation*, (E. R. Hinchee, and R. F. Olfenbuttel, eds.), Butterworth-Heinemann, Stoneham, MA, pp. 539.

Westrick, J. J., Mello, J. W., and Thomas, R. F., 1984, The groundwater supply survey, *J. Am. Water Works Assoc.* **76**:52–59.

White, D. C., Smith, G. A., Gehron, M. J., Parker, J. H., Findlay, R. H., Martz, R. F. Frederickson, H. L., 1983, The ground water aquifer microbiota: biomass, community structure, and nutritional status, *Develop. Indus. Microbiol.* **24**:189–199.

Widdel, F., 1988, Microbiology and ecology of sulfate- and sulfur-reducing bacteria, in: *Biology of Anaerobic Microorganisms*, (A. J. B. Zehnder, ed.), John Wiley & Sons, New York, pp. 469–585.

Williams, G. M., Smith, B., and Ross, C. A. M., 1991, The migration and degradation of waste organic compounds in groundwater, *Adv. Org. Geochem.* **19**:531–543.

Wilson, J. T., and Wilson, B. H., 1985, Biotransformation of trichlororethylene in soil, *Appl. Environ. Microbiol.* **49**:242–243.

Wilson, J. T., McNabb, J. F., Balkwill, D. L., and Ghiorse, W. C., 1983, Enumeration and characterization of bacteria indigenous to a shallow water-table aquifer, *Ground Water* **21**:134–142.

Wilson, B., Smith, G. B., and Rees, J. F., 1986, Biotransformation of selected alkylbenzenes and halogenated aliphatic hydrocarbons in methanogenic aquifer material: a microcosm study, *Environ. Sci. Technol.* **20**:997–1002.

Wilson, B. H., Wilson, J. T., Kampbell, D. H., Bledsoe, B. E., and Armstrong, J. M., 1990a, Biotransformation of monoaromatic and chlorinated hydrocarbons at an aviation gasoline spill site, *Geomicrobiol. J.* **8**:225–240.

Wilson, G. B., Andrews, J. N., and Bath, A. H., 1990b, Dissolved gas evidence for denitrification in the Lincolnshire groundwaters, Eastern England, *J. Hydrol.* **113**:51–60.

Wilson, J. T., Armstrong, J. M., and Rafai, H. S., 1994, A full-scale field demonstration on the use of hydrogen peroxide for *in situ* bioremediation of an aviation gasoline-contaminated aquifer, in: *Bioremediation Field Experience*, (E. P. Flathman, E. D. Ferger, and H. J. Exner, eds.), Lewis Publishers, Boca Raton, pp. 333–359.

Woods, N. R., and Murrell, J. C., 1989, The metabolism of propane in *Rhodococcus rhodochrous* PNKb1, *J. Gen. Microbiol.* **135**: 2335–2344.

Wyndham, R. C., Nakatsu, C., Peel, M., Cashore, A., Ng, J., and Szilagyi, F., 1994, Distribution of the catabolic transposon Tn5271 in a groundwater bioremediation system, *Appl. Environ. Microbiol.* **60**:86–93.

Zeyer, J., Kuhn, E. P., and Schwarzenbach, R. P., 1986, Rapid microbial mineralization of toluene and 1,3-dimethylbenzene in the absence of molecular oxygen, *Appl. Environ. Microbiol.* **52**:944–947.

Index

Adaptation, proliferation hypothesis and, 106, 169
Aggregates
   bioaggregates in chemostat, 130–131
   of marine particles, 81, 90, 97, 98
Algae, *see also* Cyanobacteria; Phytoplankton
   in degradative consortium, 132
   in lichens, 107–108
   PAM fluorescence studies, 16
Altruistic behavior, 107, 110
Ammonia oxidizers, halogenated organics and, 322–323
Amoebobacter purpureus: *see* Phototrophic sulfur bacteria
Anaerobic degradation, in groundwater, 294, 301–306
   aromatic hydrocarbons, 315–319
   halogenated organics, 323–325
   metals, 329
   nitrate, 326–327
Anaerobic digestor granules, 162–165; *see also* Sludge communities
Anaerobic microniches, 154
Anaerobic systems, effectiveness of, 106
Antimicrobial agents
   agar wedge technique, 136
   biofilm studies, 62, 135, 152–153
   in microbial community regulation, 108
   rumen microflora and, 154
Aquifers: *see* Groundwater bioremediation
Aromatic hydrocarbons, biodegradation of
   aerobic, 307–315
      chemical mechanisms, 307–308
      intrinsic bioremediation, 307
      microbial adaptations, 309–312, 314–315
      microbial interactions, 312
      microbial numbers, 312–313
      polycyclic compounds, 307, 308, 311–312, 313
      rate measurements, 308–310
      stimulation methods, 312–315, 318

Aromatic hydrocarbons, biodegradation of (*cont.*)
   anaerobic, 315–319
      electron acceptors involved, 315–317
      limits on benzene degradation, 317–319
      polycyclic compounds, 319
      stimulation methods, 317–319
   compounds represented in groundwater, 295–301, 307–308
Arsenic, reduction in aquifers, 330
Ascomycetes, spore dispersal by, 219–220
Associations
   definition of, 166, 169
   proliferation by, 106, 112
Autopoiesis, *see also* Self-organization
   defined community and, 117, 166
   definition of, 169
   evolution and, 171
   systems theory and, 39

Bacterial plates: *see* Phototrophic sulfur bacteria
Bacterial production, submicron particles and, 87–88, 90
Bacteriocins, 108
Behavioral level of analysis, 50–53
Benzene: *see* BTEX, biodegradation of
Biliproteins, UV-B damage to, 13
Bioaggregates, in chemostat, 130–131
Biodegradation: *see* Degradative consortia; Groundwater bioremediation
Biofilms, *see also* Communities, microbial; Degradative consortia
   in continuous culture systems, 133–139
      dual-dilution, 133
      microstat, 135–139, 174
      rototorque bioreactor, 133–134
      slide cultures, 134–135
   examples of, 107, 135
   experimental designs with, 54–56
      perturbations within system, 58–62
      variations in environment, 56–58

Biofilms (cont.)
  formation of, 134–135
  identification of organisms, 146, 154, 155
  on metal surfaces, 155
  scanning confocal microscopy and, 140–141, 171
  sloughing from, 59, 134–135
  system boundaries, 42–43
  system identity, 69
  variables studied
    cell distribution, 141
    charge distribution, 145
    diffusion coefficients, 142–145
    pH distribution, 141, 146–147
    substrate utilization profiles, 156–157
    viability, 147–152
  on water distribution pipes, 327
Biological control, by fungal entomopathogens, 234–235
BIOLOG system, 156
Bioremediation: see Groundwater bioremediation
Biosorption, 328
Brownian motion, in colloid degradation, 91, 95, 97
BTEX, biodegradation of
  aerobic, 307–308, 309–310, 312–313, 314
  anaerobic, 315–319

Carbon cycles, see also Dissolved organic matter
  global, 4
  marine colloids in, 81–82, 85–86, 98
  of meromictic lake, 269–272, 273, 274–278
  phytoplankton in, 4, 18
Carboxyfluorescein, as pH probe, 146–147
Carotenoids, of phototrophic sulfur bacteria, 256, 258, 260, 282–283
CASE: see Complex adaptive systems
Case study, as methodology, 32, 34–35
Cell death, probes for, 147–152
Cell distribution, by fluorescence exclusion, 141
Cell line: see Clone
Cell membrane integrity, tests of, 148–151
Change, environmental, 169
Chaos, as potential information, 168, 173
Charge distribution, in biofilms, 145
Chemostat, 128–133, 169–170
  mixed cultures in, 111, 119, 130–133, 170
Chlorinated hydrocarbons: see Degradative consortia; Halogenated organics, in groundwater

Chlorination in pipelines, biofilms and, 148
Chromium, bioremediation of, 329
Clone
  definition of, 170, 173
  vs. pure culture, 116, 174
Cobalt-EDTA, bioremediation of, 329
Cold shock, 57
Colloids, freshwater, 83, 95
Colloids, marine
  abundance, 81, 84–85
  aggregation, 81, 90, 97, 98
  in carbon cycles, 81–82, 85–86, 98
  definition of, 83–84
  degradation of
    by bacteria, 90–94, 97–98
    by metazoans, 96, 97, 98
    by protozoans, 94–96, 98
  distribution, 85
  hypothetical roles, 81
  nitrogen in, 86, 87, 89
  physical characteristics, 84–85, 86, 98
  size spectrum, 83
  sources, 88–90
  turnover, 86–88, 93–94, 98
Co-metabolism, in bioremediation of halogenated organics, 320–323
Communality, 166
Communication, in microbial communities, 48–49
Communities, microbial, see also Biofilms; Complex adaptive systems; Degradative consortia; Fluorescent probes
  as complex adaptive systems, 29
  interactions with environment, 45–47
  internal coordination, 48–49
  structural information, 47–48, 49, 51
  criteria for characterization of, 29, 126–128, 165–167
  cultivation methods, 128–139
    chemostat, 111, 119, 128–133, 169–170
    continuous-flow slide culture, 128, 134–135, 170–171
    dual-dilution, 133
    microstat, 128, 135–139, 174
    rototorque bioreactor, 133–134
  evolution and, 105–109, 110–115, 119–120, 122–126
  examples of, 107
  germ theory and, 109–110, 115–118

Communities, microbial (*cont.*)
  identification of organisms in, 28, 47
    fatty acid profiles, 155–156
    genetic probes, 28, 44–45, 153–155
    polyclonal antibodies, 146
    substrate utilization profiles, 156–157
  microelectrode techniques, 28, 43, 146, 157
  reductionism and, 109–110, 120–122
  selective enrichment theory and, 109–110, 118–120, 127, 128, 172
  spectroscopic techniques, 157
Communitization, 111–112, 118
  definition of, 170
Community, definition of, 170
Community cultures
  definition of, 170
  vs. pure cultures, 116–117, 121–122, 165–167
Competition
  in microbial communities, 48, 61
  vs. proliferation hypothesis, 106–107, 110–111, 170, 173
Complex adaptive systems, 29, 38; *see also* SCIO model
  emerging paradigm for, 29–32
    analysis, 35–36, 37, 72
    focus, 33–34, 37, 71
    methodology, 34–35, 37, 71–72
    scientific control, 36–38, 72–73
    subjectivity, 32–33, 34, 36, 37, 70–71
    techniques, 72
  open systems theory and, 39–41
    identity in, 62–64, 67–69, 70
    system boundaries, 42–43
Confocal laser microscopy: *see* Scanning confocal laser microscopy
Consortia, *see also* Communities, microbial; Degradative consortia
  definition of, 170
Constructivism
  vs. realism, 31, 32, 33, 35, 36, 38
  research designs and, 70
  systems theory and, 40
Contextual understanding, 33, 34, 36, 37, 38
  of complex adaptive systems, 64, 66–69
  systems theory and, 40
Continuous-flow slide cultures, 128, 134–135, 170–171
Cooperative behavior, 48, 106–107, 110
Coordination parameter, 41–42, 48–49, 50–54, 64–69

Corrosion biofilms, 107, 146, 157
Coupling, in complex adaptive systems, 64–69
CTC, as redox probe, 147–148
Cyanelles, 131
Cyanobacteria
  gas vacuoles, loss in laboratory, 116–117
  in lichens, 107–108
  nitrogen fixation by, 12, 13, 108
  with sulfate-reducers, in mats, 269–270
  UV-B radiation and, 10, 12–13, 16, 17

Degradative consortia, 122–123, 157–162; *see also* Diclofop methyl degrading consortia
  chemostat studies, 128–129, 131–132
  chlorinated hydrocarbons and, 112, 114
    anaerobic reduction, 324, 325
  in enrichment cultures, 119
  mobile genetic elements in, 110, 112, 114, 119, 123
  sludge-degrading, 141, 145, 154, 162–165
Denaturing gradient gel electrophoresis (DGGE), 154
Denitrification, 326–327
Density-dependent plating effect, 125–126
Dental plaque, 107
Deuteromycetes, 194, 195
  conidial viability, 204, 223, 224
  host species specificity, 230
  infection processes, 206–207, 209, 211–212
  for insect control, 234–235
  life cycle, 196
  pathogenicity, 200
  sporulation, 218, 219–220
  temperature optima, 199
  toxins of, 215
Dextrans, fluorescent, in biofilm studies, 142–145
DGGE (denaturing gradient gel electrophoresis), 154
Diclofop methyl degrading consortia, 117, 132, 133; *see also* Degradative consortia
  in continuous-flow slide culture, 158–162
  exopolysaccharides of, 145–146, 159, 161
  herbicide binding sites, 145, 159, 161–162
  in microstat, 136
  substrate variations, 156–157, 158, 162
Diffusion coefficients, in biofilms, 142–145
Diffusion gradients: *see* Microstat
Digestor granules, anaerobic, 162–165; *see also* Sludge communities

Digital image analysis, of flow cells, 134, 135
Dinoflagellates, vertical migration of, 7, 9
Dissolved organic matter, *see also* Carbon cycles
   bacterial degradation of, 90–94
   colloidal fraction, 81, 85–86
   definition of, 83–84
   sources, 82, 88–90
   turnover rates, 82, 86–88, 98
   vertical transport, 87
Dissolved organic nitrogen, 86, 87, 89
Distinctiveness, in complex adaptive systems, 67–69
DNA
   in microbial identification
      reassociation kinetics, 44
      reverse sample genome probes, 154–155
   as physiological indicator, 46
   UV radiation and, 16–17
DNA-gyrase inhibitor, 152–153
Dual-dilution continuous culture, 133

Ecogram: *see* Environmental response surface
Ecology
   definitions of, 171
   inconsistent versions of, 105–106
Ecosystems
   definition of, 171
   ecological theory and, 106
Ecotones, 126–127, 166, 171–172
Emergent design, 32, 35
Endosymbiosis, 110–111, 114–115, 123
Enrichment cultures, 109–110, 118–120, 127, 128, 172
Entomophthorales: *see* Fungal entomopathogens, terrestrial
Environmental response surface
   definition of, 171, 172
   of microstat, 138, 174
Environmental selection, 172; *see also* Selection
Euglena gracilis, UV-B radiation and, 9–10
Eukaryotes, evolution from prokaryotes, 107, 110–111, 114–115, 123
Euphotic zone, 6
Evolution, *see also* Selection
   community-level, 105–109, 110–115, 119–120, 122–126
   as self-organization, 172–173
Exopolymers, *see also* Polysaccharides
   diffusion in biofilms, 142–145
   marine, 86, 88, 95, 96, 97

Fatty acid profiles, 155–156
Fecal pellets, of flagellates, 89, 90, 98
FITC-conjugated probes, 142–146; *see also* Fluorescent probes
Flow cells
   continuous-flow slide culture, 128, 134–135, 170–171
   microstat, 128, 135–139, 174
   multi-channel, 135
Flow cytometry, 45
Fluorescein, pH dependence of, 141, 150–151
Fluorescence exclusion, 141
Fluorescence microscopy, for single-cell identification, 45, 154
Fluorescence quenching analysis, of photosynthetic apparatus, 15–16
Fluorescence recovery after photobleaching, 138, 142
Fluorescent probes, with confocal laser microscopy, 105, 121, 139–141, 171
   of biocidal activity, 152–153
   of cell distribution, 141
   of cellular growth rate, 153
   of charge distribution, 145
   of diffusion coefficients, 142–145
   of hydrogen ion distribution, 146–147
   for identification, 146
   of metal ion distribution, 146
   in microstat, 135, 136, 138
   of oligopolysaccharide distribution, 145–146
   other potential probes, 152–153
   of viability, 147–152
Focus, in naturalistic inquiry, 32, 33–34, 37, 71
Food webs
   dissolved organic matter and, 87–88
   phytoplankton and, 6
Fungal entomopathogens, terrestrial, 193–235
   appressoria of, 210–211
   community-level studies, 228–234
   conidial types, 196–198
      capilliconidia, 198, 206, 209, 224
      germ conidia, 199, 201–203
      infectivity of, 209
      secondary, 198, 206, 208, 209, 224
   epizootiology, 225–234
      co-infections, 233–234
      definition of, 225
      determining factors, 225–228
      host food and, 228–230
      host species specificity, 230–231

Fungal entomopathogens, terrestrial (*cont.*)
epizootiology (*cont.*)
host susceptibility, 231
parasitoids and, 231–233
predators and, 231–233
history of research, 193
hosts
death of, 215–216, 218
defenses of, 212–215, 228
food plants, 228–230
population density, 226, 227, 230, 231
predators and parasitoids as, 231–233
species, 194–195
species specificity, 230–231
susceptibility variations, 231
transmission of fungus and, 225, 227–228
infective cycle, 195–199
conidial adhesion, 198, 206–208
conidial dispersal, 218–222, 233
conidial germination, 208–209
conidial production, 216–218
conidial survival and transmission, 222–225
death of host, 215–216, 218
growth within host, 212–215
penetration, 211–212
pre-penetration events, 209–211
routes of infection, 206
for insect control, 193, 234–235
literature reviews, 195
species implicated, 193–195
temperature optima, 199
toxins of, 215
in unfavorable periods, 199–206
persistence in soil, 199–200, 204–206
resting spores, 198–199, 200–204
saprophytic growth, 200, 204, 205–206
Fungi
in aquifers, 291, 292
of lichens, 107–108
Fungi Imperfecti: *see* Deuteromycetes

Gaia hypothesis, 124, 125–126, 127
Genetic probes, 28, 44–45, 46, 110, 153–155
Genetic rearrangement, adaptation and, 169
Gene transfer, in degradative consortia, 110, 112, 114, 119, 123
Geotaxis
in entomopathogenic fungus dispersal, 216
in phytoplankton, 7, 8–9

Germ theory, 109–110, 115–118
Gradostat, 132–133
Greenhouse effect
marine colloids and, 82
phytoplankton and, 4, 18
Grounded theory, 32, 36
Groundwater bioremediation, 289–331; *see also*
Aromatic hydrocarbons; Halogenated organics
contaminants, 295–301, 306–307
aromatics, aerobic degradation, 307–315
aromatics, anaerobic degradation, 315–319
halogenated organics, 319–325
metals and metalloids, 327–330
nitrate, 326–327
engineered, 289–290
aromatic hydrocarbons, 312–315, 317–319
halogenated organics, 321–323, 324–325
metals, 328
nitrate, 326–327
intrinsic, 289
BTEX, 307
mercury, 330
nitrate, 326
microorganisms in aquifers
aerobic processes of, 294, 301, 304, 305
anaerobic processes of, 301–306
degradative species, 295–301
numbers and distribution, 291–294
oligotrophic environment, 290–291
types of, 291
Group selection, 109, 114–115, 172
Gypsy moth, 195; *see also* Fungal entomopathogens, terrestrial

Habitat range, synergy and, 173
Halogenated organics
degradative consortia and, 112, 114
anaerobic reduction, 324, 325
in enrichment cultures, 119
in groundwater, 319–325
abiotic transformations, 320
biodegradation, 320–325
identified compounds, 319–320
Heat shock, 57
Herbicides: *see* Diclofop methyl degrading consortia
Homeostatic communities, 29, 167
Homoserine lactones, 49

Humic acids, benzene degradation and, 318–319
Hydrocarbons: see Aromatic hydrocarbons; Groundwater bioremediation; Halogenated organics

Identification of organisms, 28, 47
  fatty acid profiles, 155–156
  genetic probes, 28, 44–45, 153–155
  polyclonal antibodies, 146
  substrate utilization profiles, 156–157
Identity, in open systems, 62–64, 67–69, 70
Idiographic interpretation, 32, 35–36, 37, 38
Image analysis, digital, 134, 135
Immune responses, to fungal entomopathogens, 212–214
Individual, definition of, 173
Inductive data analysis, 32, 35, 36
Information, 168–169, 173
Insects: see Fungal entomopathogens, terrestrial
In situ studies, 27–29; see also Communities, microbial
  of biodegradation in groundwater, 310
  emerging paradigm for, 29–30, 32, 33, 71
Interaction parameter, 41, 42, 45–47, 50–54, 64–69
Iron
  in groundwater
    bioremediation of, 328
    as electron acceptor, 294, 302–306, 315–316, 318–319
    regeneration, by marine protozoans, 96
Iron chelators, benzene degradation and, 318–319
Isolated cell line: see Clone

Koch's postulates, 115–118
k-selection, 120

Laws of nature, vs. multiple realities, 35, 36
Lectins, fluorescent, 145–146
Lichens, 107–108
Life, definition of, 174
Liposomes, as marine particles, 90, 93, 96, 98
Luciferase, bacterial, 46

Mahoney Lake: see Phototrophic sulfur bacteria
Manganese, in groundwater
  bioremediation of, 328
  as electron acceptor, 294, 302–304, 315

Marine environments: see Colloids, marine; Phytoplankton
Mecoprop, 131, 132
Mercury, bioremediation of, 330
Meromictic lakes: see Phototrophic sulfur bacteria, in Mahoney Lake
Metals, in groundwater
  bioremediation of, 327–330
  as electron acceptors, 294, 301–306, 315–319
Methanogenesis
  in groundwater, 294, 301–305, 315, 316–317
  in sludge communities, 162–165
Microcosm technique, for biodegradation studies, 310
Microecosystem, definition of, 174
Microelectrode techniques, 28, 43, 146, 157
Microstat, 128, 135–139, 174
Mini gaia hypothesis, 124, 125–126, 127
Mixed culture, definition of, 174
Mobile genetic elements, in degradative consortia, 110, 112, 114, 119, 123
Motility
  of bacteria
    on surfaces, 135
    swarming, 49, 109, 135
    swimming, 91–92
  of phytoplankton, 7–10, 17
    cyanobacteria, 12, 17
Multi-channel flow cells, 135

Naphthalene: see Polycyclic aromatic hydrocarbons
Naturalistic inquiry, 31–32
  analysis in, 35–36, 37, 72
  focus in, 33–34, 37, 71
  methodology in, 34–35, 37, 71–72
  open systems and, 40–41
  scientific control in, 36–38, 72–73
  subjectivity in, 32–33, 34, 36, 37, 70–71
  techniques in, 72
Natural selection: see Selection
Negative staining, 141
Negotiated outcomes, 36, 37
Nematodes, coinfection with entomopathogenic fungi, 233
Nitrate, in groundwater
  bioremediation of, 326–327
  as electron acceptor, 302, 304, 315, 317
Nitrogen, in marine particles, 86, 87, 89

Index

Nitrogen fixation
  by bacteroids, 123
  by cyanobacteria, 12, 13, 108
Non-destructive methods, 43–44, 53–54
Nutristat, 128, 132, 174

Ockham's razor, 107
Okenone, 256, 260, 282–283
Open systems theory, 39–41
  identity in, 62–64, 67–69, 70
  system boundaries, 42–43
Organism, definition of, 173
Organismic selection theory, 110–115, 167
Organism parameter, 41, 42, 44–45, 50–54, 64–69
Organizational level of analysis, 50–53
Orientation, in phytoplankton
  cyanobacteria, 12
  UV radiation and, 7–10
Ozone depletion, 1–3, 14, 18

PAM fluorescence, 15–16
Paradigms
  conventional, 30–31, 37, 38, 71–72, 73
  definition of, 30
  emerging: *see* Naturalistic inquiry
Paraflagellar body, UV-B radiation and, 9
Particulate organic matter, definition of, 83
PCR method, 44, 153, 154
pH, in biofilms, 141, 146–147
Phospholipid profiles, 155–156
Photosynthesis
  irradiance window for, 15
  PAM fluorescence studies, 15–16
  by phytoplankton
    productivity, 4, 18
    UV-B effects on, 13, 14, 16, 17
    vertical distribution and, 6
  pigments of purple sulfur bacteria, 256, 258, 260, 282–283
Phototaxis, in phytoplankton, 7, 9–10, 12
Phototroph/heterotroph bioaggregates, 130–131
Phototrophic sulfur bacteria, in Mahoney Lake, 251–252
  biomass accumulation, 265–266
  biomass loss, by upwelling, 266–269, 280
  buoyancy regulation, 267
  carbon cycle, 269–272, 273, 274–278
  eukaryotes and, 255–256, 280
  experimental methodology, 256

Phototrophic sulfur bacteria, in Mahoney Lake (*cont.*)
  growth-controlling factors, 260–266
    biomass, 265
    light, 251–252, 257–260, 261, 263, 265–266
    sulfide, 261–265
  heterotrophic bacteria and, 252, 260
  in oxic water layers, 278–280
  sulfate reducers, 269–272, 274–278
  sulfur reducers, 272–274
  Mahoney Lake habitat, 252–256
  paleomicrobiology, 280–284
  polysulfides and, 260, 272–274
  species composition, 256–260
  in sediment core, 282–284
Phycobilins, UV-B damage to, 13
Phytoplankton, *see also* Cyanobacteria
  in carbon cycles, 4, 18
  distribution
    global, 4–5
    vertical, 5–6, 7–10
  in food web, 6
  marine colloids derived from, 88
  in meromictic lake, 275–276, 278, 279
  UV-B effects
    aquatic ecosystems and, 18–19
    on global distribution, 4–5
    on growth, 14
    on metabolism, 14–16
    protective strategies, 17
    research methodologies, 6–7, 14–16
    solar radiation characteristics, 1–3
    target molecules, 9, 12–13, 14, 16–17
    on vertical distribution, 7–10
Picofecal pellets, 89, 90, 98
Plankton: *see* Phytoplankton; Zooplankton
Plasmids: *see* Mobile genetic elements, in degradative consortia
Pollution: *see* Groundwater bioremediation
Polyclonal antibodies, 146
Polycyclic aromatic hydrocarbons, biodegradation of
  aerobic, 307, 308, 311–312, 313
  anaerobic, 319
Polysaccharides, *see also* Exopolymers
  bacterial exopolysaccharides
    charge distributions *in situ*, 145
    in degradative consortia, 145–146, 157, 159, 161, 163, 165
    functions of, 145
    in marine colloids, 86, 88, 95, 97

Population, definition of, 174
Porins, in marine colloids, 86
Predators
　in biofilms, 60
　fungal entomopathogens of, 231–233
Prokaryotes
　community-level evolution in, 106, 107, 111, 124
　eukaryote evolution from, 107, 110–111, 114–115, 123
Proliferation hypothesis, 106–107, 168–169
Propane oxidizers, halogenated organics and, 322, 323
Protector guilds, 126
Protozoa
　in aquifers, 291, 292, 311–312
　marine colloid consumption by, 87–88, 94–96, 98
　marine colloids derived from, 89–90, 98
Pseudomonas
　hydrocarbon degradation by, 310–311, 314–315, 322, 324
　P. fluorescens, in biofilms, 146, 152–153
Pulse amplitude modulation fluorescence, 15–16
Pure cultures
　vs. community cultures, 115–118, 121
　inhomogeneity of, 116, 120, 174
Purple sulfur bacteria: *see* Phototrophic sulfur bacteria

Qualitative methods, 32, 35, 37, 72
Quasi-steady state
　biofilms in, 134–135
　definition of, 174
Quorum-sensing system, 49

Realism, vs. constructivism, 31, 32, 33, 35, 36, 38
Redox probes, for biofilms, 147–148
Reductionism, 109–110, 120–122
Reduction zones, in aquifer, 294, 301–306, 315–319
Regulation, in complex adaptive systems, 64–69
Relativity, universal, 175
Reporter genes, 46–47, 50
Reproduction, definition of, 174
Reproductive success, 106, 107, 122–124, 174
Resorufin, as redox probe, 147–148
Responsiveness, in complex adaptive systems, 67–69

Reverse sample genome probes (RSGP), 154–155
Rototorque annular bioreactor, 133–134
rRNA hybridization, 28, 44–45, 46, 110, 153–154
r-selection, 120
RSGP (reverse sample genome probes), 154–155
Rumen communities, 107, 154

Sampling, purposive, 32, 33, 34
Scanning confocal laser microscopy, *see also* Fluorescent probes
　advantages of, 28, 44, 47–48, 105, 121
　for biofilm studies, 139–141, 171
　in microstat studies, 135, 136
Scientific control, in emerging paradigm, 36–38, 72–73
SCIO model, 41–42, 49–50; *see also* Complex adaptive systems
　behavioral vs. organizational analysis, 50–54
　coupling and regulation in, 64–69
　parameter analysis in, 41–50
　　coordination, 41–42, 48–50
　　interaction, 41, 42, 45–47, 49–50
　　organisms, 41, 42, 44–45, 49–50
　　structure, 41, 43–44, 47–48, 49–50
Scytonemine, 17
Sediments
　in aquifers, 292–293
　　anaerobic processes in, 301–306, 315–319
　of meromictic lake, 251, 280–284
Selection
　organismic selection theory, 110–115, 167
　vs. proliferation, 106, 107, 172–173, 174–175
Selective enrichment: *see* Enrichment cultures
Selenium, bioremediation of, 329–330
Self-organization, *see also* Autopoiesis
　in chemostat communities, 119
　evolution as, 106, 172–173
　postulates of, 168–169
Sludge communities, 141, 145, 154, 162–165
Snow, marine, 4, 97
Social constructivism: *see* Constructivism
Solar radiation: *see* UV radiation
Speciation, 111, 175
Species, definition of, 175
Structure parameter, 41, 43–44, 47–48, 50–54, 64–69
Subjectivity, 32–33, 34, 36, 37, 70–71

Index

Submicron particles, *see also* Colloids, marine
   definition of, 84
Substrate utilization profiles, 156–157
Suicidal genetic elements, 62
Sulfate-reducing bacteria
   in groundwater, 302–306, 315, 316, 317
   in oilfields, 154–155
   in stratified lakes, 252, 260, 269–272, 274–278
Sulfur cycle, of Mahoney Lake, 252, 256, 262, 273
Surface communities: *see* Biofilms
Swarming, bacterial, 109
   homoserine lactones and, 49
   viscosity and, 135
Symbiosis, 27–28; *see also* Communities, microbial; Endosymbiosis
Synergy, in microbial communities, 166–167
Systems theory: *see* Open systems theory

Taxonomy: *see* Identification of organisms
Terminal electron-accepting processes, 294, 301–306, 315–319
Toluene: *see* BTEX, biodegradation of
Toxins, of fungal entomopathogens, 215

Unculturable organisms, 28, 44, 153
Universal relativity, 175
Uranium, bioremediation of, 328–329

UV radiation, *see also* Phytoplankton, UV-B effects
   entomopathogenic fungal spores and, 223–224
   ozone and, 1–3
   penetration into water column, 3
   UV-A
      DNA repair and, 16–17
      phytoplankton and, 8, 9, 10, 14
   UV-B, definition of, 2

Viability probes, for biofilms, 147–152
Viruses
   coinfection with entomopathogenic fungi, 233–234
   in ocean
      abundance, 84
      colloid production and, 90
      flagellate ingestion of, 95–96
Visible radiation
   DNA repair and, 16–17
   phytoplankton and, 8, 9, 10

Xylene: *see* BTEX, biodegradation of

Yeasts, lethal proteins from, 108

Zooplankton, food sources of
   marine colloids, 96, 97, 98
   phototrophic bacteria, 252, 280